Astrophysics and Space Science Proceedings

More information about this series at
http://www.springer.com/series/7395

Carlos F. Sopuerta
Editor

Gravitational Wave Astrophysics

Proceedings of the Third Session of the Sant Cugat Forum on Astrophysics

Springer

Sant Cugat Forum on Astrophysics

Editor
Carlos F. Sopuerta
Institut de Ciències de l'Espai (CSIC-IEEC)
Edifici Nexus
C/ Gran Capità, 2-4
08034 Barcelona
Spain

ISSN 1570-6591 ISSN 1570-6605 (electronic)
ISBN 978-3-319-10487-4 ISBN 978-3-319-10488-1 (eBook)
DOI 10.1007/978-3-319-10488-1
Springer Cham Heidelberg New York Dordrecht London

Library of Congress Control Number: 2014957540

Printed on acid-free paper

Springer is part of Springer Science+Business Media (www.springer.com)

Sant Cugat Forum on Astrophysics

To Alberto Lobo (1953–2012).
Many thanks for all the inspiration and leadership provided over the years.

Preface

The Sant Cugat Forum on Astrophysics[1] is a framework setup to host international meetings. Among the meetings hosted by the Forum, the most recognizable will be the Workshop sessions focusing on a specific aspect of astrophysics, like the one this book is about. These will typically be held every 2 years, with a typical attendance of up to 100 scientists, hosting many plenary talks, and having a duration of about 3–4 days. The Forum has been established under the auspices of the Sant Cugat City Hall, the Institute for Space Sciences (ICE-CSIC), the Institut d'Estudis Espacials de Catalunya (IEEC), and the Institució Catalana de Recerca i Estudis Avançats (ICREA). This book gives account of the third session of the Forum, devoted to "Gravitational Wave Astrophysics" and was held on April 22–25, 2014.

The detection and analysis of the gravitational radiation emitted by diverse astrophysical and cosmological sources promises to open a completely new window to the exploration of the Universe. The developments towards Gravitational Wave Astronomy are currently following different paths. On the one hand, second-generation ground-based interferometric detectors [LIGO (USA), VIRGO (Italy-EGO), and KAGRA (Japan)] operating in the high-frequency band ($1 - 10^4$ Hz) are going to start operations very soon and are expected to provide the first detections during this decade. There are also plans for a third-generation observatory, the Einstein Telescope, one of the *Magnificent Seven* European projects recommended by the ASPERA network for the future development of astroparticle physics in Europe. These detectors will observe general relativistic phenomena such as the coalescence and merger of stellar compact binaries containing neutron stars and/or black holes. They are also sensitive to gravitational emissions from core-collapse supernovae and to gravitational-wave backgrounds from the early universe. On the other hand, the design of space-based detectors that will cover the low-frequency band ($10^{-5} - 1$ Hz) has a long history, and in particular the *Laser Interferometer Space Antena* (LISA). At present, evolved LISA (eLISA) is the mission concept proposed in *The Gravitational Universe*, the science theme

[1]Its website is http://www.ice.csic.es/research/forum.

selected by the European Space Agency for its future L3 mission. eLISA will search for supermassive black hole binary coalescence, inspirals of stellar compact objects into supermassive black holes, thousands of galactic binaries, and stochastic backgrounds of gravitational waves. Finally, we have the Pulsar Timing Arrays (PTAs), which look at variations in the arrival times of radio emissions from millisecond pulsars due to the passage of gravitational waves. There are three PTAs [NANOGrav (USA), EPTA (Europe), and PPTA (Australia)] operating radio-telescopes that monitor sets of millisecond pulsars to detect the presence of gravitational waves in the very low frequency band ($10^{-9} - 10^{-7}$ Hz). These arrays, organized in the International PTA (IPTA) consortium, are expected to make the first detections on a timescale of decades, and are currently providing unique constraints on gravitational waves from inspirals of supermassive black holes, cosmic strings, and other sources. There is therefore a strong motivation to study the astrophysical mechanisms that create the different gravitational-wave sources (to understand their distribution and event rates) and also the physical connection between these sources and other astrophysical and cosmological processes. For instance, observations by large telescopes for electromagnetic transient signals together with gravitational-wave observations will provide rich information for high-energy phenomena such as gamma-ray bursts. Moreover, the detection of supermassive black hole mergers can help us understand the mechanisms of galaxy formation, explore the structure of black holes, and test General Relativity and other theories of gravity.

Taking into account the technological developments in gravitational-wave detectors, it is an excellent time to organize a workshop to discuss the state of the art of the main astrophysics associated with the sources of gravitational waves and the hidden universe that can be unveiled through these observations. We hope this session of the Forum and this volume provide an ample view of these topics and can encourage other scientists to join the efforts in Gravitational Wave Astronomy, an emerging area which promises great discoveries in the next years and decades.

We would like to warmly thank the Scientific Organizing Committee (SOC), who have helped shaping the content of the workshop. The SOC was formed by

Nils Andersson (University of Southampton, UK)
Pierre Binétruy (APC-Paris, France)
Monica Colpi (University of Milano-Bicocca, Italy)
Enrique García-Berro (Polytechnic University of Catalonia, Spain)
Philippe Jetzer (University of Zurich, Switzerland)
Kostas Kokkotas (University of Tübingen, Germany)
Pablo Laguna (Georgia Institute of Technology, USA)
Alicia M. Sintes (Universitat de les Illes Balears, Spain)
Carlos F. Sopuerta (IEEC-CSIC, Spain)
Alberto Vecchio (University of Birmingham, UK)

The meeting would have been impossible without the help of the local organization, which was formed by

Anna Bertolín (IEEC-CSIC)
Ivan Lloro (IEEC-CSIC)

Pilar Montes (IEEC-CSIC)
Daniel Santos Oliván (IEEC-CSIC)
Carlos F. Sopuerta (IEEC-CSIC, *chair*)
Diego F. Torres (ICREA & IEEC-CSIC, Sant Cugat Forum)

and especially, without the support of the City Hall, especially of Exmo. Madam Major Mercè Conesa i Pagès, and Esther Salat i Llorente, Major Lt. for Education, University and Family. The poster and artwork were created by Anna Pedescoll from the Sant Cugat City Hall Communication Department, to whom we are indebted. Last but not least, we acknowledge the excellent support of all the staff at the Casa de Cultura and the Museu de Sant Cugat, which hosted the different events of the meeting.

Barcelona, Spain
June 2014

Carlos F. Sopuerta

Contents

Gravitational Radiation from Compact Binary Pulsars 1
John Antoniadis
1 Introduction 1
2 Pulsars and Pulsar Timing 4
  2.1 Pulsar Timing 4
  2.2 Binary Pulsars.. 6
  2.3 Gravitational Wave Emission 7
3 The Hulse-Taylor Binary and the Double Pulsar 9
4 Beyond the Double Pulsar 10
5 Low-Mass White Dwarfs in Binaries 11
6 White Dwarf Masses.. 12
7 Constraints on Dipolar Gravitational Radiation 13
  7.1 PSR J1738+0333, the Most Stringent Constraints on Dipolar Radiation 13
  7.2 PSR J0348+0432, a Massive Pulsar in the Most Compact Binary.. 14
8 Ramifications for the Phase Evolution of Neutron Star Mergers 16
9 Constraints on Alternative Theories of Gravity 17
  9.1 Scalar-Tensor Gravity 17
  9.2 TeVeS.. 18
  9.3 Local Lorenz Invariance.. 18
10 Final Remarks... 19
References 20

The Coalescence Rates of Compact Object Binaries 23
Tomasz Bulik
1 Introduction 23
2 Observation Based Estimates.. 24
  2.1 Double Neutron Stars.. 24
  2.2 Binary Black Holes 25
  2.3 Black Hole Neutron Star Binaries 26

3 Population Synthesis Based Results ... 27
3.1 How to Make a Compact Object Binary? ... 27
3.2 Rate Density Estimates ... 29
3.3 Population III Binaries ... 30
4 Summary ... 30
References ... 31

Enhancing Gravitational Wave Astronomy with Galaxy Catalogues ... 35
Xilong Fan, Christopher Messenger, and Ik Siong Heng
1 Introduction ... 35
2 Identification of Gravitational-Wave Host Galaxies ... 38
3 Enhanced Inclination Angle Inference ... 39
Conclusions ... 40
References ... 41

X, Gamma-Rays, and Gravitational Waves Emission in a Short Gamma-Ray Burst ... 43
F.G. Oliveira, Jorge A. Rueda, and R. Ruffini
1 Introduction ... 43
2 The Short GRB 090227B ... 45
3 Gravitational Wave Emission ... 46
References ... 49

Localizing Gravitational Wave Sources with Optical Telescopes and Combining Electromagnetic and Gravitational Wave Data ... 51
Shaon Ghosh and Gijs Nelemans
1 Introduction ... 51
2 Gravitational Wave Parameter Estimation in Presence of Electromagnetic Information ... 52
3 Gravitational Wave Sky Localization for Electromagnetic Follow-Up Studies ... 55
4 Discussion ... 57
References ... 58

Advanced Models of Black Hole–Neutron Star Binaries and Their Astrophysical Impact ... 59
Zachariah B. Etienne, Vasileios Paschalidis, and Stuart L. Shapiro
1 Introduction ... 60
2 BHNS Binary Parameter Space ... 61
3 Constraining the NS EOS from BHNS Binary GWs ... 62
3.1 What Can Be Learned About the NS Equation of State from a Plunging Merger into the BH? ... 62
3.2 What Parameters Are Important for Tidal Disruption? ... 63
3.3 Tidal Disruption in BHNS Mergers: Results from Fully General Relativistic Simulations ... 64
3.4 Can We Extract NS EOS Information from BHNS Merger GWs Alone? ... 66

4 EM Counterparts to BHNS GW Signals 67
4.1 Electromagnetic Counterparts to BHNS Inspiral: Magnetospheric Phenomena 67
4.2 Electromagnetic Counterparts to BHNS Merger: Short Gamma-Ray Burst 68
4.3 Electromagnetic Counterparts to BHNS Merger: Kilonovae 70
Conclusions 71
References 72

Coincidence Searches of Gravitational Waves and Short Gamma-Ray Bursts 75
Andrea Maselli and Valeria Ferrari
1 Introduction 75
2 Selecting Candidates for Gamma-Ray Bursts Emission 76
3 The Uncertainties on the Binary Parameters 78
4 Numerical Results 80
Conclusions 81
References 82

Gravitational Waves from Rapidly Rotating Neutron Stars 85
Brynmor Haskell, Nils Andersson, Caroline D'Angelo, Nathalie Degenaar, Kostas Glampedakis, Wynn C.G. Ho, Paul D. Lasky, Andrew Melatos, Manuel Oppenoorth, Alessandro Patruno, and Maxim Priymak
1 Introduction 86
2 Gravitational Wave Emission Mechanisms 87
3 Low Mass X-Ray Binaries 88
4 The r-Mode Instability 89
5 Mountains on Neutron Stars 93
5.1 Thermal Mountains 94
5.2 Magnetic Mountains 96
6 Torque Balance Revisited 98
7 Summary 100
References 101

Supermassive Black Hole Binaries: The Search Continues 103
Tamara Bogdanović
1 Theoretical Background: Formation and Evolution of SBHBs 103
2 Observational Evidence for SBHBs 106
2.1 Direct Imaging of Double Nuclei 106
2.2 Photometric Measurements of Quasi-Periodic Variability 108
2.3 Spectroscopic Measurements of Offset Broad Emission-Lines 109
3 Future Theoretical and Observational Prospects 111
Conclusions 113
References 114

General Relativistic Simulations of Accretion Disks Around Tilted Kerr Black Holes ... 121
Vassilios Mewes, Pedro J. Montero, Nikolaos Stergioulas, Filippo Galeazzi, and José A. Font
1 Introduction ... 121
2 Results ... 124
References ... 126

Prompt Flare and Disk Formation in Tidal Disruptions by Massive Black Holes ... 129
Christopher Evans and Pablo Laguna
1 Introduction ... 129
2 Tidal Disruption Events in Brief ... 131
3 Preliminary Results ... 132
Conclusions ... 133
References ... 134

Rival Families: Waveforms from Resonant Black-Hole Binaries as Probes of Their Astrophysical Formation History ... 137
Davide Gerosa
1 Stellar-Mass Black-Hole Binaries and the Key Role of Gravitational-Wave Astronomy ... 137
2 Constraining Black-Hole Binary Formation ... 139
3 Telling the Two Resonant Families Apart ... 142
References ... 145

Pulsar Timing Arrays and the Challenge of Massive Black Hole Binary Astrophysics ... 147
A. Sesana
1 Introduction ... 147
2 Pulsar Timing Array Response to Gravitational Waves ... 148
3 The Stochastic GW Background ... 151
4 The GW Background as a Tool for SMBH Binary Astrophysics ... 152
4.1 Spectral Amplitude: The Cosmic SMBH Binary Merger Rate ... 153
4.2 Spectral Shape: The SMBH Binary-Environment Coupling ... 155
5 Resolvable Sources ... 159
5.1 Science with Resolvable PTA Sources ... 160
Conclusions ... 161
References ... 162

Gravitational Wave Verification Sources ... 167
Mukremin Kilic, Warren R. Brown, J.J. Hermes, and A. Gianninas
1 Ultra-Compact Binary Systems ... 167
2 Verification Sources ... 168
2.1 AM CVn Stars ... 168
2.2 Double WDs ... 168

3 Future Prospects 172
References 173

The Gravitational Wave Emission of White Dwarf Dynamical Interactions 175
Gabriela Aznar-Siguán, Enrique García-Berro, and Pablo Lorén-Aguilar
1 Introduction 175
2 Numerical Setup 176
3 Gravitational Wave Radiation 176
Conclusions 182
References 183

Gravitational Recoil and Astrophysical Impact 185
Ulrich Sperhake
1 Introduction and Motivation 185
2 Calculation of the Recoil 187
3 Open Questions 192
4 Spin-Orbit Resonances 193
5 Suppression of Superkicks 196
Conclusions 199
References 200

Fully Covariant and Conformal Formulation of the Z4 System Compared to the BSSN Formulation in Spherical Symmetry 203
Nicolas Sanchis-Gual, Pedro J. Montero, José A. Font, Ewald Müller, and Thomas W. Baumgarte
1 Introduction 203
2 The Fully Covariant and Conformal Z4 Formulation 204
3 Numerical Results 206
3.1 Schwarzschild Black Hole 206
3.2 Stable Spherical Relativistic Star 207
4 Summary 208
References 208

Extraction of GWs from a Numerical Simulation 209
Nigel T. Bishop and Luciano Rezzolla
1 Introduction 209
2 Quadrupole Formula (QF) 210
3 Perturbative Methods (PM) 211
4 ψ_4 (Fixed Radius) and ψ_4 (Extrapolated) 211
5 Characteristic Extraction (CE) 212
6 Comparisons 213
References 214

Quasinormal Modes Beyond Kerr 217
Aaron Zimmerman, Huan Yang, Zachary Mark, Yanbei Chen, and Luis Lehner
1 Introduction 217
2 Quasinormal Modes of Deformed Spacetimes 218
3 Parametric Instability of Rapidly Rotating Kerr Black Holes 220
4 The Quasinormal Modes of Kerr-Newman Black Holes 221
5 Outlook 222
References 223

The Scientific Potential of Space-Based Gravitational Wave Detectors 225
Jonathan R. Gair
1 Introduction 225
1.1 Detectors 227
1.2 Sources 228
2 Science Applications: Astrophysics 230
2.1 MBH Constraints on Growth of Structure 230
2.2 EMRI Measurements of the MBH Mass Function 233
2.3 Other Astrophysical Applications 234
3 Science Applications: Cosmology 235
4 Science Applications: Fundamental Physics 236
4.1 Tests of Gravitational Physics 237
4.2 Tests of the Nature and Structure of Black Holes 238
5 Summary 242
References 242

Connecting Numerical Relativity and Data Analysis of Gravitational Wave Detectors 245
Deirdre Shoemaker, Karan Jani, Lionel London, and Larne Pekowsky
1 Introduction 245
2 Historical Perspective on NR and GWs 246
3 Binary Black Hole Mergers 247
4 Role of Higher Harmonics on BBH Mergers and GWs 249
5 End State of BBH Merger 252
6 Mergers as Transient GW Events 254
Conclusions 254
References 255

Memory Effect from Spinning Unbound Binaries 259
Lorenzo De Vittori, Achamveedu Gopakumar, Anuradha Gupta, and Philippe Jetzer
1 Introduction 259
2 Waveforms for Spinning Compact Binaries on Hyperbolic Orbits 260
3 Waveforms for Non-spinning Binaries on Hyperbolic Orbits 263

Conclusions 265
References 265

The Challenges in Gravitational Wave Astronomy for Space-Based Detectors 267
Edward K. Porter
1 Introduction 267
2 Galactic Binaries 269
2.1 Verification Binaries 269
2.2 Current Status of Existing Algorithms 270
2.3 Outstanding Problems 270
3 Supermassive Black Hole Binaries 271
3.1 Algorithmic Development 272
3.2 Outstanding Problems and Questions 273
3.3 Comparable Mass Waveforms 274
4 Extreme Mass Ratio Inspirals 274
4.1 Algorithmic Development 275
4.2 Outstanding Issues 275
5 Cosmological Sources 276
5.1 Outstanding Issues 277
Conclusion 277
References 278

Investigating Binary Black Hole Mergers with Principal Component Analysis 281
J. Clark, L. Cadonati, J. Healy, I.S. Heng, J. Logue, N. Mangini, L. London, L. Pekowsky, and D. Shoemaker
1 Introduction 281
2 Binary Black Hole Merger Simulations 282
3 Principal Component Analysis and Bayesian Model Selection 283
4 Identifying Binary Black Hole Merger Phenomenology 285
References 287

Split Bregman Method for Gravitational Wave Denoising 289
Alejandro Torres, Antonio Marquina, José A. Font, and José M. Ibáñez
1 Introduction 289
2 Mathematical Background 290
3 Gravitational Wave Catalogs 291
4 Results 292
5 Summary 293
References 294

Interaction of Gravitational Waves with Charged Particles 295
Thulsi Wickramasinghe, Will Rhodes, and Mitchell Revalski
1 Introduction 295
2 Interaction of GWs with Charged Particles 296
2.1 GW of a Specific Simple Form 297
3 Results and Discussion 298
References 299

The Emission of Electromagnetic Radiation from Charges Accelerated by Gravitational Waves and Its Astrophysical Implications 301
Mitchell Revalski, Will Rhodes, and Thulsi Wickramasinghe
1 Introduction 301
2 Previous Work 302
3 Radiation from Accelerating Charges: Calculations 303
4 Additional Arguments for Radiation: Memory Effects 306
5 Astrophysical Implications 307
Conclusions 307
References 308

Contributors

Nils Andersson Mathematical Sciences and STAG Research Centre, University of Southampton, Southampton, UK

Caroline D'Angelo Leiden Observatory, Leiden University, Leiden, The Netherlands

ASTRON, The Netherlands Institute for Radio Astronomy, Dwingeloo, The Netherlands

John Antoniadis Max-Planck-Institut für Radioastronomie, Bonn, Germany

Gabriela Aznar–Siguán Departament de Física Aplicada, Universitat Politècnica de Catalunya, Castelldefels, Spain

T.W. Baumgarte Department of Physics and Astronomy, Bowdoin College, Brunswick, ME, USA

Nigel T. Bishop Department of Mathematics (Pure & Applied), Rhodes University, Grahamstown, South Africa

Tamara Bogdanović Center for Relativistic Astrophysics, School of Physics, Georgia Institute of Technology, Atlanta, GA, USA

Warren R. Brown Smithsonian Astrophysical Observatory, Cambridge, MA, USA

Tomasz Bulik Astronomical Observatory, University of Warsaw, Warsaw, Poland

Laura Cadonati University of Massachusetts Amherst, Amherst, MA, USA

Cardiff University, Cardiff, UK

Yanbei Chen Theoretical Astrophysics 350-17, California Institute of Technology, Pasadena, CA, USA

James Clark University of Massachusetts Amherst, Amherst, MA, USA

Nathalie Degenaar Department of Astronomy, University of Michigan, Ann Arbor, MI, USA

Lorenzo De Vittori Physik-Institut, Universität Zürich, Zürich, Switzerland

Zachariah B. Etienne Department of Mathematics, West Virginia University, Morgantown, WV, USA

Christopher Evans Center for Relativistic Astrophysics, School of Physics, Georgia Institute of Technology, Atlanta, GA, USA

Xilong Fan School of Physics and Electronics Information, Hubei University of Education, Wuhan, China

SUPA, School of Physics and Astronomy, University of Glasgow, Glasgow, UK

Valeria Ferrari Dipartimento di Fisica, Università di Roma "La Sapienza" & Sezione INFN Roma1, Roma, Italy

José A. Font Departamento de Astronomía y Astrofísica, Universitat de València, Burjassot, València, Spain

Filippo Galeazzi Institut für Theoretische Physik, J.W. Goethe - Universität, Frankfurt am Main, Germany

Jonathan R. Gair Institute of Astronomy, University of Cambridge, Cambridge, UK

Enrique García–Berro Departament de Física Aplicada, Universitat Politècnica de Catalunya, Castelldefels, Spain

Institute for Space Studies of Catalonia, Barcelona, Spain

Davide Gerosa Department of Applied Mathematics and Theoretical Physics, Centre for Mathematical Sciences, University of Cambridge, Cambridge, UK

Shaon Ghosh Radboud University, Nijmegen and Nikhef, The Netherlands

Alexandros Gianninas Department of Physics and Astronomy, University of Oklahoma, Norman, OK, USA

Kostas Glampedakis Departamento de Fisica, Universidad de Murcia, Murcia, Spain

Achamveedu Gopakumar Department of Astrophysics, Tata Institute of Fundamental Research, Mumbai, India

Anuradha Gupta Department of Astrophysics, Tata Institute of Fundamental Research, Mumbai, India

Brynmor Haskell School of Physics, The University of Melbourne, Parkville, VIC, Australia

James Healy Center for Relativistic Astrophysics, Georgia Institute of Technology, Atlanta, GA, USA

Ik Siong Heng SUPA, School of Physics and Astronomy, University of Glasgow, Glasgow, UK

James J. Hermes Department of Physics, University of Warwick, Coventry, UK

Wynn C.G. Ho Mathematical Sciences and STAG Research Centre, University of Southampton, Southampton, UK

José María Ibáñez Departamento de Astronomía y Astrofísica, Universitat de València, Burjassot, València, Spain

Karan Jani Center for Relativistic Astrophysics, School of Physics, Georgia Institute of Technology, Atlanta, GA, USA

Philippe Jetzer Physik-Institut, Universität Zürich, Zürich, Switzerland

Mukremin Kilic Department of Physics and Astronomy, University of Oklahoma, Norman, OK, USA

Pablo Laguna Center for Relativistic Astrophysics, School of Physics, Georgia Institute of Technology, Atlanta, GA, USA

Paul D. Lasky School of Physics, The University of Melbourne, Parkville, VIC, Australia

Luis Lehner Perimeter Institute for Theoretical Physics, Waterloo, ON, Canada

Josh Logue SUPA, School of Physics and Astronomy, University of Glasgow, Glasgow, UK

Lionel London Center for Relativistic Astrophysics, School of Physics, Georgia Institute of Technology, Atlanta, GA, USA

Pablo Lorén–Aguilar School of Physics, University of Exeter, Exeter, UK

Nicholas Mangini University of Massachusetts Amherst, Amherst, MA, USA

Zachary R. Mark Oberlin College, Oberlin, OH, USA

Antonio Marquina Departamento de Matemática Aplicada, Universitat de València, Burjassot, València, Spain

Andrea Maselli Dipartimento di Fisica, Università di Roma "La Sapienza" & Sezione INFN Roma1, Roma, Italy

Andrew Melatos School of Physics, The University of Melbourne, Parkville, VIC, Australia

Christopher Messenger SUPA, School of Physics and Astronomy, University of Glasgow, Glasgow, UK

Vassilios Mewes Departamento de Astronomía y Astrofísica, Universitat de València, Burjassot, València, Spain

Pedro J. Montero Max-Planck-Institute für Astrophysik, Garching bei München, Germany

Ewald Müller Max-Planck-Institute für Astrophysik, Garching bei München, Germany

Gijs Nelemans Radboud University, Nijmegen and Nikhef, The Netherlands

F.G. Oliveira Dipartimento di Fisica and ICRA, Sapienza Università di Roma, Rome, Italy

Manuel Oppenoorth Department of Earth Sciences, Utrecht University, Utrecht, The Netherlands

Vasileios Paschalidis Department of Physics, University of Illinois, Urbana, IL, USA

Alessandro Patruno Leiden Observatory, Leiden University, Leiden, The Netherlands

ASTRON, The Netherlands Institute for Radio Astronomy, Dwingeloo, The Netherlands

Larne Pekowsky Center for Relativistic Astrophysics, School of Physics, Georgia Institute of Technology, Atlanta, GA, USA

Edward K. Porter François Arago Center, APC, Université Paris Diderot, CNRS/IN2P3, CEA/Ifru, Obs. de Paris, Sorbonne Paris Cité, Paris Cedex 13, France

Maxim Priymak School of Physics, The University of Melbourne, Parkville, VIC, Australia

Mitchell Revalski The College of NJ, Ewing, NJ, USA

Luciano Rezzolla Institut fur Theoretische Physik, Frankfurt am Main, Germany

Will Rhodes The College of NJ, Ewing, NJ, USA

Jorge A. Rueda Dipartimento di Fisica and ICRA, Sapienza Università di Roma, Rome, Italy

Remo Ruffini Dipartimento di Fisica and ICRA, Sapienza Università di Roma, Rome, Italy

N. Sanchis-Gual Departamento de Astronomía y Astrofísica, Universitat de València, Burjassot, València, Spain

Alberto Sesana Max-Planck-Institut für Gravitationsphysik, Albert Einstein Institut, Golm, Germany

Stuart L. Shapiro Departments of Physics and Astronomy, University of Illinois, Urbana, IL, USA

Deirdre Shoemaker Center for Relativistic Astrophysics, School of Physics, Georgia Institute of Technology, Atlanta, GA, USA

Ulrich Sperhake Department of Applied Mathematics and Theoretical Physics, University of Cambridge, Cambridge, UK

California Institute of Technology, Pasadena, CA, USA

Department of Physics and Astronomy, The University of Mississippi, University, MS, USA

Nikolaos Stergioulas Department of Physics, Aristotle University of Thessaloniki, Thessaloniki, Greece

Alejandro Torres Departamento de Astronomía y Astrofísica, Universitat de València, Burjassot, València, Spain

Thulsi Wickramasinghe The College of NJ, Ewing, NJ, USA

Huan Yang Perimeter Institute for Theoretical Physics, Waterloo, ON, Canada

Aaron Zimmerman Canadian Institute for Theoretical Astrophysics, Toronto, ON, Canada

Gravitational Radiation from Compact Binary Pulsars

John Antoniadis

Abstract An outstanding question in modern Physics is whether general relativity (GR) is a complete description of gravity among bodies at macroscopic scales. Currently, the best experiments supporting this hypothesis are based on high-precision timing of radio pulsars. This chapter reviews recent advances in the field with a focus on compact binary millisecond pulsars with white-dwarf (WD) companions. These systems—if modeled properly—provide an unparalleled test ground for physically motivated alternatives to GR that deviate significantly in the strong-field regime. Recent improvements in observational techniques and advances in our understanding of WD interiors have allowed for a series of precise mass measurements is such systems. These masses, combined with high-precision radio timing of the pulsars, result to stringent constraints on the radiative properties of gravity, qualitatively very different from what was available in the past.

1 Introduction

The centennial of Einstein's General Theory of Relativity (GR) will also mark the beginning of a new era in experimental gravity. Ground-based antennas such as advanced LIGO and VIRGO will be put in use to observe mergers of compact objects through their gravitational-wave (GW) emission. This new endeavor builds on the remarkable success of GR in describing gravitational phenomena and its wide-spread use in contemporary astrophysics. Following a hibernation period after Einstein's original publication (Einstein 1915b, 1916), discoveries such as quasars, X-ray binaries and neutron stars (NSs), as well as the emergence of observational Cosmology made GR a vital part of astronomers' toolbox. At the same time however, it was realized that GR, being a non-renormalizable field theory, was difficult to reconcile with the Standard Model (SM) of Particle Physics (Feynman et al. 1995). Furthermore, some key features of modern Cosmology like dark matter and dark energy forced scientists to consider the possibility that the theory is only

J. Antoniadis (✉)
Max-Planck-Institut für Radioastronomie, Auf dem Hügel 69, 53121, Bonn, Germany
e-mail: jantoniadis@mpifr-bonn.mpg.de

C.F. Sopuerta (ed.), *Gravitational Wave Astrophysics*, Astrophysics and Space Science Proceedings, DOI 10.1007/978-3-319-10488-1_1

valid for a limited range of energy scales. This motivated the investigation of numerous alternative theories, inevitably leading to the need for high-precision tests of gravity (Will 1994, 2006).

During the past 60 years GR has been tested in numerous different settings and emerged victorious in every single case. An important chapter in precision tests opened with the discovery of the first binary pulsar by Russell Hulse and Joe Taylor in 1974 (Hulse and Taylor 1975). For the first time, this system allowed to test the self-consistency of GR for orbiting strongly self gravitating objects. Most importantly, it led to the first experimental proof of gravitational radiation through the detection of the system's orbital decay (Taylor et al. 1979; Weisberg and Taylor 1981; Taylor and Weisberg 1982; Weisberg and Taylor 1984; Weisberg et al. 2010).

Precision tests of gravity in the Solar System include:

- The perihelion-advance of Mercury (Einstein 1915a; Verma et al. 2014),
- Light deflection in Sun's and Jupiter's gravitational fields (Dyson et al. 1920; Lambert and Le Poncin-Lafitte 2011),
- Strong Equivalence Principle (SEP) tests and the de-Sitter precession of the Moon's orbit (Hofmann et al. 2010; Merkowitz 2010),
- Lens-Thirring effect in satellite orbits (Ciufolini and Pavlis 2004),
- Relativistic Spin Precession of gyroscopes (Everitt et al. 2011),
- the Shapiro delay due to the Sun's gravitational field (Shapiro 1964; Bertotti et al. 2003).

Most of these tests have now reached a precision of 10^{-4} or better. For a thorough review see (Will 1994, 2006).

Presently this effect has been measured in ten more binary pulsars and a double white-dwarf (WD) system (Table 1). These binaries allow for a range of strong-field and radiative tests of gravity in a regime where many alternatives predict significant deviations from the GR predictions. This chapter summarizes recent advances in the field focusing on compact binary pulsars with WD companions which, in the context of many alternative theories of gravity should radiate significant amounts of dipolar GWs. This chapter is a brief review of a few recent studies from the point of view of an experimentalist. The text is split in two major parts, background theory (Sects. 2–6) and results (Sects. 7–9). For a more thorough review on theoretical aspects the reader is referred to (Wex 2014; Will 2014).

Table 1 Compact binaries in the Galaxy with observed orbital decays

Name	Comp. type	P_{spin} (ms)	P_b (h)	m_p^a ($M_\odot$)	m_c^a ($M_\odot$)	e	$\dot{P}_b^{GR}$ ($\times 10^{-13}$)	$\dot{P}_b^{obs,b}$ ($\times 10^{-13}$)	d^c (kpc)	Ref.
J0737−3039	PSR	22.7	2.5	1.3381(7)	1.2489(7)	0.08	−1.24787(13)	−1.252(17)	1.15(22)	Kramer et al. (2006) and Deller et al. (2009)
B1534+12	NS	37.9	10.1	1.3330(4)	1.3455(4)	0.27	−0.1366(3)	−0.19244(5)	0.7	Stairs et al. (1998) and Fonseca et al. (2014)
J1756−2251	NS	28.5	7.7	1.312(17)	1.258(17)	0.18	−0.22(1)	−0.21(3)	2.5	Faulkner et al. (2005) and Ferdman et al. (2014)
J1906+0746	NS	144	3.98	1.323(11)	1.290(11)	0.08	−0.52(2)	−0.565(6)	5.4	Kasian (2012)
B1913+16	NS	59.0	7.8	1.4398(2)	1.3886(2)	0.61	−2.402531(14)	−2.396(5)	9.9	Weisberg et al. (2010)
B2127+11C	NS	30.5	8.0	1.358(10)	1.354(10)	0.18	−3.95(13)	−3.961(2)	10.3(4)	Jacoby et al. (2006) and Kirsten et al. (2014)
J0348+0432	WD	39.1	2.5	2.01(4)	0.172(3)	10^{-6}	−0.258(11)	−0.273(45)	2.1(2)	Antoniadis et al. (2013)
J0751+1807	WD	3.4	6.3	1.26(14)	0.13(2)	10^{-6}	–	−0.031(14)[d]	2.0	Lundgren et al. (1995) and Nice et al. (2008)
J1012+5307	WD	5.2	14.4	1.64(22)	0.16(2)	10^{-6}	−0.11(2)	−0.15(15)	0.836(80)	Lazaridis et al. (2009) and Callanan et al. (1998)
J1161−6545	WD	394	4.8	1.27(1)	1.02(1)	0.17	−0.403(25)	−0.401(25)	3.7	Bhat et al. (2008), Ord et al. (2002), and Antoniadis et al. (2011)
J1738+0333	WD	5.9	8.5	1.46(6)	0.181(7)	10^{-7}	−0.028(2)	−0.0259(32)	1.47(10)	Antoniadis et al. (2012) and Freire et al. (2012b)
WD J0651+2844	WD	–	0.212	0.26(4)	0.50(4)	0	−8.2(17)	−9.8(28)	1.0	Hermes et al. (2012) and Brown et al. (2011)

2 Pulsars and Pulsar Timing

Pulsars play an important role in precision tests of gravity for two reasons: Firstly, they contain extremely strong gravity fields (surface strengths of order $GM/c^2R \simeq 0.2$, compared to 10^{-6} for the Sun). Secondly, they possess clock-like properties that enable the detection of tiny orbital effects with high precision.

The regularity of spin periods for some of the known pulsars rivals the best atomic clocks on Earth. However, radio pulses *do not* arrive at a constant rate at our telescopes and this is exactly what allows us to measure most of their properties. Figure 1 displays the measured $P_{\rm spin}$ of all known 2,200 pulsars in the Galaxy (Hobbs et al. 2005) against the most common change in the spin period—the first spin-period derivative $\dot{P}_{\rm spin}$ (Lorimer and Kramer 2005). Pulsars tend to cluster in two distinct regions of the $P_{\rm spin} - \dot{P}_{\rm spin}$ diagram. There are "normal" pulsars with $(P_{\rm spin}, \dot{P}_{\rm spin}) \sim (1\,{\rm s}, 10^{-15}\,{\rm s\,s^{-1}})$ and "millisecond" pulsars with $(P_{\rm spin}, \dot{P}_{\rm spin}) \sim (3\,{\rm ms}, 10^{-20}\,{\rm s\,s^{-1}})$. The latter are usually the most stable, i.e. there is no other intrinsic rotation change other than the secular variation described above. Conveniently, millisecond pulsars are also more likely to be found in a binary system, a consequence of their evolution, which is influenced by accretion of mass from a binary companion. Most systems discussed here belong to this second category. Because of their intrinsic stability, millisecond pulsars can be used as experimental clocks to infer the properties of spacetime around them. The technique used to achieve this is called pulsar timing and is briefly described below.

2.1 *Pulsar Timing*

For most astronomical objects, the relative motion of the source along the line of sight is determined from Doppler measurements of the emitted signals. For pulsars however, one can do much better. Here, the signal of interest is the beamed electromagnetic radiation sweeping the telescope once per rotation. The pulse profile—especially for millisecond pulsars—occupies a very small fraction of the rotational phase, typically few percent. This allows to determine a precise pulse time of arrival (TOA) relative to a fiducial point on the profile (e.g. its peak), which corresponds to a constant physical latitude on the NS. Thus, since the time difference between any pair of TOAs corresponds to an integer number of rotations, the parametric model describing deviations from constant spin can be optimized to take into account all pulsar rotations between two measurements. This phase-coherent approach of pulsar timing enables inference of rotational and orbital parameters orders of magnitude more precise than traditional Doppler techniques.

The timing model that relates observed TOAs with proper emission times has to account for the accelerating reference frame of the observer, propagation delays

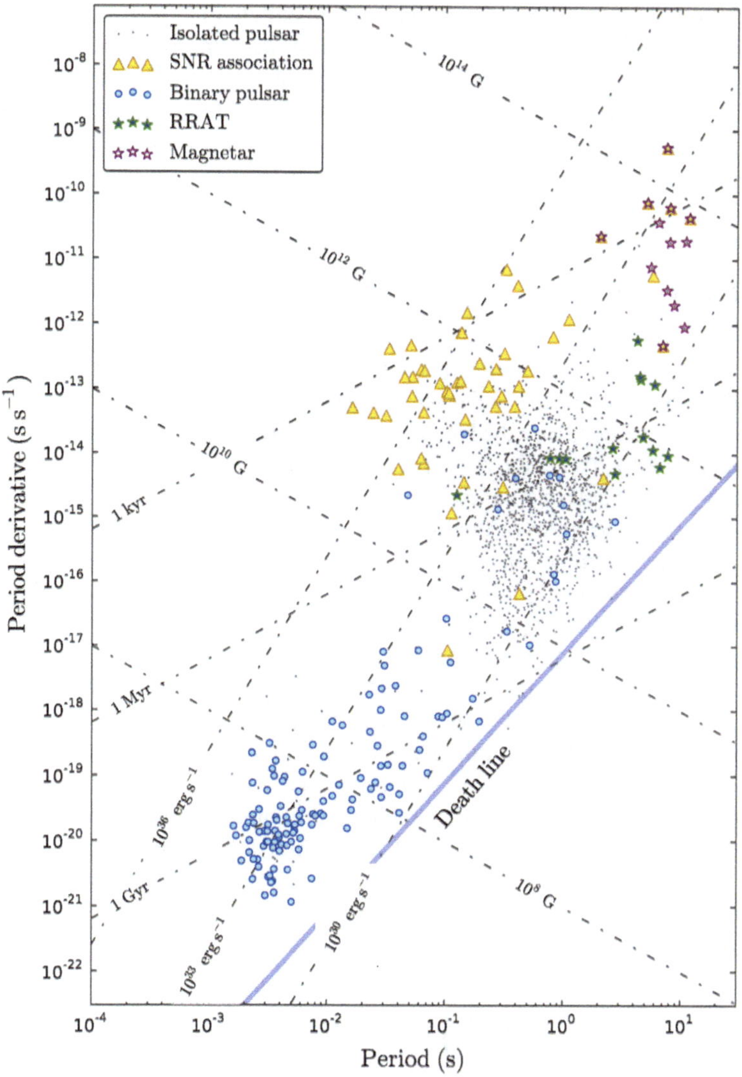

Fig. 1 A $P_{\rm spin} - \dot{P}_{\rm spin}$ diagram of the known radio pulsars in the Galactic disk (adapted from Ewan 2013). The secular change of the spin period is caused by energy emission from the star's rotating electromagnetic field. *Dotted lines* depict representative values for the spin-down energy and the magnetic field strength. The *blue line* corresponds to a critical value of the spin-down energy bellow which electromagnetic emission is switched off. For more details see Lorimer and Kramer (2005) (Color figure online)

caused by the interstellar medium and the motion of the pulsar and the radio telescope relative to the Solar System Barycentre (SSB):

$$t_0 = t_{\rm telesope} + D/f^2 - \Delta_{\rm R} - \Delta_{\rm E} - \Delta_{\rm S} + \Delta_{\rm PSR} \tag{1}$$

Here, the first term accounts for frequency-dependent delays of the signal caused by the interstellar medium. The first three Δ terms account for the Roemer, Einstein and Shapiro delays in the Solar System and transform the TOA at the non-inertial terrestrial reference frame of the observer to a TOA at the inertial SSB frame. The last term of the equation encodes information about the pulsar's orbital motion and is what gives access to most gravity tests.

2.2 *Binary Pulsars*

For binary pulsars, the time delay due to the orbital motion can be modeled by means of the Keplerian parameters and a set of post-Keplerian (pK) terms that account for any possible deviation from the Newtonian orbit (Damour and Taylor 1991; Lorimer and Kramer 2005). Typically, these pK parameters include secular and higher-order variations of the Keplerian parameters ($\dot{e}$, $\dot{\omega}$, $\dot{x}$ etc.), two "Shapiro delay" terms, r and s that model variations of the light-travel time due to the curvature of spacetime around the binary companion and an "Einstein delay" term γ, that takes into account changes of the apparent spin rate due to the motion of the pulsar and the varying gravitational redshift along the orbit. For boost-invariant gravity theories, in the weak-field, slow motion approximation, the pK parameters become functions of the orbital elements, the inertial masses of the system and the equation-of-state (EoS) describing the properties of the stellar interior (Damour and Deruelle 1985, 1986).

In GR, even thought the coupling of the gravitational field to matter inside the star can be extremely strong, the field equations satisfy exactly the strong equivalence principle (SEP) leading to an effacement of the internal stellar structure. In other words, internal coupling terms can be absorbed in redefinitions of the masses, so that the pK parameters are only functions of the masses and orbital elements. In the first post-Newtonian approximation level $\mathcal{O}(v^2/c^2)$, one finds the following deviations from the Keplerian orbit:

$$\dot{\omega} = 3\left(\frac{P_{\rm b}}{2\pi}\right)^{-5/3} (T_{\odot}M)^{2/3}\left(1-e^2\right)^{-1}, \tag{2}$$

$$r = T_{\odot}m_{\rm c} \tag{3}$$

$$s = \sin i = x\left(\frac{P_{\rm b}}{2\pi}\right)^{-2/3} T_{\odot}^{-1/3}M^{2/3}m_{\rm c}^{-1} \tag{4}$$

$$\gamma = e\left(\frac{P_{\rm b}}{2\pi}\right)^{1/3} T_{\odot}^{2/3}M^{-4/3}m_{\rm c}\left(m_{\rm p}+2m_{\rm c}\right), \tag{5}$$

where, $T_{\odot} = GM_{\odot}/c^3$ and $M = m_1 + m_2$ is the total mass of the binary. We note that the former set of equations refers to the intrinsic pK effects that can be extracted from the measured values after taking into account kinematic corrections.

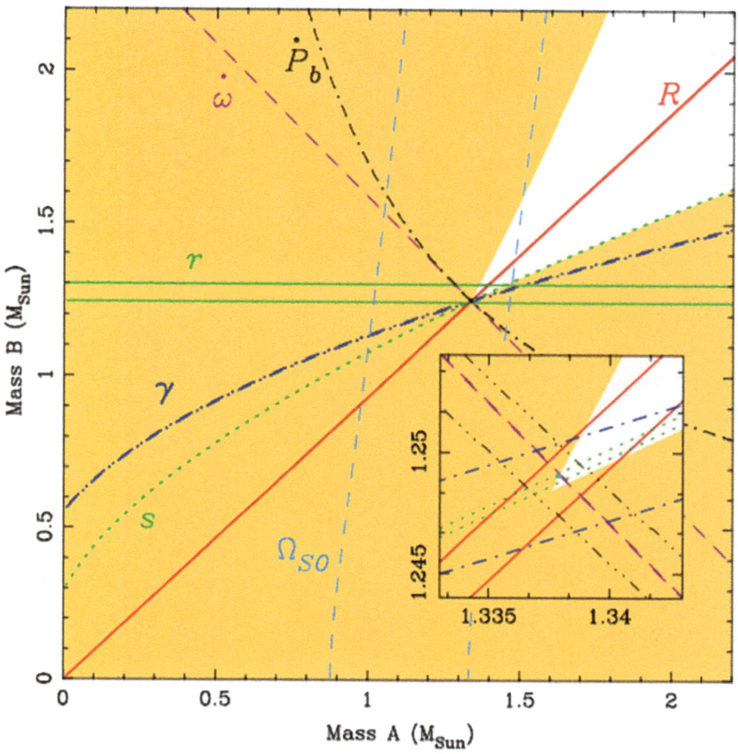

Fig. 2 Post-Keplerian effects and corresponding 1σ uncertainties for the double pulsar system PSR J0737−3039 as of 2011 (courtesy of Michael Kramer)

Equations (2)–(5) relate observable, model-independent quantities with the a priori unknown masses. Whenever two pK parameters become measurable the masses are uniquely determined, of course withing the framework of a particular gravity theory. Thus, additional pK parameters become tests of the gravity theory, since their values are fixed from the already determined masses (see Fig. 2 for a graphical example in the case of the double pulsar).

2.3 *Gravitational Wave Emission*

Gravitational wave emission from accelerating masses is expected for theories in which gravity propagates with a finite speed. As we shall see below, detection and characterization of GWs is of fundamental importance for gravity tests as the radiative properties are a sensitive probe of the underlying gravity theory. In GR, the loss of energy to GWs first enters the binary motion as a quadrupole term at the

2.5 post-Newtonian level ($\mathcal{O}(v^5/c^5)$). The associated orbital-period decay, averaged over one orbit reads:

$$\dot{P}_{\rm b} = -\frac{192\pi}{5}\left(1+\frac{73}{24}e^2+\frac{37}{96}e^4\right)\left(1-e^2\right)^{-7/2}\left(\frac{2\pi\mathcal{M}}{P_{\rm b}}\right)^{5/3}, \tag{6}$$

where $\mathcal{M} = (m_{\rm p}m_{\rm c})^{3/5}(m_{\rm p}+m_{\rm c})^{-1/5}$ is the so-called chirp mass of the system.

In addition to the emission of GWs, the observed orbital decay of a binary pulsar can generally be affected by several "classical" terms. These effects are either apparent (kinematic) or intrinsic to the system.

Kinematic effects include a secular change due to the Galactic gravitational potential and the proper motion of the pulsar μ on the sky:

$$\delta\dot{P}_{\rm b}^{\rm kin} = \frac{P_{\rm b}}{c}\left[\hat{\mathbf{K}}_0\left(\mathbf{g}_{\rm PSR}-\mathbf{g}_{\odot}\right)+\mu d^2\right], \tag{7}$$

where $\hat{\mathbf{K}}_0$ is the unit vector pointing from the Earth to the pulsar.

Intrinsic orbital-period variations due to "classical" effects may arise from:

1. Changes in the companion star's quadrupole moment (Lazaridis et al. 2011; Archibald et al. 2013) that become apparent due to spin-orbit coupling,
2. mass losses either from the pulsar or the companion, $\dot{P}_{\rm b}^{\dot{M}} = 2\dot{M}/M$, where $M = m_{\rm p} + m_{\rm c}$ (Antoniadis et al. 2013) and
3. an orbital decay caused by tides:

$$\dot{P}_{\rm b}^{\rm T} = \frac{3k\Omega_{\rm c}}{2\pi q(q+1)}\left(\frac{R_{\rm c}P_{\rm b}\sin i}{xc}\right)^2\frac{1}{\tau_{\rm s}} \tag{8}$$

where $\Omega_{\rm c}$ is the angular velocity of the companion, $\tau_{\rm s} = \dot{\Omega}_{\rm c}/\Omega_{\rm c}$ is the synchronization timescale and k is related to the companion's moment of inertia, $k \equiv I_{\rm c}/(m_{\rm c}R_{\rm c}^2)$.

For the systems discussed here these terms can either be taken into account or are much bellow the measurement uncertainties.

From Eq. (6) we see that the back-reaction to the orbit depends sensitively on the orbital period and less so on the eccentricity and the stellar masses. As of today, $\dot{P}_{\rm b}^{\rm intrinsic}$ has been detected in 11 binary pulsars and a double-WD system (Table 1). All of these binaries have orbital periods $\leq$1 day and span a broad range of masses and eccentricities. While all systems are interesting in their own regard, only few systems stand out as laboratories for strong gravity. The remaining of this text focuses on those, following a somewhat historical order.

3 The Hulse-Taylor Binary and the Double Pulsar

The orbital decay due to emission of GWs was first detected in PSR B1913+16, the "original" binary pulsar discovered by Russel Hulse and Joe Taylor in the summer of 1974. The pulsar orbits an unseen NS every 17 h with an eccentricity of $e = 0.68$. Soon after the discovery, the system's periastron was found to change by $0.1\,\mathrm{deg\,yr^{-1}}$. Assuming the validity of GR, this change corresponds to a total mass of the system, $M = 2.7\,\mathrm{M}_\odot$. Together with γ, also measured soon after, the masses of the pulsar and the companion are constrained to $\sim$1.33 and $1.34\,\mathrm{M}_\odot$ respectively. The detection of the orbital period decay was announced a few years later by Weisberg and Taylor (1981). The measured value was in agreement with Einstein's quadrupole formula, providing direct experimental evidence for the existence of GWs. Perhaps not surprisingly, this result earned Hulse and Taylor a ticket to Stockholm in 1993.

The most recent value of $\dot{P}_\mathrm{b}$ for the system agrees with GR at the 1 % level. While the experimental accuracy continues to increase with time, the precision of this radiative test is not expected to improve significantly. The main reason is the unknown distance to the binary which limits the ability to correct the observed $\dot{P}_\mathrm{b}$ for the apparent kinematic contributions.

Today we know of a more "extreme" system, that outshines the Hulse-Taylor binary in all aspects. PSR J0737−3039A/B, also known as the double pulsar, was discovered in a survey of the southern sky with the Parkes telescope in 2003 (Burgay et al. 2003; Lyne et al. 2004). The system has an orbital period of 2.45 h and an eccentricity of $e = 0.088$ and it is viewed at a remarkably edge-on inclination of $i = 88.7°$. Furthermore, both NSs are seen as pulsars allowing to infer the theory-independent mass ratio from the observed Roemer delays. In total six pK parameters are measured; five in the timing data of pulsar A (Kramer et al. 2006) and an additional from the effects of geodesic precession of pulsar B (Breton et al. 2008). These quantities, together with the mass-ratio allow for $7 - 2 = 5$ independent tests of gravity, which all happen to agree with GR.

Because of its proximity to Earth, its distance is directly measured with Very-Long Baseline interferometry (Deller et al. 2009). This allows to correct the observed $\dot{P}_\mathrm{b}$ for kinematic contributions with much better precision compared to the Hulse-Taylor binary. The latest published value (Kramer et al. 2006) agrees with GR at the 1.4 % level, while the current unpublished result reaches a precision well bellow 0.1 %. These results will be published in an upcoming publication by Kramer et al. The double pulsar provides by far the best verification of Einstein's quadrupole formula. Furthermore, the high-precision of the experiment rules out a range of alternatives such as Beckenstein's TeVeS (see Sect. 9).

4 Beyond the Double Pulsar

The orbital decay of J0737−3039 leaves very little room for deviations from Einstein's quadruple formula. Unfortunately, this doesn't necessarily guarantee the absolute validity of GR for the energy scales probed by orbiting NSs—many alternatives to GR *also* predict the same rate of GW emission, even if NSs behave very differently. In many (metric) alternatives, the gravitational interaction is mediated not only by a tensor field but also by additional fields, for example scalar or vector. These fields usually break local symmetries leading to a SEP violation and/or a breaking of the Local Lorenz Invariance for the gravity sector. Such effects lead to a number of modifications to the binary orbit. For the class of compact binaries discussed here the most significant effect is the emission of *dipolar* gravitational radiation, on top of the quadrupole and higher-order terms, also present in GR. The associated orbital decay is a $\mathcal{O}(v^3/c^3)$ pK effect that reads (Damour and Esposito-Farese 1996a; Wex 2014):

$$\dot{P}_{\rm b}^{\rm dipole} = -\frac{4\pi^2 G_*}{c^3 P_{\rm b}} \frac{m_{\rm p} m_{\rm c}}{m_{\rm p} + m_{\rm c}} (\alpha_{\rm p} - \alpha_{\rm c})^2 \tag{9}$$

where G_* is the effective gravitational constant and other symbols have their usual meaning. The terms $\alpha_{\rm p}$ and $\alpha_{\rm c}$, often referred to as "charges", depend on the coupling strength of the extra fields to matter inside the stars and usually scale non-linearly with the binding energy. Obviously in GR, $\alpha_{\rm p} = \alpha_{\rm c} = 0$, but in some alternative theories they can reach unity (Damour and Esposito-Farese 1993; Esposito-Farese 2004). Similarly to the quadrupole term, dipole radiation depends sensitively on the orbital period. A fundamental difference however is that the associated orbital decay depends on $\alpha_{\rm p} - \alpha_{\rm c}$, in other words, the difference between the binding energies of the pulsar and its companion. It should now be obvious why double NSs make poor laboratories for testing the existence of this effect. In systems like J0737−3039, the compositions of the two stars are almost identical, meaning that $(\alpha_{\rm p} - \alpha_{\rm c})^2 \simeq 0$, even if $\alpha_{\rm p}$ is significantly larger that zero. Therefore, dipole radiation is not expected for DNSs, even if there is an underlying modification to gravity that forces stars to respond differently to gravity compared to GR.

Evidently, the best laboratories to test the predictions of these theories are tight binaries where the companion star is expected to have $\alpha_{\rm p} \neq \alpha_{\rm c}$. Such asymmetric systems are in fact not rare. The advent of recent radio surveys has reveled a rich ensemble of binary pulsars with main-sequence, semi-degenerate and WD companions. Most of these systems however either have long orbital separations or their orbital motion is affected by classical phenomena, such as movement of material due to irradiation, tides and quadrupole-moment variations. A fortunate exception to this picture are pulsars with WD companions. WDs have extremely small gravitational binding energies compared to NSs, but yet they are "compact" enough to be treated as point masses. Presently we know five pulsar—WD binaries with orbital periods less than a day. These systems are the end-point of evolution of stars that survived a common envelope phase, followed by a long period of stable mass transfer. As we shall see bellow their formation—which is not yet fully understood—gives them a set of salient orbital and intrinsic properties that make them exquisite labs for strong-field gravity. We should also note that the *ideal* system to test dipolar GW emission would be a pulsar orbited by a black hole. Unfortunately, such a system is yet to be found.

5 Low-Mass White Dwarfs in Binaries

Low-mass white dwarfs (LMWDs) represent the final evolutionary product of stars with masses $\leq 0.4\,M_{\odot}$ that cannot evolve beyond the helium flash. Single stars with such low mass evolve extremely slow and have in fact not yet settled on the WD cooling track. However, formation of LMWDs is still possible in binaries where higher-mass stars can loose significant amounts of matterial through their life. LMWDs are the most common counterparts to millisecond pulsars, a fact that strongly supports that the high-angular momentum of these NSs is a consequence of mass and angular momentum transfer from the LMWD progenitors. These binaries possess numerous interesting properties which are the subject of extensive studies over the past few decades (e.g. Driebe et al. 1998; Tauris and Savonije 1999; Serenelli et al. 2002; Panei et al. 2007). The properties most to gravity tests are the following:

1. **LMWDs are not exactly WDs.** Textbook WDs are objects where all nuclear-fusion reactions have ceased and the entire stellar luminosity is provided by the latent heat of the core. For most LMWDs, this is not the case. When these stars settle on their cooling tracks they may still have an extensive amount of hydrogen surrounding their degenerate helium cores (Driebe et al. 1998). In most cases the pressure provided by the envelope leads to conditions that allow residual hydrogen burning close to the degenerate core. Nuclear burning becomes the dominant energy and allows the stars to evolve on nuclear rather than thermal timescales, which is the case for higher-mass WDs. For gravity tests this is important because it allows to study some of these stars spectroscopically and infer their masses.
2. **LMWDs flash.** The initial envelope mass depends mainly on the orbital separation during detachment, i.e. it is a strong function of the stellar mass. The less massive the WD the larger the fractional size of the envelope. Therefore, because the size of the nuclear burning region also depends on the envelope size, there exists a range of masses for which nuclear burning is unstable. This thermal instability develops because the shell is not thick enough to compensate sudden changes in temperature by regulating its density, like normal stars do. For a shell of thickness $d = r - r_0 \ll R$, fractional changes in temperature and density are related via:

$$\frac{\delta T}{T} = \frac{1}{\chi_T}\left(4\frac{d}{r} - \chi_\rho\right)\frac{\delta\rho}{\rho} \tag{10}$$

where $\chi_T > 0$ and χ_ρ are the logarithmic pressure gradients under constant temperature and density respectively. From Eq. (10) it follows that a shell is thermally unstable when $d < r\chi_\rho/4$. In LMWDs the most severe flashes are those induced by unstable CNO burning (rate $\propto T^{18}$) shortly before the star enters the cooling track. CNO flashes may consume large amounts of hydrogen resulting to complete cessation of nuclear reactions afterwards. Furthermore, during a flash episode, the radius of the star may expand beyond its Roche lobe causing further decrease of its mass. Modern stellar evolution calculations place the lower mass limit for CNO-induced flashes between 0.17 and 0.21 $M_\odot$, however the exact value depends on the envelope metallicity, rate of gravitational settling and treatment of microphysics. Unfortunately knowledge of the critical mass value is crucial for determining LMWD masses and is one of the main limiting factors for gravity tests. Recent discoveries of high-temperature LMWDs with masses <0.2 $M_\odot$ strongly suggest that the threshold for flashes is indeed above this value (see e.g. Brown et al. 2010). This is further supported by observations of the system PSR J1909−3744 for which the mass and radius are measured independently (Antoniadis 2013). Hence, assuming that the mass threshold for flashes is universal, it is probably safe to conclude that all LMWDs with $m_c < 0.2\,M_\odot$—have thick hydrogen envelopes.
3. **Pulsars with LMWD companions have extremely circular orbits.** A final remark is that tidal dissipation during the mass-transfer phase circularizes the orbit on very short timescales. For gravity tests this becomes an issue because it renders most pK effects non-measurable. Thus, masses can be inferred from timing only for orbits sufficiently edge on, which is not the case for any of the known binaries. Consequently alternative mass-determination methods should be applied, some of which are further discussed in the following section.

6 White Dwarf Masses

Strong-field gravity tests with binary pulsars require knowledge of the stellar masses independently of any particular theory of gravity. For compact binaries with low eccentricities the only way to achieve this is through timing is by measuring the Shapiro-delay signature of the pulsar signal as it passes through the (weak) gravitational potential of the companion. Since, this requires a sufficiently edge-on viewing, it can only be applied to very limited cases. In fact, non of the known relativistic binaries has a sufficiently edge-on inclination to allow for a clear measurement of the Shapiro delay. However, because LMWDs stay hot and bright for a significant fraction of their lifetimes, one can use methods based on optical observations to infer their masses.

The most straightforward technique involves phase-resolved spectroscopy of the WD. Doppler measurements of the WD's Balmer lines along the orbit, trace the orbital motion, the amplitude of which is directly proportional to the size of the WD orbit. Because the size of the pulsar's orbit is also measured via radio timing, the ratio of the two quantities gives the mass ratio, q of the system. Furthermore, one can also compare the observed spectrum with atmospheric models. Atmospheric Balmer lines have shapes and depths that can be modeled using only two parameters, the effective temperature $T_{\rm eff}$ and surface gravity $g = Gm_{\rm c}/R_{\rm WD}^2$ (Koester 2008). Once these two quantities are determined the WD mass and radius can be inferred using an appropriate finite-temperature mass-radius relation (see Sect. 6). An additional side-product is the measurement of the systemic radial velocity γ, which together with the proper motion of the system and the distance, make possible to infer the 3D motion of the binary in the Galaxy. A drawback, apart from the dependence on theoretical mass-radius relations, is the usage of model atmospheres. Although LMWDs are extremely simple, atmospheric models based on 1D calculations are known to yield biased values for the surface gravity if the outer envelope is convective (Antoniadis 2013). However, errors can be accounted for with appropriate 3D models (Tremblay et al. 2013) and vanish completely for hot, radiative atmospheres.

Another method that does not depend on atmospheric models relies on the parallax measurement of the system (Antoniadis et al. 2012). Once the distance is known with sufficient precision, one can infer the absolute stellar luminosity. Together with the temperature this yields the WD radius, which again combined with a mass-radius relation, give the WD mass. This method has not yet yield any precise mass measurements, but it may be proven useful in the near future.

The first compact binary studied with spectroscopy was PSR J1012+5307, a 5.2 ms pulsar in a 14.4 h orbit. Because of the large uncertainties in atmospheric models at that time, the mass was pinned down with a precision of only ~12 % (van Kerkwijk et al. 1996; Callanan et al. 1998). Nevertheless, together with a ~1 % detection of the orbital period decay, which is agreement with the prediction of GR (Lazaridis et al. 2009), the system places a constraint on dipole radiation of $\dot{P}_{\rm b}^{\rm dipole} = (-0.2 \pm 1.6) \times 10^{-14}$. Presently the system is superseded by two more compact systems, discussed in the remaining of this chapter.

7 Constraints on Dipolar Gravitational Radiation

7.1 PSR J1738+0333, the Most Stringent Constraints on Dipolar Radiation

PSR J1738+0333 is a 5.9 ms pulsar in a 8.5 h circular orbit around a LMWD. It was discovered in 2001 with the Parkes telescope in Australia and is being timed ever since with Arecibo, Effelsberg and Westerborg telescopes, on a monthly basis. The long baseline of precise timing observations yields highly precise measurements of the spin, kinematic and orbital parameters, including the parallax and systemic proper motion (Freire et al. 2012b). These quantities allow to compute the Shklovskii contribution to the orbital decay, $\dot{P}_{\rm b}^{\rm Shk} = 8.3^{+0.6}_{-0.5}\,{\rm fs\,s^{-1}}$. Together with the measured orbital decay $\dot{P}_b = -17.0 \pm 3.3\,\mathrm{fs\,s^{-1}}$, this implies that the intrinsic orbital period decay is $\dot{P}_b^{\rm int} = -25.9 \pm 3.2\,\mathrm{fs\,s^{-1}}$.

Optical observations yield a mass ratio of $q = 8.1 \pm 0.2$, a WD mass of $m_c = 0.181^{+0.007}_{-0.005}\,\mathrm{M_\odot}$ and a pulsar mass of $1.43\,\mathrm{M_\odot}$ (Antoniadis et al. 2012). The latter is based on the assumption that the WD envelope is thick, which is most likely the case, given the high temperature and the slightly lower mass compared to the WD in PSR J1909−3744. Given the masses, the rate of orbital decay according to GR (Eq. (6)) should be $-27.7^{+1.5}_{-1.9}\,\mathrm{fs\,s^{-1}}$ which is in agreement with the observed value. Since both stars can be treated as point particles (Freire et al. 2012b), the "excess" orbital decay of $+2.0^{+3.7}_{3.6}$,$\mathrm{fs\,s^{-1}}$ directly translates to a constraint on deviations from the quadruple formula. Apart from dipole radiation, these may already enter at the Newtonian level, due to a varying gravitational constant $\dot{G}$. Lunar Laser Ranging experiments yield a limit of $\dot{G}/G = (0.7 \pm 3.8) \times 10^{-13}\,\mathrm{yr^{-1}}$, thus $\dot{P}_b^{\dot{G}} = -2P_b\dot{G}/G = (+0.14 \pm 0.74)\,\mathrm{fs\,s^{-1}}$. Therefore, the limit on dipole radiation from this test is $\dot{P}_b^{\rm dipole} = 1.9^{+3.8}_{-3.7}\,\mathrm{fs\,s^{-1}}$, the most stringent obtained so far. Since GR predicts $\dot{P}_b^{\rm dipole} = 0$, the theory passes the test. As we shall see below, this is not the case for most alternative theories.

7.2 *PSR J0348+0432, a Massive Pulsar in the Most Compact Binary*

This extraordinary system was discovered in a drift-scan survey (Boyles et al. 2012; Lynch et al. 2012) conducted with the Green-bank Telescope in West Virginia. The pulsar appears to be "mildly recycled", with a spin period of 39 ms and relatively strong surface magnetic field of $B = 3 \times 10^9$ Gauss. The NS orbits a LMWD companion every 2.45 h. The orbital period—by far the shortest among this type of systems—is only 15 s longer than that of the double pulsar. Consequently, the orbital period decays relatively fast and it is easier to measure precisely on short timescales.

Spectroscopic observations of the WD companion (Antoniadis et al. 2013) were conducted with the FORS2 instrument of the VLT in Chile. The data show an effective temperature of ~10,100 K and a surface gravity of $\log g \sim 6.0$ dex. Independently of any model, the high temperature constraint the WD mass to a narrow interval: the star cannot be less massive than $\sim 0.165\,\mathrm{M_\odot}$ as WDs with lower masses never reach such high temperatures on the cooing track. Furthermore, if its hydrogen envelope were thin, it would have cooled to this temperature in only a few Myrs, which would imply that there is a very high abundance of these stars, given the 13 Gyrs age of the Universe. Using appropriate cooling models with thick envelopes the WD mass is constrained to 0.172 ± 0.003. Combined with the measured mass ratio, $q = 10$, the inferred pulsar mass is 2.01 ± 0.04, the highest NS mass measured until the time of writing.

High NS masses are important for constraining the EoS of supra-nuclear matter, as many models place the dividing line between NSs and black holes at lower values. A two-solar mass NS is also important for gravity as it has a gravitational binding energy ~50 % higher than lower-mass NS like PSR J1738+0333. As the magnitude of deviations from GR can depend non-linearly on the binding energy, these stars can be used to probe a strong-field gravity regime which is qualitatively very different compared to other binary-pulsar experiments (Fig. 3).

Since April 2011 the system is being monitored monthly with Arecibo and Effelsberg. To match the arrival times, the orbital solution requires a significant rate of orbital decay, $\dot{P}_{\rm b}^{\rm obs} = (-273 \pm 45)\,{\rm fs\,s^{-1}}$. The former is largely unaffected by kinematic contributions which are an order of magnitude bellow the measurement error. The GR prediction given the masses is $-258^{+8}_{-11}\,{\rm fs\,s^{-1}}$, in agreement with the observed rate. The 20 % precision of this GR test is quantitatively poor compared to the remarkable precision achieved for e.g. the double pulsar. However, because of the high pulsar mass, the resulting constraints on some alternative theories are infinitely more stringent.

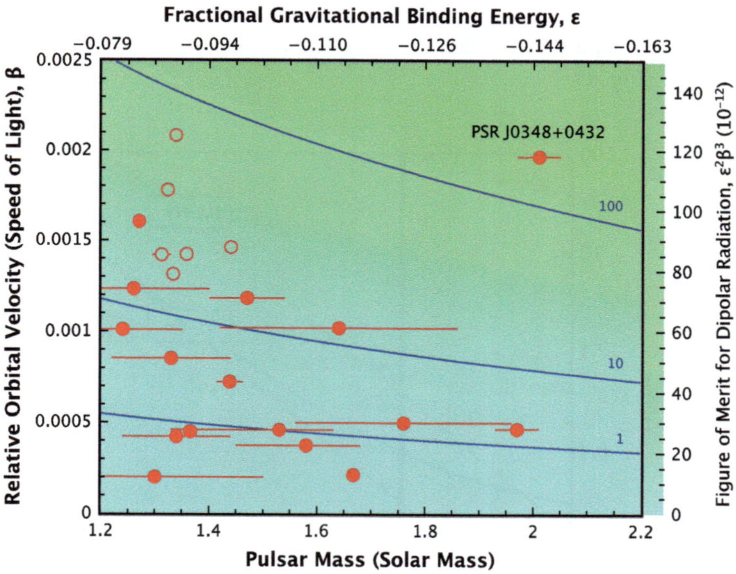

Fig. 3 Mean orbital velocity as a function of NS mass for the compact binary pulsars of Table 1. *Hollow circles* represent double NS systems where no significant dipolar emission is expected. PSR J0348+0432 stands out as the most "luminous" emitter

8 Ramifications for the Phase Evolution of Neutron Star Mergers

In-spiralling compact binaries will be the main source of GW detectable by the forthcoming ground-based detectors. The GW signal from merger events will be deeply buried in the detectors' noise, requiring matched filtering techniques for extraction. Ideally, theoretical templates used for detection should track the phase evolution of the merger with a precision better than 1 cycle over the $\sim 10^4$ orbits covering the detectors' band. Presently, models calculated by various groups have reached a precision of $(v/c)^7$, with the current forefront of research being pushing towards the fourth post-Newtonian order (Blanchet 2014). However if the properties of gravity differ from the GR prediction, loss of energy to dipole GWs may drive the merger many cycles away from the templates, jeopardizing detection. This is true even for double NSs that have small differences on their binding energies. The limits on dipolar radiation imposed by PSRs J1738+0333 and J0348+0432 imply $|\alpha_p - \alpha_0| \leq 0.004$ for the hole range of NSs observed in nature. Consequently, mergers cannot loose more than half a cycle compared to GR, supporting the use of GR templates. We should note that this may not be true if the partner particle to graviton is massive, although PSR J0348+0432 already constraints its mass above $\sim 10^{-13}\,\mathrm{eV/c^2}$ (Fig. 4).

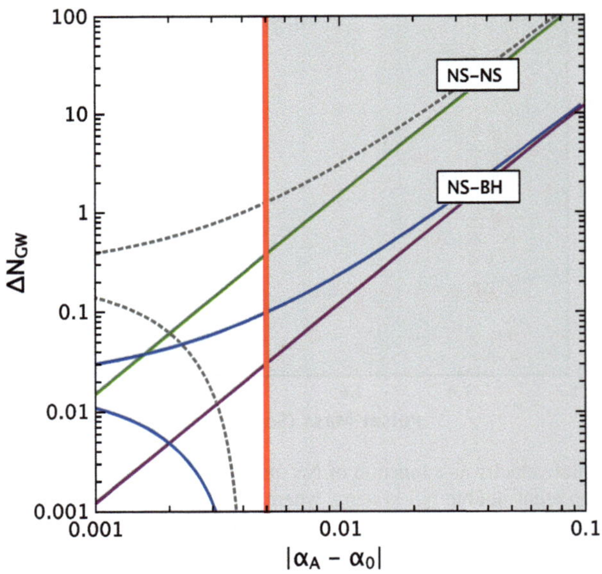

Fig. 4 Maximum phase offset in the band 20 Hz to few kHz between the GR template and the true in-spiral in the presence of dipolar radiation, as a function of the scalar coupling strength. The two examples show the merger of a $2\,M_\odot$ NS with a $1.25\,M_\odot$ NS (*green*) and a $2\,M_\odot$ NS with a $10\,M_\odot$ black hole. The *gray area* is excluded by the limits from binary pulsars (for details see Antoniadis et al. 2013) (Color figure online)

9 Constraints on Alternative Theories of Gravity

9.1 Scalar-Tensor Gravity

Scalar-Tensor theories of gravity are the most natural extensions of GR, where the gravitational interaction involves an additional scalar-field contribution ϕ, in addition to the spin-2 graviton. In these theories, matter is assumed to be coupled to a physical metric $\hat{g}_{\mu\nu} \equiv A^2(\phi)g_{\mu\nu}$, where $A(\phi)$ is an generic function of the matter-scalar coupling. Since $A(\phi)$ is a-priori unknown, it is useful to expand it around the cosmologically imposed scalar field ϕ_0 as $\ln A(\phi) = \ln A(\phi_0) + \alpha_0(\phi - \phi_0) + 1/2\beta_0(\phi - \phi_0)^2 + \ldots$ (Damour and Esposito-Farese 1992, 1993, 1996a, b, 1998; Esposito-Farese 1999, 2004). Here, α_0 and β_0 denote the linear and quadratic coupling terms respectively and higher-order terms are neglected. This notation allows to embed GR and other well-studied theories in the space of scalar-tensor alternatives. For example in GR, $\alpha_0 = \beta_0 = 0$ and in the Jordan, Fierz, Brans-Dicke theory, $a_0^2 = 1/(2\omega_{\rm BD} + 3)$, $\beta_0 = 0$ (Wex 2014). In the strong-gravity potential of a NS, α_0 and β_0 are modified by strong-gravity effects and become body-dependent "charges" $\alpha_{\rm p,c}$ which can be computed by means of numerical integration of the modified hydrostatic equation, assuming an EoS. For binary pulsars with WD companions $\alpha_{\rm c} \simeq \alpha_0$ which is constrained below $\sim 4 \times 10^{-4}$ by Solar-System experiments. Therefore, limits on dipolar radiation can be viewed as direct constraints on $\alpha_{\rm p}$ which can rich unity for specific values of α_0, β_0 due to non-perturbative effects (Esposito-Farese 2004).

Figure 5 shows the limits on α_0, β_0 imposed by binary pulsar experiments and various Solar-System tests. For most of the parameter space, PSR J1738+0333

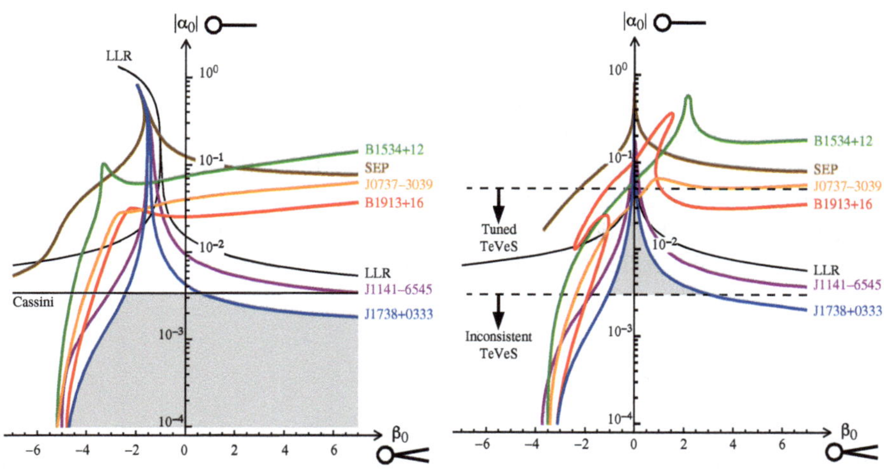

Fig. 5 *Left:* Constraints on the linear and quadrupole scalar-matter coupling in Scalar-Tensor Gravity. *Coloured lines* show various binary pulsar and Solar-System experiments. The *shaded area* is the regime still consistent with all experiments. *Right:* Constraints on scalar-matter coupling in the TeVeS framework (Color figure online)

posses the most significant constraints. This is especially true for negative values of β_0, where non-perturbative effects are more prominent. For the original Brance-Dicke scalar-tensor theory the Cassini bound is still more constraining by a factor of $\sim$2, but this may change as more data for PSR J1738+0333 become available.

Because of its high mass, PSR J0348+0432 provides a complementary constraint for negative values of β_0 still consistent with PSR J1738+0333. Due to the high gravitational binging energy inside the NS, α_p may assume order-unity values even for background values of α_0, orders of magnitude below the constraints that can be obtained in the Solar system. The timing data impose a limit of $|\alpha_p - \alpha_c| < 5\times10^{-3}$, therefore excluding "spontaneous scalarization" even for 2 $M_\odot$ NSs.

Similar constrains are obtained for certain $f(\mathbf{R})$ theories that have very similar behaviour (De Laurentis and De Martino 2013a, b).

9.2 TeVeS

Tensor-Vector-Scalar (TeVeS) gravity is a relativistic formulation of Modified Newtonian Dynamics (MOND) that seeks to explain galaxy rotation curves and week lensing without invoking Dark Matter (Milgrom 1983). In MOND this is achieved by modifying Newton's equation below a fundamental acceleration scale $\alpha_0 \simeq 10^{-10}\,\mathrm{m\,s}^{-10}$. In TeVeS gravity is mediated by tensor, vector and scalar fields. The scalar-field kinetic term is an unknown non-linear function, which has a natural shape only if $|\alpha_0| > 0.05$, otherwise the theory would be inconsistent with Solar-System constraints (Freire et al. 2012b). Below this value the function becomes highly fine-tuned and completely unnatural (bi-valued) for $|\alpha_0| < 0.003$. If one assumes a generalized scalar-matter coupling just like in scalar-tensor theories, binary pulsar experiments impose the limits shown in the right panel of Fig. 5 (Freire et al. 2012b). For most of the parameter space, the theory predicts significant amounts of dipolar radiation and therefore PSR J1738+0333 posses the most stringent constraints. For the original theory however where $A(\phi) = \exp(\alpha_0\phi)$ there is no low-order modification to the gravitational radiation damping. However, modifications still enter the orbital dynamics through other pK effects, thus the much more precise (published) double-pulsar data pose the most stringent constraints. As it can be seen in Fig. 5, TeVeS is already confined within the "unnatural" regime.

9.3 Local Lorenz Invariance

Local Lorenz Invariance violations (LLIV) in the gravity sector would introduce a preferred frame at each spacetime point, that would alter the orbital dynamics of binary pulsars. In the parametrized post-Newtonian framework, LIV is characterized by two parameters $\alpha_{1,2}$. In the general case where the binary moves relative to the preferred frame with velocity $\mathbf{w}$, α_1 would induce a "rotating" eccentricity and α_2 a precession of the projected semi-major axis, x around $\mathbf{w}$ (Shao and Wex 2012). Assuming that the preferred frame coincides with the Cosmic-Microwave background, PSRs J1738+0333 and J1012+5307 placed a stringent limit of $\alpha_1 = 0.4^{+3.7}_{-3.1} \times 10^{-5}$ (Shao and Wex 2012), which represents a significant improvement over Solar-System tests. The same binaries limit $|\alpha_2|$ below 1.8×10^{-4} (Shao and Wex 2012). This parameter is in fact better constrained from studies of the profile stability in solidary pulsars ($|\alpha_2| < 1.6 \times 10^{-9}$, Shao et al. 2013). LIV would also result to emission of dipolar waves. Several studies have used the limits from the aforementioned binaries to place stringent constraints on LLIV theories such as Einstein-Æther and khronometric gravity (Yagi et al. 2014a, b).

10 Final Remarks

Present timing experiments have reached remarkable sensitivities and place stringent constraints on a wide range of possible deviations from GR. Binary pulsars are often regarded as tests of gravity at the binary-separation scale. While this is certainly true for theories like GR, a wide-range of possible strong-field deviations in the non-linear environment of the NS interior could propagate over large distances, making binary pulsars sensitive (and thus far unique) laboratories for strong gravity. This is especially true when considering GW emission which was the focus of this chapter.

In view of the stringent limits summarized here and the anticipated operation of advanced ground-based detectors, it may be tempting to ask whether if it is necessary to continue searching for modified gravity in pulsar experiments. In the author's view, the answer is positive, both for philosophical and practical reasons. GR, is one of the fundamental pillars of modern physics and therefore deserved to be tested in every way possible. Since the magnitude and scale of possible deviations is a-priori unknown, it makes sense to both continue improving the sensitivity of current experiments and also seek additional systems able to provide complementary tests. For example, improving the precision of the double pulsar experiment may provide tests of the orbital dynamics at higher pK orders. Similarly, systems like a pulsar-black hole binary or the newly discovered triple pulsar (Freire et al. 2012a; Ransom et al. 2014) may provide a many-fold sensitivity improvement in the search for SEP violations and novel tests of other fundamental predictions like the no-hair theorem (Liu et al. 2012).

Acknowledgements I am grateful to the BEACON project (ERC Grant: FP7-IDEAS-ERC) for financial support throughout this research. I also wish to thank Paulo Freire, Norbert Wex and Michael Kramer for thorough discussions, insightful comments and ideas.

References

J. Antoniadis, C.G. Bassa, N. Wex, M. Kramer, R. Napiwotzki, MNRAS **412**, 580 (2011). doi:10.1111/j.1365-2966.2010.17929.x

J. Antoniadis, M.H. van Kerkwijk, D. Koester, P.C.C. Freire, N. Wex, T.M. Tauris, M. Kramer, C.G. Bassa, MNRAS **423**, 3316 (2012)

J. Antoniadis, P.C.C. Freire, N. Wex, T.M. Tauris, R.S. Lynch, M.H. van Kerkwijk, M. Kramer, C. Bassa, V.S. Dhillon, T. Driebe, J.W.T. Hessels, V.M. Kaspi, V.I. Kondratiev, N. Langer, T.R. Marsh, M.A. McLaughlin, T.T. Pennucci, S.M. Ransom, I.H. Stairs, J. van Leeuwen, J.P.W. Verbiest, D.G. Whelan, Science **340**, 448 (2013). doi:10.1126/science.1233232

J. Antoniadis, Multi-wavelength studies of pulsars and their companions. Ph.D. thesis, University of Bonn, 2013

A.M. Archibald, V.M. Kaspi, J.W.T. Hessels, B. Stappers, G. Janssen, A. Lyne, ArXiv e-prints (2013)

B. Bertotti, L. Iess, P. Tortora, Nature **425**, 374 (2003). doi:10.1038/nature01997

N.D.R. Bhat, M. Bailes, J.P.W. Verbiest, Phys. Rev. D **77**(12), 124017 (2008)

L. Blanchet, Living Rev. Relat. **17**, 2 (2014). doi:10.12942/lrr-2014-2

J. Boyles, R.S. Lynch, S.M. Ransom, I.H. Stairs, D.R. Lorimer, M.A. McLaughlin, J.W.T. Hessels, V.M. Kaspi, V.I. Kondratiev, A. Archibald, A. Berndsen, R.F. Cardoso, A. Cherry, C.R. Epstein, C. Karako-Argaman, C.A. McPhee, ArXiv: 1209.4293 (2012)

R.P. Breton, V.M. Kaspi, M. Kramer, M.A. McLaughlin, M. Lyutikov, S.M. Ransom, I.H. Stairs, R.D. Ferdman, F. Camilo, A. Possenti, Science **321**, 104 (2008). doi:10.1126/science.1159295

W.R. Brown, M. Kilic, C. Allende Prieto, S.J. Kenyon, ApJ **723**, 1072 (2010). doi:10.1088/0004-637X/723/2/1072

W.R. Brown, M. Kilic, J.J. Hermes, C. Allende Prieto, S.J. Kenyon, D.E. Winget, ApJ **737**, L23 (2011)

M. Burgay, N. D'Amico, A. Possenti, R.N. Manchester, A.G. Lyne, B.C. Joshi, M.A. McLaughlin, M. Kramer, J.M. Sarkissian, F. Camilo, V. Kalogera, C. Kim, D.R. Lorimer, Nature **426**, 531 (2003). DOI 10.1038/nature02124

P.J. Callanan, P.M. Garnavich, D. Koester, MNRAS **298**, 207 (1998). doi:10.1046/j.1365-8711.1998.01634.x

I. Ciufolini, E.C. Pavlis, Nature **431**, 958 (2004). doi:10.1038/nature03007

T. Damour, N. Deruelle, Ann. Inst. H. Poincare (Physique Theorique) **43**, 107 (1985)

T. Damour, N. Deruelle, Ann. Inst. H. Poincare (Physique Theorique) **44**, 263 (1986)

T. Damour, G. Esposito-Farese, Class. Quantum Gravity **9**, 2093 (1992)

T. Damour, G. Esposito-Farese, Phys. Rev. Lett. **70**, 2220 (1993). doi 10.1103/PhysRevLett.70.2220

T. Damour, G. Esposito-Farese, Phys. Rev. D **53**, 5541 (1996a). doi 10.1103/PhysRevD.53.5541

T. Damour, G. Esposito-Farese, Phys. Rev. D **54**, 1474 (1996b)

T. Damour, G. Esposito-Farese, Phys. Rev. D **58**(042001), 1 (1998)

T. Damour, J.H. Taylor, ApJ **366**, 501 (1991)

M. De Laurentis, I. De Martino, astrp-ph:1310.0711 (2013)

M. De Laurentis, I. De Martino, MNRAS **431**, 741 (2013). doi:10.1093/mnras/stt216

A.T. Deller, M. Bailes, S.J. Tingay, Science **323**, 1327 (2009). doi:10.1126/science.1167969

T. Driebe, D. Schoenberner, T. Bloecker, F. Herwig, A&A **339**, 123 (1998)

F.W. Dyson, A.S. Eddington, C. Davidson, R. Soc. Lond. Philos. Trans. Ser. A **220**, 291 (1920). doi:10.1098/rsta.1920.0009

A. Einstein, Sitzungsber. preuss. Akad. Wiss. **47**(2), 831–839 (1915a)

A. Einstein, *Sitzungsberichte der Königlich Preußischen Akademie der Wissenschaften (Berlin)* [Seite 844–847] (1915b), pp. 844–847

A. Einstein, Ann. Phys. **354**, 769 (1916). doi:10.1002/andp.19163540702

G. Esposito-Farese, Binary-pulsar tests of strong-field gravity and gravitational radiation damping, in *Contribution to 10th Marcel Grossmann Meeting* (2004) [gr-qc/0402007]

G. Esposito-Farese. Binary-pulsar tests of strong-field gravity
C.W.F. Everitt, D.B. Debra, B.W. Parkinson, J.P. Turneaure, J.W. Conklin, M.I. Heifetz, G.M. Keiser, A.S. Silbergleit, T. Holmes, J. Kolodziejczak, M. Al-Meshari, J.C. Mester, B. Muhlfelder, V.G. Solomonik, K. Stahl, P.W. Worden Jr., W. Bencze, S. Buchman, B. Clarke, A. Al-Jadaan, H. Al-Jibreen, J. Li, J.A. Lipa, J.M. Lockhart, B. Al-Suwaidan, M. Taber, S. Wang, Phys. Rev. Lett. **106**(22), 221101 (2011). doi:10.1103/PhysRevLett.106.221101
B. Ewan, Searching for pulsars with the effelsberg telescope. Ph.D. thesis, University of Bonn, 2013
A.J. Faulkner, M. Kramer, A.G. Lyne, R.N. Manchester, M.A. McLaughlin, I.H. Stairs, G. Hobbs, A. Possenti, D.R. Lorimer, N. D'Amico, F. Camilo, M. Burgay, ApJ **618**, L119 (2005)
R.D. Ferdman, I.H. Stairs, M. Kramer, G.H. Janssen, C.G. Bassa, B.W. Stappers, P.B. Demorest, I. Cognard, G. Desvignes, G. Theureau, M. Burgay, A.G. Lyne, R.N. Manchester, A. Possenti, astro-ph:1406.5507 (2014)
R.P. Feynman, F.B. Morinigo, W.G. Wagner, *Feynman Lectures on Gravitation* (1995)
E. Fonseca, I.H. Stairs, S.E. Thorsett, ApJ **787**, 82 (2014). doi:10.1088/0004-637X/787/1/82
P.C.C. Freire, M. Kramer, N. Wex, Class. Quantum Gravity **29**(18), 184007 (2012a). doi:10.1088/0264-9381/29/18/184007
P.C.C. Freire, N. Wex, G. Esposito-Farèse, J.P.W. Verbiest, M. Bailes, B.A. Jacoby, M. Kramer, I.H. Stairs, J. Antoniadis, G.H. Janssen, MNRAS **423**, 3328 (2012b)
J.J. Hermes, M. Kilic, W.R. Brown, D.E. Winget, C. Allende Prieto, A. Gianninas, A.S. Mukadam, A. Cabrera-Lavers, S.J. Kenyon, ApJ **757**, L21 (2012)
G. Hobbs, R. Manchester, A. Teoh, M. Hobbs, AJ
F. Hofmann, J. Müller, L. Biskupek, A&A **522**, L5 (2010). doi:10.1051/0004-6361/201015659
R.A. Hulse, J.H. Taylor, ApJ **195**, L51 (1975)
B.A. Jacoby, P.B. Cameron, F.A. Jenet, S.B. Anderson, R.N. Murty, S.R. Kulkarni, ApJ **644**, L113 (2006). doi:10.1086/505742
L.E. Kasian, Radio observations of two binary pulsars. Ph.D. thesis, The University of British Columbia, 2012
F. Kirsten, W. Vlemmings, P. Freire, M. Kramer, H. Rottmann, R.M. Campbell, A&A **565**, A43 (2014). doi:10.1051/0004-6361/201323239
D. Koester, ArXiv:0812.0482 (2008)
M. Kramer, I.H. Stairs, R.N. Manchester, M.A. McLaughlin, A.G. Lyne, R.D. Ferdman, M. Burgay, D.R. Lorimer, A. Possenti, N. D'Amico, J.M. Sarkissian, J.E. Reynolds, B.C. Joshi, P.C.C. Freire, F. Camilo, (2006) (submitted)
S.B. Lambert, C. Le Poncin-Lafitte, A&A **529**, A70 (2011). doi:10.1051/0004-6361/201016370
K. Lazaridis, N. Wex, A. Jessner, M. Kramer, B.W. Stappers, G.H. Janssen, G. Desvignes, M.B. Purver, I. Cognard, G. Theureau, A.G. Lyne, C.A. Jordan, J.A. Zensus, MNRAS **400**, 805 (2009). doi:10.1111/j.1365-2966.2009.15481.x
K. Lazaridis, J.P.W. Verbiest, T.M. Tauris, B.W. Stappers, M. Kramer, N. Wex, A. Jessner, I. Cognard, G. Desvignes, G.H. Janssen, M.B. Purver, G. Theureau, C.G. Bassa, R. Smits, MNRAS **414**, 3134 (2011). doi:10.1111/j.1365-2966.2011.18610.x
K. Liu, N. Wex, M. Kramer, J.M. Cordes, T.J.W. Lazio, ApJ **747**, 1 (2012). doi:10.1088/0004-637X/747/1/1
D.R. Lorimer, M. Kramer, *Handbook of Pulsar Astronomy* (Cambridge University Press, Cambridge, 2005)
S.C. Lundgren, A.F. Zepka, J.M. Cordes, ApJ **453**, 419 (1995)
R.S. Lynch, J. Boyles, S.M. Ransom, I.H. Stairs, D.R. Lorimer, M.A. McLaughlin, J.W.T. Hessels, V.M. Kaspi, V.I. Kondratiev, A.M. Archibald, A. Berndsen, R.F. Cardoso, A. Cherry, C.R. Epstein, ArXiv: 1209.4296 (2012)
A.G. Lyne, M. Burgay, M. Kramer, A. Possenti, R.N. Manchester, F. Camilo, M.A. McLaughlin, D.R. Lorimer, N. D'Amico, B.C. Joshi, J. Reynolds, P.C.C. Freire, Science **303**, 1153 (2004). DOI 10.1126/science.1094645
S.M. Merkowitz, Living Reviews in Relativity **13**(7) (2010). doi:10.12942/lrr-2010-7. URL http://www.livingreviews.org/lrr-2010-7

M. Milgrom, ApJ **270**, 365 (1983)
D.J. Nice, I.H. Stairs, L.E. Kasian, in *40 Years of Pulsars: Millisecond Pulsars, Magnetars and More*, vol. 983, ed. by C. Bassa, Z. Wang, A. Cumming, V.M. Kaspi (American Institute of Physics Conference Series, 2008), pp. 453–458
S.M. Ord, M. Bailes, W. van Straten, MNRAS **337**, 409 (2002)
J.A. Panei, L.G. Althaus, X. Chen, Z. Han, MNRAS **382**, 779 (2007)
S.M. Ransom, I.H. Stairs, A.M. Archibald, J.W.T. Hessels, D.L. Kaplan, M.H. van Kerkwijk, J. Boyles, A.T. Deller, S. Chatterjee, A. Schechtman-Rook, A. Berndsen, R.S. Lynch, D.R. Lorimer, C. Karako-Argaman, V.M. Kaspi, V.I. Kondratiev, M.A. McLaughlin, J. van Leeuwen, R. Rosen, M.S.E. Roberts, K. Stovall, Nature **505**, 520 (2014). doi:10.1038/nature12917
A.M. Serenelli, L.G. Althaus, R.D. Rohrmann, O.G. Benvenuto, MNRAS **337**, 1091 (2002)
L. Shao, N. Wex, Class. Quantum Gravity **29**(21), 215018 (2012). doi:10.1088/0264-9381/29/21/215018
L. Shao, R.N. Caballero, M. Kramer, N. Wex, D.J. Champion, A. Jessner, Class. Quantum Gravity **30**(16), 165019 (2013). doi:10.1088/0264-9381/30/16/165019
I.I. Shapiro, Phys. Rev. Lett. **13**, 789 (1964). doi:10.1103/PhysRevLett.13.789
I.H. Stairs, Z. Arzoumanian, F. Camilo, A.G. Lyne, D.J. Nice, J.H. Taylor, S.E. Thorsett, A. Wolszczan, ApJ **505**, 352 (1998)
T.M. Tauris, G.J. Savonije, A&A **350**, 928 (1999)
J.H. Taylor, L.A. Fowler, P.M. McCulloch, Nature **277**, 437 (1979). doi:10.1038/277437a0
J.H. Taylor, J.M. Weisberg, ApJ **253**, 908 (1982)
P.E. Tremblay, H.G. Ludwig, M. Steffen, B. Freytag, ArXiv:1302.2013 (2013)
M.H. van Kerkwijk, P. Bergeron, S.R. Kulkarni, ApJ **467**, L89 (1996). doi:10.1086/310209
A.K. Verma, A. Fienga, J. Laskar, H. Manche, M. Gastineau, A&A **561**, A115 (2014). doi:10.1051/0004-6361/201322124
J. Weisberg, J. Taylor, Gen. Relat. Gravit. **13**, 1 (1981)
J.M. Weisberg, J.H. Taylor, Phys. Rev. Lett. **52**, 1348 (1984)
J.M. Weisberg, D.J. Nice, J.H. Taylor, ApJ **722**, 1030 (2010)
N. Wex, astro-ph:1402.5594 (2014)
C.M. Will, Phys. Rev. D **50**, 6058 (1994)
C.M. Will, Living Rev. Relat. **9**(3) (2006). doi:10.12942/lrr-2006-3. http://www.livingreviews.org/lrr-2006-3
C.M. Will, Living Reviews in Relativity **17**, 4 (2014). doi:10.12942/lrr-2014-4
K. Yagi, D. Blas, N. Yunes, E. Barausse, Phys. Rev. Lett. **112**(16), 161101 (2014a). doi:10.1103/PhysRevLett.112.161101
K. Yagi, D. Blas, E. Barausse, N. Yunes, Phys. Rev. D **89**(8), 084067 (2014b). doi:10.1103/PhysRevD.89.084067

The Coalescence Rates of Compact Object Binaries

Tomasz Bulik

Abstract Mergers of compact object binaries have been considered the most promising sources of gravitational waves for ground based interferometric detectors. Investigating their properties was hampered by their elusive nature: only a handful are known. In this paper I review the estimates of the compact object merger rate density. This rate density can be estimated based directly on observations of objects that will merge, or using elaborate computer simulations of binary evolution. The two methods give consistent results, yet the uncertainties are still quite large. Nevertheless the current estimates point that the probability of detection of gravitational waves in the coming years is quite high.

1 Introduction

The estimate of the compact object binary coalescence rates has been investigated by many authors. The subject is mentioned in several fields of astrophysics: from the binary pulsar studies, through gamma ray-bursts, and finally in the gravitational wave astrophysics.

Coalescences of compact object binaries are the most promising sources of gravitational waves to be detected by the interferometric detectors. Currently LIGO and VIRGO are being reassembled to achieve the sensitivity roughly ten times better then in the previous configurations. KAGRA is in the construction phase, and LIGO India is in the early stage. Thus the coming years are bound to bring a number of observatories, which raises expectations for detection of gravitational wave sources.

It is therefore important to investigate and review our knowledge of properties of compact object binaries and the constraints that can be imposed on their coalescence rates. This paper does not aim at a comprehensive review of the subject. We must note that a comprehensive review of the rates has been recently compiled (Abadie et al. 2010).

T. Bulik (✉)
Astronomical Observatory, University of Warsaw, Al. Ujazdowskie 4, 00-478, Warsaw, Poland
e-mail: tb@astrouw.edu.pl

C.F. Sopuerta (ed.), *Gravitational Wave Astrophysics*, Astrophysics and Space Science Proceedings, DOI 10.1007/978-3-319-10488-1_2

In the literature there are several conventions for the units of the coalescence rates. Some papers quote the number per unit time in the Milky Way. This has been to found to be too Earth centered, and basically irrelevant for the gravitational wave searches since we do not expect to see a coalescence of compact binary in the Milky Way, neither in our lifetime nor probably even in the lifetime of our civilisation. Thus people started using the units of the coalescence rate per MWEG, which stands for Milky Way Equivalent Galaxy. However, the MWEG is an imaginary galaxy, a copy of the Milky Way distributed uniformly in the Universe with the density of $0.02\,\mathrm{Mpc}^{-3}$. Thus it is a unit of volume albeit quite strange. In this paper I will there fore use the units that are quite common and easy to understand i.e. number of coalescences per year per megaparsec cube.

The outline of the paper is as follows. In Sect. 2 we describe and review the observation based estimates. Section 3 contains the outline of population synthesis results. Finally in Sect. 4 we present the summary and discussion of uncertainties.

2 Observation Based Estimates

2.1 Double Neutron Stars

There are currently ten double neutron star binaries known, out of which four have coalescence times smaller than the Hubble time. The list of known binary neutron stars is shown in Table 1. The calculation of the coalescence rate of double neutron star systems can be described quite easily. Let us take into focus the only known real double pulsar: J0737-3039, where both neutron stars have been detected as radio pulsars. Given its radio luminosity the pulsar J0737-3039 could be detected in about 10 % of the volume of the Milky Way. Moreover given the width of the beam we can

Table 1 Pulsars in binary neutron star systems according to ATNF database (Manchester et al. 2005)

	NAME	P (s)	$\dot{P}$	P_b (d)
1	J0737-3039A	0.022699	1.76×10^{-18}	0.1023
1	J0737-3039B	2.773461	8.92×10^{-16}	
2	J1518+4904	0.040935	$2.72e \times 10^{-20}$	8.6340
3	B1534+12	0.037904	2.42×10^{-18}	0.4207
4	J1753-2240	0.095138	7.90×10^{-19}	13.6376
5	J1756-2251	0.028462	1.02×10^{-18}	0.3196
6	J1811-1736	0.104182	9.01×10^{-19}	18.7792
7	J1829+2456	0.041010	5.25×10^{-20}	1.1760
8	J1906+0746	0.144072	2.03×10^{-14}	0.1660
9	B1913+16	0.059030	8.63×10^{-18}	0.3230
10	B2127+11C	0.030529	4.99×10^{-18}	0.3353

calculate that only about 3 % of the sky is covered by the radio emission. Thus there are about 300 not detected binaries like J0737-3039 for each one that we can see. The lifetime and the age of the pulsar have been determined by Piran and Shaviv (2005) and Willems et al. (2006). The pulsar will merge in about 80 Myrs, and its current age is approximately 90 Myrs, giving a total lifetime as a binary pulsar of 170 Myrs. Thus, we expect a coalescence of a binary pulsar like J0737-3039 every:

$$\delta t \approx \frac{170\,\mathrm{Myrs}}{300} \approx 5.6 \times 10^5\,\mathrm{yrs} \tag{1}$$

This means that the Galactic merger rate is $1.8 \times 10^{-6}\,\mathrm{yr}^{-1}$. In order to present this number as the merger rate density we need to use the average galaxy density in the Universe $\rho_G = 0.0\,\mathrm{Mpc}^{-3}$:

$$\mathscr{R} = \rho_G \delta t^{-1} \approx 4 \times 10^{-8}\,\mathrm{Mpc}^{-3}\,\mathrm{yr}^{-1} \tag{2}$$

A more detailed calculation includes detailed models of the pulsar luminosity function, neutron star binaries distribution in the Galaxy, detectability as a function of position in the sky by various surveys, see e.g. Kim et al. (2010).

Estimating the properties of the binaries containing BHs is not straightforward, as as neither BHNS nor BHBH binary has been detected directly. It must be mentioned however that there is a potential BHBH candidate detected by the OGLE microlensing experiment (Dong et al. 2007). The microlensing event OGLE-2005-SMC-001, can be interpreted as a binary BH with the likely masses of 3 and 7 $\mathrm{M}_\odot$, with an orbital separation of 4.7 AU. Such a system has merger time much longer than the Hubble time and there fore does not contribute to the rate of observable gravitational wave systems. Nevertheless it is interesting to see that there is evidence for existence of binary BHs.

2.2 Binary Black Holes

The observational estimates of the binary BH coalescence rates have been based on the analysis of X-ray binaries and their future evolution. The first object to found to be candidate progenitor of the coalescing BH binary is IC10 X-1 (Bulik and Belczynski 2010). The X-ray binary IC10 X-1 is a member of a small class of binaries which host a Wolf-Rayet donor transferring mass onto a BH. The orbit is quite tight—the orbital period is $\approx$30 h. The mass of the BH is 23–33 $\mathrm{M}_\odot$, and the mass of the WR star is 17–35 $\mathrm{M}_\odot$. The lifetime of the WR star is only about $1-2\times 10^5$ years. After this time the mass of the WR star will decrease to primarily due the wind mass loss, and will explode as a supernova creating a BH with very little mass loss. This will lead to formation of the binary BH on a tight orbit. The potential natal kick of the BH is unlikely to disrupt the system as it is already very tight and the orbital velocity is $\approx$600 km s^{-1}. The binary BH that will form will have the time to merger of about 2–3 Gyrs.

The estimate of the merger rate of binary BH based on the IC10 X-1 is pretty straightforward. Since the merger time is shorter than the Hubble time we can assume that the merger rate is equal to the birth rate of such systems. The birth rate can can be estimated by considering the observability of IC10 X-1. There are two key key points that determine this: the volume in which one obtain the spectroscopy required to find the masses, and the time that IC10 X-1 is detectable as an X-ray source. The distance out to which one can obtain spectroscopy of a WR star is about $R = 2\,\text{Mpc}$, while the source is X-ray active for about the lifetime of the WR star, i.e. up to $t_D = 200\,\text{kyr}$. Thus the formation, and merger rate is:

$$\mathscr{R}_{BBH} = \left(\frac{4\pi R^3}{3}\right)^{-1} t_D^{-1} \approx 1.6 \times 10^{-7}\,\text{Mpc}^{-3}\,\text{yr}^{-1} \quad (3)$$

This is a surprisingly high rate, comparing to the one obtained for the binary neutron stars above.

A more detailed calculation has been done with the inclusion of another WR binary—NGC300 X-1 (Bulik et al. 2011). This paper contains a much more detailed derivation of the probability distribution of the expected detection rate by the detectors like VIRGO and LIGO. In Fig. 1 we present the analysis of the probability distribution of the merger rate density of binary BHs. Recently another object, a Wolf Rayet binary in NGC 253 was added to the list of binaries that may lead to formation of the binary BH (Maccarone et al. 2014). However this object needs still a more detailed spectroscopic study before its contribution to the merger rate density can be estimated.

2.3 Black Hole Neutron Star Binaries

A number of X-ray binaries have been analysed in search for progenitors of black hole neutron star binaries. It has been found that Cyg X-3 is a potential progenitor of such a merging binary (Belczynski et al. 2013). Cyg X-3 hosts a 2–4.5 $M_\odot$ black hole and 7.5–14.2 $M_\odot$ Wolf-Rayet star on 4.8hr orbit. The most likely fate of this system is it will be disrupted by a supernova. However, if the Wolf-Rayet star mass is large it may end up as a merging binary BH. In the case the Wolf-Rayet star mass is in the lower end of the possible range the system may end as a BHNS binary. The probability of such an outcome is quite low, and hence the expected merger rate density is of the order of

$$\mathscr{R}_{BHNS} \approx 10^{-8}\,\text{Mpc}^{-3}\,\text{yr}^{-1} \quad (4)$$

This estimate is highly uncertain. The derivation of this number carried a number of assumptions on the binary evolution, especially the wind mass loss, and the natal

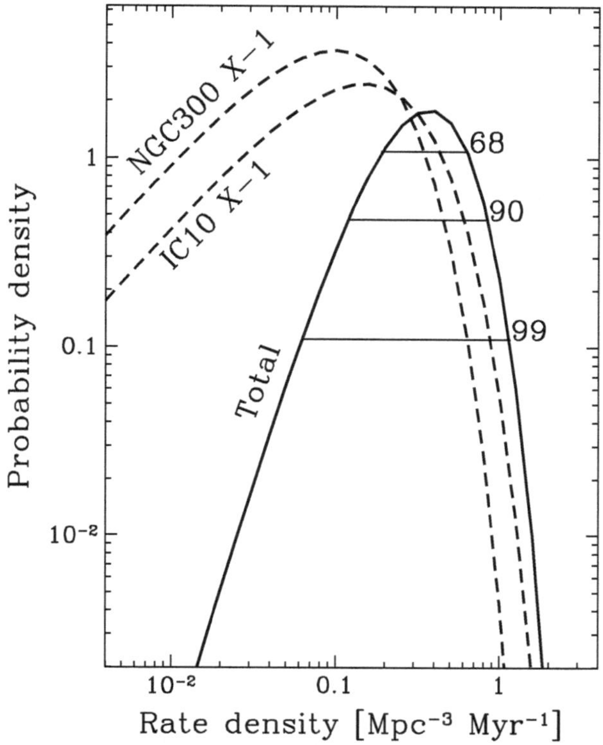

Fig. 1 The probability distribution of the merger rate of binary BHs. We present the contributions of the IC10 X-1 and NGC300 X-1 separately as well as the combined value. The *horizontal bars* correspond to the regions that contain, 68, 90, and 98 % of probability

kicks imparted on the newly formed compact object. It is based on a single object but the problem of small number statistics is a plague of the observational estimates of the compact object coalescence rates. It must be stressed that this is the lower limit on the BHNS merger rate density as the estimate has not taken into account the selection effects that lead to non detection of a some fraction of Cyg X-3 like binaries in the Milky Way

3 Population Synthesis Based Results

3.1 How to Make a Compact Object Binary?

Formation of compact object binaries has been known for a long time see, e.g. Lipunov et al. (1995). A short description of an evolutionary scenario leading to formation of a merging compact object binary is described below. The main problem is that we need form a merging binary, with the orbital separation much smaller that the size of the initial stars. This requires that there are processes that tighten the orbit.

The general scenario goes along the following steps:

1. We start we two massive stars at an orbit separation a few hundred solar radii. The masses must be in excess of $20\,M_{\odot}$ so that each of the components can form a compact object.
2. The initially more massive star evolves first, fills its Roche lobe and initiates mass transfer. The system losses some mass the orbit is not changed a lot. The initially more massive star looses its envelope and becomes a Wolf Rayet star, the mass ratio is inverted.
3. The Wolf Rayet star undergoes a supernova explosion, and forms the first compact object. the system may be disrupted by the natal kick imparted on the newly formed compact object.
4. The initially less massive star starts to evolve, becomes a giant, and fills its Roche lobe. The mass transfer is unstable and the common envelope phase starts. The orbit shrinks and the giant losses its envelope. The system may end up in a merger at this time forming a Thorne-Zytkow object.
5. If the system survives the common envelope it consists of a Wolf-Rayet star and a compact object on a tight orbit. It can be seen as an X-ray binary accreting either from the wind or through the Roche lobe overflow.
6. The Wolf-Rayet star ends up its life as supernova and forms a second compact object. The system can again be disrupted by the natal kick.
7. If the system survives formation of the second compact object it is now a tight binary consisting of two compact objects.
8. The system tightens emitting gravitational waves and merges.

This scenario has been published in a many papers (Lipunov et al. 1997; Belczynski et al. 2002) and was a basis to the belief that binary black hole formation is quite well understood. It was also believed that the binary BH binaries are likely to be the most common sources of gravitational waves.

This belief ran into trouble about 8 years ago when it was found that making compact object binaries containing black holes showed some problems. In particular,surviving stage 4. above was difficult as most systems simulated would not survive the common envelope and merge forming a Thorne-Zytkow object. This lead us to a pessimistic conclusion that the number of binaries containing black holes will be very small, published under a title "On the Rarity of Double Black Hole Binaries: Consequences for Gravitational Wave Detection" (Belczynski et al. 2007).

This pessimistic view has changed with the discovery and interpretation of the two Wolf-Rayet binaries described in Sect. 2. These are the system that have survived the difficult stage 4. of the evolutionary scenario above and nothing can prevent them from becoming merging compact object binaries containing black holes. What made them special is that they originate in the low metallicity dwarf galaxies: the metallicity of IC10 is $0.3\,Z_{\odot}$, and that of NGC300 is $0.6\,Z_{\odot}$. This lead to realization that a crucial factor in the formation of compact object binaries containing black holes is played by metallicity.

Metallicity affects the stellar evolution quite significantly. It influences the stellar winds and mass loss. It also changes the sizes of stars because of strong dependence of the opacity on metallicity. These two factors play a major role in allowing formation of the compact object binaries. The stellar mass loss determines the stellar masses at the moment they collapse and form a compact object. Thus they affect the mass spectrum of the compact objects. In fact the maximum mass of a stellar black hole in a low metallicity environment ($0.1\,Z_\odot$) rises to about $30\,M_\odot$, and for the very low metallicity like the one in globular clusters it can reach even $100\,M_\odot$ (Belczynski et al. 2010a). The second factor is connected with stellar structure. The low metallicity giants have smaller radii and therefore the survival of the common envelope phase may be easier. These factors cause a very strong dependence of the compact object formation rate on metallicity, first shown in Belczynski et al. (2010b).

3.2 Rate Density Estimates

The population synthesis of compact objects has been performed by several groups and several population synthesis codes are now in use in the astrophysical community (Portegies Zwart and Yungelson 1998; Bethe and Brown 1998; Hurley et al. 2002; Belczynski et al. 2002; Tutukov and Yungelson 2002; Voss and Tauris 2003).

The estimate of the rate density must take into account three inputs: the population synthesis results, the star formation history and the cosmology model. As a result one obtains the merger rate density as a function of the redshift and masses of the binaries. Let us assume that the population synthesis provides an estimate of the distribution of masses and merger times of compact object binaries $\frac{d^2N}{dMdt}$, where t is the delay time and M is the mass (or masses) of the compact object binary. The star formation rate dependence on the redshift is $SFR(z)$. In general the star formation can also depend on the metallicity. The differential coalescence rate is then

$$\frac{d^2R}{dzdM} = \int dt' \frac{d^2N}{dMdt'} SFR(z')\delta(z' - z(t(z) + t')) \tag{5}$$

where $z(t)$ is the dependence of redshift on cosmic time, and $t(z)$ is the inverse relation. The cosmology enters through the $z(t)$ relation.

The recent modelling of the compact binary formation has been done in a series of papers (Dominik et al. 2012, 2013). This work includes the metallicity dependence of the compact binary formation, as well as the metallicity evolution in the Universe. For each metallicity there is a different star formation rate history. The calculation tales into account various time delays between the formation of each

Table 2 The summary of local merger rate density of compact object binaries from Dominik et al. (2012)

Binary type	NS-NS	BH-NS	BH-BH
Rate density ($\mathrm{Mpc}^{-3}\,\mathrm{yr}^{-1}$)			
Solar metallicity	2.3×10^{-7}	1.6×10^{-8}	8.2×10^{-8}
Subsolar metallicity	8.1×10^{-8}	3.4×10^{-8}	7.3×10^{-7}

binary and its coalescence. The rate density range found in this study is summarized in Table 2.

The uncertainty of the rate densities presented in Table 2 is quite high, i.e up to a factor of ten up or down.

3.3 *Population III Binaries*

The extreme case of the low metallicity evolution corresponds to the very first stars in the Universe—Population III stars. The properties of Population III stars are highly uncertain: as yet we have not detected even a single such star. We need to relay on simulations to find their properties. Thus the initial mass function and the binary fraction of Population III stars are only known very approximately. It is however believed that Population III stars can lead to formation of massive black holes—with masses reaching 500–1,000 $M_\odot$, provided that the initial mass function reaches that high.

An early analysis of the detectability of Population III binaries (Belczynski et al. 2004) has shown that they can dominate the observed rate in the detectors. This work has used quite optimistic assumptions about the number of binaries, and their initial mass function. Since this paper was published a number of detailed studies investigated the process of collapse of metal free clouds. The possibility of formation of binaries has been confirmed by several studies (Stacy et al. 2010; Machida et al. 2008; Saigo et al. 2004). Recently a very detailed study has shown that the Population III binaries will dominate the rate above the total mass of 30 $M_\odot$ (Kinugawa et al. 2014).

4 Summary

I have presented several approaches to find the merger rate density of the compact object binaries. The list and the discussion are meant to provide an outline of the problem and it is by no means to be considered as an exhaustive description of the subject.

A relatively clear picture emerges based on these various leads. First, there exists a population of potential coalescing compact object binaries that will become sources of gravitational waves. The evidence comes from double pulsars as well as Wolf Rayet binaries that will evolve to form binary black holes soon. Thus the existence of such sources can not be denied. The estimate of their coalescence rate is much tougher. Nevertheless one can obtain several estimates of the compact object binary merger rate density. At the top end of the estimates we obtain the values reaching $\approx 10^{-6}\,\mathrm{Mpc}^{-3}\,\mathrm{yr}^{-1}$, and at the low end they are two orders of magnitude lower $\approx 10^{-8}\,\mathrm{Mpc}^{-3}\,\mathrm{yr}^{-1}$. It is expected that the DNS and BBH merger rates densities are close to the upper value, while the BHNS merger rates are rather low and close to the lower bound. There are some models that quench formation of compact object binaries, however at this point they seem quite unlikely. This rates can be used to calculate the detector range require to obtain a single detection in 1 year. For the optimistic rates the required range is 62 Mpc, and for the very pessimistic ones it is 287 Mpc. Note that I have not specified the types of the binaries and their masses. The expected angle and orientation averaged range for detection of a binary neutron star with the Advanced detectors will reach $\approx$200 Mpc. Thus, even for the pessimistic rates we expect the advanced detectors to see merger of a coalescing compact object binary.

In the discussion above I have not touched on the subject of the dynamically formed coalescing binaries. This has been investigated in a number of papers (Grindlay et al. 2006; Hopman et al. 2006; Sadowski et al. 2008; Lee et al. 2010; Downing et al. 2010, 2011). The evolution in clusters may lead to formation of a significant number of merging compact object binaries, increasing it average merger density. However, the studies of these rates is more complicated than in the case of solitary evolution. Consideration of the dynamically formed binaries will only increase the expected merger rate density. A different question is whether the dynamically formed binaries can be distinguished from those in the filed by gravitational wave observations only.

In summary the merger rate density estimates are quite solid and based on multiple astrophysical arguments. The uncertainty in the rate is quite large, however it seems that the direct detection of gravitational waves from a compact object merger is possible in the coming years.

Acknowledgement This work was supported by the National Science Center, Poland with the grant number DEC-2013/01/ASPERA/ST9/00001.

References

J. Abadie, B.P. Abbott, R. Abbott, M. Abernathy, T. Accadia et al., Topical review: Predictions for the rates of compact binary coale scences observable by ground-based gravitational-wave detectors. Class. Quantum Gravity **27**(17), 173001 (2010). doi:10.1088/0264-9381/27/17/173001

K. Belczynski, V. Kalogera, T. Bulik, A comprehensive study of binary compact objects as gravitational wave sources: Evolutionary channels, rates, and physical properties. ApJ **572**, 407–431 (2002) doi:10.1086/340304

K. Belczynski, T. Bulik, B. Rudak, The first stellar binary black holes: The strongest gravitational wave burst sources. ApJ **608**, L45–L48 (2004). doi:10.1086/422172

K. Belczynski, R.E. Taam, V. Kalogera, F.A. Rasio, T. Bulik, On the rarity of double black hole binaries: Consequences for gravitational wave detection. ApJ **662**, 504–511 (2007) doi:10.1086/513562

K. Belczynski, T. Bulik, C.L. Fryer, A. Ruiter, F. Valsecchi, J.S. Vink, J.R. Hurley, On the maximum mass of stellar black holes. ApJ **714**, 1217–1226 (2010a). doi:10.1088/0004-637X/714/2/1217

K. Belczynski, M. Dominik, T. Bulik, R. O'Shaughnessy, C. Fryer, D.E. Holz, The effect of metallicity on the detection prospects for gravitational waves. ApJ **715**, L138–L141 (2010b). doi:10.1088/2041-8205/715/2/L138

K. Belczynski, T. Bulik, I. Mandel, B.S. Sathyaprakash, A.A. Zdziarski, J. Mikołajewska, Cyg X-3: A galactic double black hole or black-hole-neutron-star P rogenitor. ApJ **764**, 96 (2013). doi:10.1088/0004-637X/764/1/96

H.A. Bethe, G.E. Brown, Evolution of binary compact objects that merge. ApJ **506**, 780–789 (1998). doi:10.1086/306265

T. Bulik, K. Belczynski, Observational evidence for stellar mass binary black holes and their coalescence rate. Mem. Soc. Astron. Italiana **81**, 302 (2010)

T. Bulik, K. Belczynski, A. Prestwich, IC10 X-1/NGC300 X-1: The very immediate progenitors of BH-BH binaries. ApJ **730**, 140 (2011). doi:10.1088/0004-637X/730/2/140

M. Dominik, K. Belczynski, C. Fryer, D.E. Holz, E. Berti, T. Bulik, I. Mandel, R. O'Shaughnessy, Double compact objects, I. The significance of the common envelope on merger rates. ApJ **759**, 52 (2012). doi:10.1088/0004-637X/759/1/52

M. Dominik, K. Belczynski, C. Fryer, D.E. Holz, E. Berti, T. Bulik, I. Mandel, R. O'Shaughnessy, Double compact objects, II. Cosmological merger rates. ApJ **779**, 72 (2013). doi:10.1088/0004-637X/779/1/72

S. Dong, A. Udalski, A. Gould, W.T. Reach, G.W. Christie et al., First space-based microlens parallax measurement: Spitzer Observations of OGLE-2005-SMC-001. ApJ **664**, 862–878 (2007) doi:10.1086/518536

J.M.B. Downing, M.J. Benacquista, M. Giersz, R. Spurzem, Compact binaries in star clusters - I. Black hole binaries inside globular clusters. MNRAS **407**, 1946–1962 (2010). doi:10.1111/j.1365-2966.2010.17040.x

J.M.B. Downing, M.J. Benacquista, M. Giersz, R. Spurzem, Compact binaries in star clusters - II. Escapers and detection rates. MNRAS **416**, 133–147 (2011). doi:10.1111/j.1365-2966.2011.19023.x

J. Grindlay, S. Portegies Zwart, S. McMillan, Short gamma-ray bursts from binary neutron star mergers in globular clusters. Nat. Phys. **2**, 116–119 (2006). doi:10.1038/nphys214

C. Hopman, D. Guetta, E. Waxman, S. Portegies Zwart, The redshift distribution of short gamma-ray bursts from dynamically formed neutron star binaries. ApJ **643**, L91–L94 (2006). doi:10.1086/505141

J.R. Hurley, C.A. Tout, O.R. Pols, Evolution of binary stars and the effect of tides on binary popula tions. MNRAS **329**, 897–928 (2002). doi:10.1046/j.1365-8711.2002.05038.x

C. Kim, V. Kalogera, D. Lorimer, The effect of PSR J0737-3039 on the DNS merger rate and implications for gravity-wave detection. New A Rev. **54**, 148–151 (2010). doi:10.1016/j.newar.2010.09.010

T. Kinugawa, K. Inayoshi, K. Hotokezaka, D. Nakauch i, T. Nakamura, *Possible Indirect Confirmation of the Existence of Pop III Massive Stars by Gravitational Wave* (2014) [ArXiv e-prints]

W.H. Lee, E. Ramirez-Ruiz, G. van de Ven, Short gamma-ray bursts from dynamically assembled compact binaries in globular clusters: Pathways, rates, hydrodynamics, and cosmological setting. ApJ **720**, 953–975 (2010). doi:10.1088/0004-637X/720/1/953

V.M. Lipunov, K.A. Postnov, M.E. Prokhorov, I.E. Panchenko, H.E. Jorgensen, Evolution of the double neutron star merging rate and the cosmological origin of gamma-ray burst sources. ApJ **454**, 593 (1995). doi:10.1086/176512

V.M. Lipunov, K.A. Postnov, M.E. Prokhorov, First LIGO events: Binary black holes mergings. New A **2**, 43–52 (1997). doi:10.1016/S1384-1076(97)00007-9

T.J. Maccarone, B.D. Lehmer, J.C. Leyder, V. Antoniou, A. Hornschemeier, A. Ptak, D. Wik, A. Zezas, A new candidate Wolf-Rayet X-ray binary in NGC 253. MNRAS **439**, 3064–3072 (2014) doi:10.1093/mnras/stu167

M.N. Machida, K. Omukai, T. Matsumoto, S.I. Inutsuka, Conditions for the formation of first-star binaries. ApJ **677**, 813–827 (2008). doi:10.1086/533434

R.N. Manchester, G.B. Hobbs, A. Teoh, M. Hobbs, The australia telescope national facility pulsar catalogue. AJ **129**, 1993–2006 (2005). doi:10.1086/428488

T. Piran, N.J. Shaviv, Origin of the binary pulsar J0737-3039B. Phys. Rev. Lett. **94**(5), 051102 (2005). doi:10.1103/PhysRevLett.94.051102

S.F. Portegies Zwart, L.R. Yungelson, Formation and evolution of binary neutron stars. A&A **332**, 173–188 (1998)

A. Sadowski, K. Belczynski, T. Bulik, N. Ivanova, F.A. Rasio, R. O'Shaughnessy, The total merger rate of compact object binaries in the local universe. ApJ **676**, 1162–1169 (2008). doi:10.1086/528932

K. Saigo, T. Matsumoto, M. Umemura, The formation of population III binaries. ApJ **615**, L65–L68 (2004). doi:10.1086/426389

A. Stacy, T.H. Greif, V. Bromm, The first stars: Formation of binaries and small multiple systems. MNRAS **403**, 45–60 (2010). doi:10.1111/j.1365-2966.2009.16113.x

A.V. Tutukov, L.R. Yungelson, A model for the population of binary stars in the galaxy. Astron. Rep. **46**, 667–683 (2002). doi:10.1134/1.1502227

R. Voss, T.M. Tauris, Galactic distribution of merging neutron stars and black holes - prospects for short gamma-ray burst progenitors and LIGO/VIRGO. MNRAS **342**, 1169–1184 (2003) doi:10.1046/j.1365-8711.2003.06616.x

B. Willems, J. Kaplan, T. Fragos, V. Kalogera, K. Belczynski, Formation and progenitor of PSR J0737-3039: New constraints on the supernova explosion forming pulsar B. Phys. Rev. D **74**(4), 043003 (2006). doi:10.1103/PhysRevD.74.043003

Enhancing Gravitational Wave Astronomy with Galaxy Catalogues

Xilong Fan, Christopher Messenger, and Ik Siong Heng

Abstract Joint gravitational wave (GW) and electromagnetic (EM) observations, as a key research direction in *multi-messenger astronomy*, will provide deep insight into the astrophysics of a vast range of astronomical phenomena. Uncertainties in the source sky location estimate from gravitational wave observations mean follow-up observatories must scan large portions of the sky for a potential companion signal. A general frame of joint GW-EM observations is presented by a multi-messenger observational triangle. Using a Bayesian approach to multi-messenger astronomy, we investigate the use of galaxy catalogue and host galaxy information to reduce the sky region over which follow-up observatories must scan, as well as study its use for improving the inclination angle estimates for coalescing binary compact objects. We demonstrate our method using a simulated neutron stars inspiral signal injected into simulated Advanced detectors noise and estimate the injected signal sky location and inclination angle using the Gravitational Wave Galaxy Catalogue. In this case study, the top three candidates in rank have 72 %, 15 % and 8 % posterior probability of being the host galaxy, respectively. The standard deviation of cosine inclination angle (0.001) of the neutron stars binary using gravitational wave-galaxy information is much smaller than that (0.02) using only gravitational wave posterior samples.

1 Introduction

The detection of Gravitational waves (GW) will herald a new era of astronomy, with Advanced LIGO (Harry 2010) and Advanced Virgo (The Virgo Collaboration 2009) expected to make the first detection of GWs within the next few years. *Multi-*

X. Fan (✉)
School of Physics and Electronics Information, Hubei University of Education, 430205 Wuhan, China

SUPA, School of Physics and Astronomy, University of Glasgow, Glasgow G12 8QQ, UK
e-mail: Xilong.Fan@glasgow.ac.uk

C. Messenger • I.S. Heng
SUPA, School of Physics and Astronomy, University of Glasgow, Glasgow G12 8QQ, UK
e-mail: Christopher.Messenger@glasgow.ac.uk; Ik.Heng@glasgow.ac.uk

C.F. Sopuerta (ed.), *Gravitational Wave Astrophysics*, Astrophysics and Space Science Proceedings, DOI 10.1007/978-3-319-10488-1_3

messenger astronomy involves the joint observation of astrophysical phenomena by a combination of GW, electromagnetic (EM), and astroparticles observatories. Examples of multi-messenger astronomy involving GWs include EM and neutrinos observatories (e.g. Kanner et al. 2012; Evans et al. 2012; Aasi et al. 2013; Blackburn et al. 2013; Abadie et al. 2012a; Ando et al. 2013; Aasi et al. 2014; Abadie et al. 2012b, 2010; Nissanke et al. 2013; Kanner et al. 2008). Galaxies, as the hosts of GW and/or EM sources, are examples of common observables in multi-messenger astronomy. For example, searches for GWs in association with gamma-ray bursts (GRB) GRB070201 (Abbott et al. 2008) and GRB051103 (Abadie et al. 2012c) have ruled out the possibility that their progenitors are binary neutron stars (NS) sources in Andromeda galaxy (M31) and M81, respectively. The basic assumption of this research is that there are common parameters among observations of GWs, GRBs and their host galaxy: sky location and distance. Besides of these common parameters for GW sources, EM sources and host galaxy, there are physical links between each other depending on the nature of sources. For example, merging NSs or NS-black hole systems and the collapse of supermassive stars, as potential sources of inspiral and burst GW, are thought to be the most likely progenitors for short and long GRBs respectively (e.g. Paczynski 1986; Eichler et al. 1989; Fox et al. 2005; Woosley and Bloom 2006). Nearby ($z < 1$) long-GRBs are more likely to be observed in irregular galaxies or the outermost regions of spiral disks (Fan et al. 2010), which are usually metal poor, while short-GRBs are observed in all types of galaxy (e.g. Fong et al. 2013). This general picture is summarized by the *multi-messenger astronomy observational triangle* shown in Fig. 1.

The broad sky location estimates ($>10\,\text{deg}^2$, e.g. Wen et al. 2008; Wen and Chen 2010; Veitch and Vecchio 2010; Schutz 2011; Fairhurst 2011; Sidery et al. 2014) from GW observations is a challenge for joint EM-GW observation. In addition to exploiting common parameters, the link between GWs and galaxies is used to improve the efficiency of searching for an EM counterpart of a GW. With the help of a galaxy catalog, Nuttall and Sutton (2010) proposed a ranking statistic to identify the most likely GW host galaxy based on galaxy distance and luminosity and the

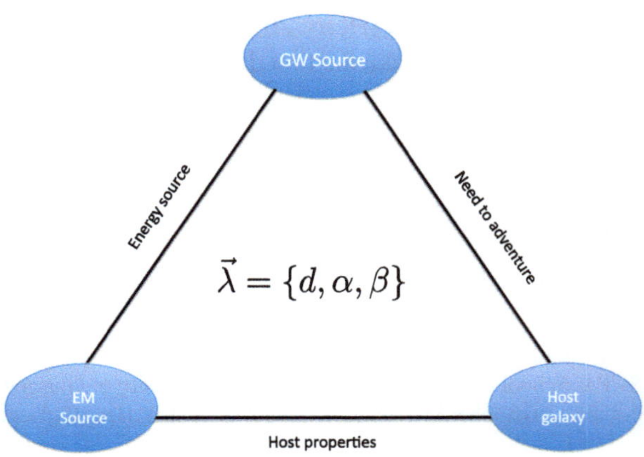

Fig. 1 Multi-messenger observational triangle. The common parameters ($\boldsymbol{\lambda}$) of GW source, EM source and host galaxy include: distance (d), sky location (α, β). Models and data have suggest the physical links between EM source and their host galaxies. The physical links between EM and GW source have been investigated by theoretical studies, while it is not clear for the link between GW source and galaxies

sky position error box. The scan area needed by an EM followup team could be reduced to the regions of sky most likely associated with the GW host galaxy. In past searches, blue band luminosity, which is assuming to be the tracer of the star formation rate of a galaxy, is used to estimated the rate of compact binary coalescences (CBC) (e.g. Evans et al. 2012; Aasi et al. 2014; Nuttall and Sutton 2010). Several other properties, such as morphology and metallicity, of galaxies are suggested to have effects on the CBC event rate in a galaxy via stellar population synthesis studies (e.g. Belczynski et al. 2010; Fryer et al. 2012; O'Shaughnessy et al. 2010). Amongst the many existing galaxy catalogues, the Gravitational Wave Galaxy Catalogue (GWGC, White et al. 2011) has been specifically compiled for current follow-up searches of optical counterparts from GW triggers. GWGC is $\sim$100 % and $\sim$60 % complete out to about 40 and 100 Mpc respectively, estimated by blue band luminosity function (see discussion in White et al. 2011). Hanna et al. (2014) estimated that an average of $\sim$500 galaxies are located in a typical GW sky location error box for Advanced LIGO-Vigo network ($\sim$20 deg^2), up to range of 200 Mpc. The distance limitation of GWGC (or any other galaxy catalogs) comparing with the GW detection horizon leads to a completeness issue for catalog based multi-messenger astronomy studies. An ongoing study by Messenger et al. (2014) aims to address the completeness issue.

Fan et al. (2014) have proposed a Bayesian approach to multi-messenger astronomy. An example of this approach is the joint research of GW and it's galaxy host with a GW-galaxy relation model (multi-messenger prior function in Fan et al. 2014). One of merits of this Bayesian approach is that posterior probability of a galaxy hosting the GW source is estimated, as well as it's ranking in host candidates. The probabilities of GW host candidates can be used to guide EM follow-up observations to focus on the particular galaxy with very high posterior probability. When the first few galaxies in the rank have similar posterior probabilities, the absolute rank becomes of less importance since these galaxies should be considered candidates of similar importance. Moreover, better constraints on the distance of a galaxy-GW observations can benefit the inference on the inclination angle of a CBC event. The better inclination angle estimation could help with understanding the nature of EM counterparts. For example, a set of inclination angles, combined with GRB observations, might constrain the beaming factor of GRBs, therefore the dynamics and the energetics of GRB jets and so on (e.g. Arun et al. 2014).

In this article, we present a case study of the Bayesian approach designed for joint EM and GW observations, in particular, the galaxy catalog (GWGC) and NS-NS coalescence events, with a blue band luminosity based multi-messenger prior function. The aim of this research is to identify the GW source host galaxy (see Sect. 2) and provide better inclination angle estimates (see Sect. 3).

2 Identification of Gravitational-Wave Host Galaxies

We present a case study of simulated GW signals from NS-NS inspirals injected into simulated noise from the advanced LIGO-Virgo network. The sky location of the injected NS-NS signal is randomly chosen from the locations of galaxies within the GWGC. In Fig. 2, we plot a 12 deg^2 region around the sky location where

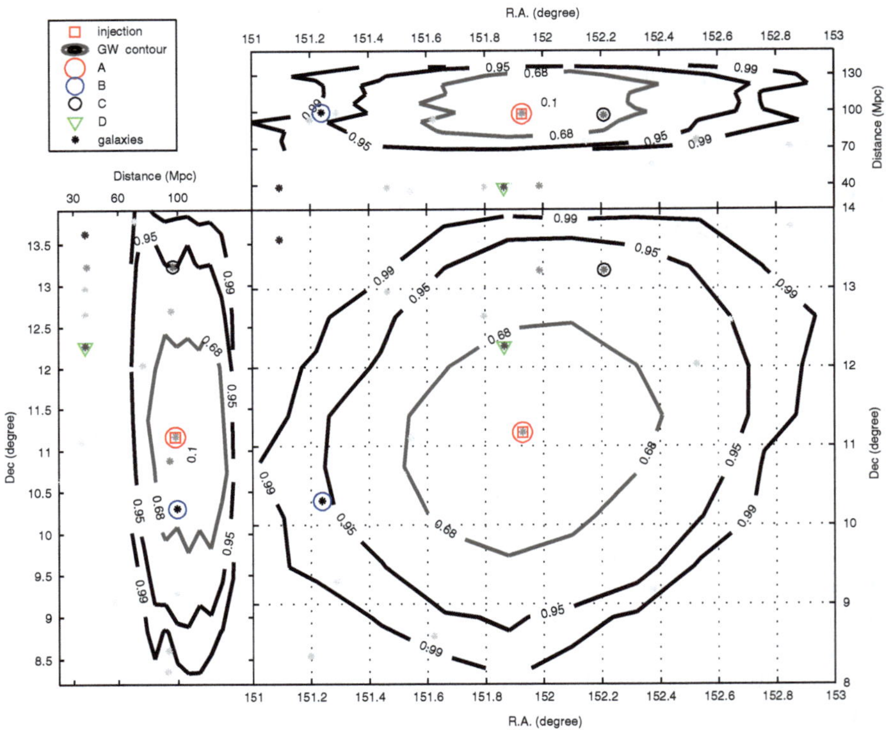

Fig. 2 Sky localisation for a single BNS coalescence signal. The contours map out the 68, 95 and 99 % confidence regions on the estimate of the sky location of the signal progenitor obtained using only GW observations. Also plotted are circle markers corresponding to the first (*red circle*), second (*blue circle*) and third (*black circle*) ranked host galaxy candidates, labelled *A*, *B* and *C* respectively, as determined using our Bayesian approach to multi-messenger astronomy. The posterior probability of *A*, *B*, *C* hosting the GW source are 72 %, 15 % and 8 %, respectively. Marker *D* (*green downward-pointing triangle*) has a lower probability because of its distance from the GW sky location estimate. Additionally, *grayscale asterisk markers* are for all galaxies in this sky region, with the shade of the markers corresponding to each galaxies B-band luminosity. The *darkest markers* are the most luminous in the B-band. The BNS coalescence signal (*red square*) was injected at a sky location and distance and in this case it corresponds to the top ranking galaxy candidate. The simulated signal has an optimal network SNR 27.89 (Color figure online)

a simulated GW signal was injected. The galaxies in this region of the sky are plotted with grayscale asterisk markers. Bolder markers are galaxies with strong B-band luminosities. The top three galaxy candidates are marked by circles and the top galaxy marked by the largest circle is the injection. It is interesting to note that galaxy marked as D in Fig. 2, which is within the $1\,\sigma$ skymap error area, is automatically excluded by the incompatible distance estimate provided by the GW signal analysis. One of merits of this Bayesian approach is that the galaxy candidates are presented by ranking *and* posterior probability. In this case, the top three candidates in rank have 72 %, 15 % and 8 % posterior probability of being the host galaxy, respectively. Therefore, EM follow-up teams could, with relative confidence, focus only on the top candidate in rank.

3 Enhanced Inclination Angle Inference

Once the top ranked galaxies are identified, the distance of the candidates can be used to provide an improved estimate on the inclination angle (ι) of the CBC progenitor. For example, each candidate in the GW signal error region is assigned an estimate of the posterior probability of hosting the GW signal. Only the GW posterior samples within certain distance ranges should have non-zero probability. Galaxies with non-zero probability tend to be at the center of a distance range with 10 % of that distance as the radius of that range. Therefore, the GW posterior distribution combined with the probabilities of the galaxy candidates hosting the GW source at certain distances can lead to a better estimation on inclination angle.

In Fig. 3, for the same simulation shown in Sect. 2, we see that the posterior probability peaks strongly around the correct value of inclination. Furthermore, the inclination angle (ι) posterior distribution is now much narrower given the additional distance information provided by the galaxy candidates. In this example, the standard deviation of $\cos(\iota)$ (0.001) obtained from joint GW-galaxy information is much smaller than that (0.02) obtained via a GW analysis alone. It is worth noting that the host galaxy for the GW signal does not need to be the top ranked galaxy for the inclination angle inference to improve. Even the true host galaxy is not be identified due to a cluster of galaxies within the GW location estimate, the additional distance information provided by all galaxies within the GW location estimate will still reduce the width of the inclination angle posterior.

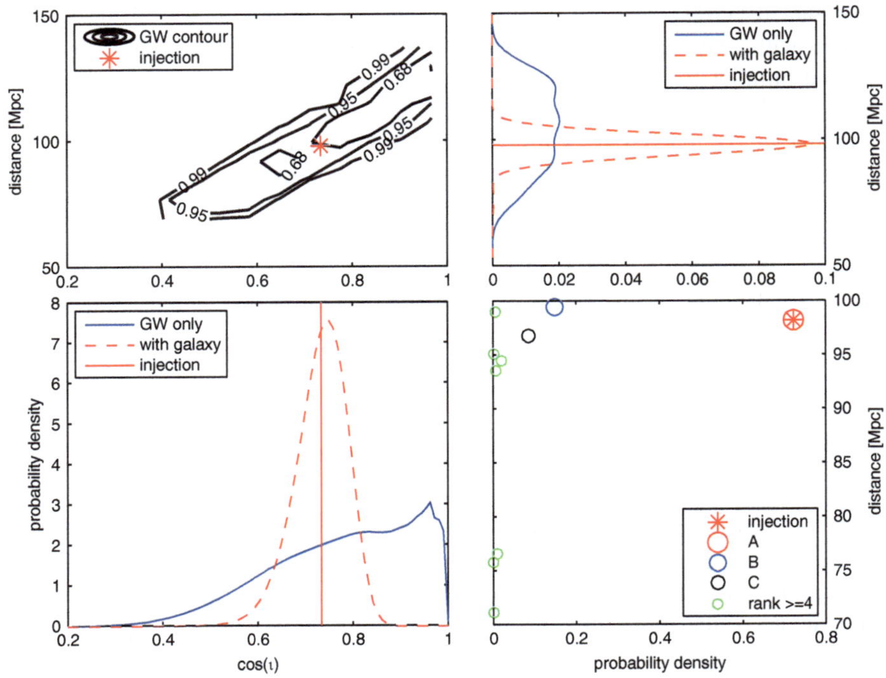

Fig. 3 An example showing the reduction in the degeneracy between distance and inclination angle ι. *Top-left panel* correspond to GW posterior samples contour. *Bottom-right panel* correspond to posterior probability of host galaxy, and plotted are *circle markers* corresponding to the first (*red*), second (*blue*) and third (*black*) ranked host galaxy candidates, labelled *A*, *B* and *C* respectively, as determined using our Bayesian approach to multi-messenger astronomy. *Top-right* and *bottom-left panels* correspond to probability density function of distance and $\cos(\iota)$ estimated by Kernel smoothing function estimate, respectively. *Blue lines* and *red dished lines* correspond to the probability density function using only GW posterior samples and GW-galaxy information, respectively. Injection value of distance and $\cos(\iota)$ are shown in *red lines*. Using GW-galaxy information, the standard deviation of $\cos(\iota)$ (0.001) is much smaller than that (0.02) using only GW posterior samples. The simulated signal is the same one used in Fig. 2 (Color figure online)

Conclusions

The implication of a Bayesian approach to multi-messenger astronomy is presented using a NS-NS inspiral GW signal injected into simulated advanced detectors noise assuming sky and distance information drawn from the GWGC. A merit of this approach is that both the rank and the posterior probability of a galaxy hosting the GW source are estimated with the help of a GW-galaxy relation model (blue band luminosity based multi-messenger prior function in this case study). With this information, EM follow-up observation could focus on the first galaxy in the rank with a very high probability. Once

(continued)

the host galaxy is identified, the additional distance information could benefit the inference of the inclination angle of CBC, shown in our example as the reduction of the variance in the estimation on inclination angle. In this case study, the posterior probabilities of the top three candidates being the host galaxy in rank are 72 %, 15 % and 8 %, respectively. The standard deviation of cosine inclination angle (0.001) using GW-galaxy information is much smaller than that (0.02) using only GW posterior samples. The results reported in this research depend on various selection effects, such as the dependence of the EM data on the common parameter set ($\boldsymbol{\lambda}$) (e.g. multi-messenger prior function, see discussion in Fan et al. 2014), the completeness of the EM data. The limited range of the GWGC with respect to the expect range for Advanced detectors highlights the need to take into account selection effects introduced by the catalogue and elsewhere. We will address the completeness issue in next work (Messenger et al. 2014).

Acknowledgements We would like to acknowledge valuable input from J. Kanner. The authors gratefully acknowledge the support of this research by the Royal Society, the Scottish Funding Council, the Scottish Universities Physics Alliance and the Science and Technology Facilities Council of the United Kingdom. XF acknowledges financial support from National Natural Science Foundation of China (grant No. 11303009). XF is a Newton Fellow supported by the Royal Society and CM is a Lord Kelvin Adam Smith Fellow supported by the University of Glasgow.

References

J. Aasi, J. Abadie, B.P. Abbott, R. Abbott, T. Abbott, M.R. Abernathy, T. Accadia, F. Acernese, C. Adams, T. Adams et al., Phys. Rev. D **88**(12), 122004 (2013). doi:10.1103/PhysRevD.88.122004

J. Aasi, J. Abadie, B.P. Abbott, R. Abbott, T. Abbott, M.R. Abernathy, T. Accadia, F. Acernese, C. Adams, T. Adams et al., ApJS **211**, 7 (2014). doi:10.1088/0067-0049/211/1/7

J. Abadie et al., Class. Quantum Gravity **27**, 173001 (2010). doi:10.1088/0264-9381/27/17/173001

J. Abadie, B.P. Abbott, R. Abbott, T.D. Abbott, M. Abernathy, T. Accadia, F. Acernese, C. Adams et al., A&A **539**, A124 (2012a). doi:10.1051/0004-6361/201118219

J. Abadie, B.P. Abbott, R. Abbott, T.D. Abbott, M. Abernathy, T. Accadia, F. Acernese, C. Adams, R.X. Adhikari, C. Affeldt et al., ApJ **760**, 12 (2012b). doi:10.1088/0004-637X/760/1/12

J. Abadie, B.P. Abbott, T.D. Abbott, R. Abbott, M. Abernathy, C. Adams, R. Adhikari, C. Affeldt, P. Ajith, B. Allen et al., ApJ **755**, 2 (2012c). doi:10.1088/0004-637X/755/1/2

B. Abbott, R. Abbott, R. Adhikari, J. Agresti, P. Ajith, B. Allen, R. Amin, S.B. Anderson, W.G. Anderson, M. Arain et al., ApJ **681**, 1419 (2008). doi:10.1086/587954

S. Ando, B. Baret, I. Bartos, B. Bouhou, E. Chassande-Mottin, A. Corsi, I. Di Palma, A. Dietz, C. Donzaud, D. Eichler, C. Finley, D. Guetta, F. Halzen, G. Jones, S. Kandhasamy, K. Kotake, A. Kouchner, V. Mandic, S. Márka, Z. Márka, L. Moscoso, M.A. Papa, T. Piran, T. Pradier, G.E. Romero, P. Sutton, E. Thrane, V. Van Elewyck, E. Waxman, Rev. Mod. Phys. **85**, 1401 (2013). doi:10.1103/RevModPhys.85.1401. http://link.aps.org/doi/10.1103/RevModPhys.85.1401

K.G. Arun, H. Tagoshi, C. Kant Mishra, A. Pai, Phys. Rev. D **90**, 024060 (2014). doi:10.1103/PhysRevD.90.024060
K. Belczynski, M. Dominik, T. Bulik, R. O'Shaughnessy, C. Fryer, D.E. Holz, ApJ **715**, L138 (2010). doi:10.1088/2041-8205/715/2/L138
L. Blackburn, M.S. Briggs, J. Camp, N. Christensen, V. Connaughton, P. Jenke, J. Veitch, ArXiv e-prints:1303.2174 (2013)
D. Eichler, M. Livio, T. Piran, D.N. Schramm, Nature **340**, 126 (1989). doi:10.1038/340126a0
P.A. Evans, J.K. Fridriksson, N. Gehrels, J. Homan, J.P. Osborne, M. Siegel, A. Beardmore, P. Handbauer, J. Gelbord, J.A. Kennea et al., ApJS **203**, 28 (2012). doi:10.1088/0067-0049/203/2/28
S. Fairhurst, Class. Quantum Gravity **28**(10), 105021 (2011). doi:10.1088/0264-9381/28/10/105021
X.L. Fan, J. Yin, F. Matteucci, A&A **521**, A73 (2010). doi:10.1051/0004-6361/201015293
X. Fan, C. Messenger, I.S. Heng, ApJ **795**, 43 (2014). doi:10.1088/0004-637X/795/1/43
W. Fong, E. Berger, R. Chornock, R. Margutti, A.J. Levan, N.R. Tanvir, R.L. Tunnicliffe, I. Czekala, D.B. Fox, D.A. Perley, S.B. Cenko, B.A. Zauderer, T. Laskar, S.E. Persson, A.J. Monson, D.D. Kelson, C. Birk, D. Murphy, M. Servillat, G. Anglada, ApJ **769**, 56 (2013). doi:10.1088/0004-637X/769/1/56
D.B. Fox, D.A. Frail, P.A. Price, S.R. Kulkarni, E. Berger, T. Piran, A.M. Soderberg, S.B. Cenko, P.B. Cameron, A. Gal-Yam, M.M. Kasliwal, D.S. Moon, F.A. Harrison, E. Nakar, B.P. Schmidt, B. Penprase, R.A. Chevalier, P. Kumar, K. Roth, D. Watson, B.L. Lee, S. Shectman, M.M. Phillips, M. Roth, P.J. McCarthy, M. Rauch, L. Cowie, B.A. Peterson, J. Rich, N. Kawai, K. Aoki, G. Kosugi, T. Totani, H.S. Park, A. MacFadyen, K.C. Hurley, Nature **437**, 845 (2005). doi:10.1038/nature04189
C.L. Fryer, K. Belczynski, G. Wiktorowicz, M. Dominik, V. Kalogera, D.E. Holz, ApJ **749**, 91 (2012). doi:10.1088/0004-637X/749/1/91
C. Hanna, I. Mandel, W. Vousden, ApJ **784**, 8 (2014). doi:10.1088/0004-637X/784/1/8
G.M. Harry, LIGO scientific collaboration. Class. Quantum Gravity **27**, 084006 (2010). doi:10.1088/0264-9381/27/8/084006
J. Kanner, T.L. Huard, S. Márka, D.C. Murphy, J. Piscionere, M. Reed, P. Shawhan, Class. Quantum Gravity **25**(18), 184034 (2008). doi:10.1088/0264-9381/25/18/184034
J. Kanner, J. Camp, J. Racusin, N. Gehrels, D. White, ApJ **759**, 22 (2012). doi:10.1088/0004-637X/759/1/22
C. Messenger, X. Fan, I.K. Heng, Galaxy catalogues as gravitational-wave priors: Selection bias and completeness (2014) (in preparation)
S. Nissanke, M. Kasliwal, A. Georgieva, ApJ **767**, 124 (2013). doi:10.1088/0004-637X/767/2/124
L.K. Nuttall, P.J. Sutton, Phys. Rev. D **82**(10), 102002 (2010). doi:10.1103/PhysRevD.82.102002
R. O'Shaughnessy, V. Kalogera, K. Belczynski, ApJ **716**, 615 (2010). doi:10.1088/0004-637X/716/1/615
B. Paczynski, ApJ **308**, L43 (1986). doi:10.1086/184740
B.F. Schutz, Class. Quantum Gravity **28**(12), 125023 (2011). doi:10.1088/0264-9381/28/12/125023
T. Sidery, B. Aylott, N. Christensen, B. Farr, W. Farr, F. Feroz, J. Gair, K. Grover, P. Graff, C. Hanna, V. Kalogera, I. Mandel, R. O'Shaughnessy, M. Pitkin, L. Price, V. Raymond, C. Röver, L. Singer, M. van der Sluys, R.J.E. Smith, A. Vecchio, J. Veitch, S. Vitale, Phys. Rev. D **89**(8), 084060 (2014). doi:10.1103/PhysRevD.89.084060
The Virgo Collaboration, Technical Report VIR-0027A-09 (2009)
J. Veitch, A. Vecchio, Phys. Rev. D **81**(6), 062003 (2010). doi:10.1103/PhysRevD.81.062003
L. Wen, Y. Chen, Phys. Rev. D **81**(8), 082001 (2010). doi:10.1103/PhysRevD.81.082001
L. Wen, X. Fan, Y. Chen, J. Phys. Conf. Ser. **122**(1), 012038 (2008). doi:10.1088/1742-6596/122/1/012038
D.J. White, E.J. Daw, V.S. Dhillon, Class. Quantum Gravity **28**(8), 085016 (2011). doi:10.1088/0264-9381/28/8/085016
S.E. Woosley, J.S. Bloom, ARA&A **44**, 507 (2006). doi:10.1146/annurev.astro.43.072103.150558

X, Gamma-Rays, and Gravitational Waves Emission in a Short Gamma-Ray Burst

F.G. Oliveira, Jorge A. Rueda, and R. Ruffini

Abstract The recent progress in the understanding the physical nature of neutron stars (NSs) and the first observational evidence of a genuinely short gamma-ray burst (GRB), GRB 090227B, allow to give an estimate of the gravitational waves versus the X and gamma-rays emission in a short GRB. NS binaries represent good candidates for the detection of gravitational waves emitted during the spiraling-in and final merging phase of the system that leads to the short GRB emission. The data analysis of the GRB 090227B by Muccino et al. (2013) have been shown to be consistent with a NS binary progenitor with masses $M_1 = M_2 = 1.34\,M_\odot$, radii $R_1 = R_2 = 12.2$ km, and a crust thickness $\Delta r \approx 0.47$ km, obtained from the new mass-radius relation by Belvedere et al. (2012) of NSs fulfilling global charge neutrality. Muccino et al. (2013) estimated that GRB 090227B is located at redshift $z \approx 1.6$, corresponding to a luminosity distance $d_L \approx 12.2$ Gpc. We assess the detectability of this source by the Advanced LIGO interferometer computing the signal-to-noise ratio (SNR) averaged over all polarizations and possible positions of the source with respect to the interferometer. We simulate the dynamics of the binary up to the contact point using the effective one-body formalism (EOB) in the fourth post-Newtonian approximation. We find that the gravitational waves signal would have been produced an SNR = 0.32 for a redshift $z = 1.61$. We find that, instead, this GRB would have been detected with an SNR = 8 if it would have been located at a redshift $z \approx 0.05$, or $d_L \approx 200$ Mpc.

1 Introduction

The connection between short GRB and gravitational waves signals as a coincidence of the same event would allow in principle to understand more about the origin of short GRBs (Kobayashi and Meszaros 2003). The first observational evidence of a genuinely short GRB, GRB 090227B (Muccino et al. 2013), offers the possibility of making the first joint analysis of the electromagnetic (X and gamma-rays) and

F.G. Oliveira (✉) • J.A. Rueda • R. Ruffini
Dipartimento di Fisica and ICRA, Sapienza Università di Roma, P.le Aldo Moro 5, 00185 Rome, Italy
e-mail: fe.fisica@gmail.com; jorge.rueda@icra.it; ruffini@icra.it

C.F. Sopuerta (ed.), *Gravitational Wave Astrophysics*, Astrophysics and Space Science Proceedings, DOI 10.1007/978-3-319-10488-1_4

gravitational waves emission in a short GRB. Genuine short GRBs have been theoretically predicted by the Fireshell model (Ruffini et al. 2001, 2002) as bursts with the same inner engine as the long bursts but endowed with a severely low value of the baryon load $B = M_B c^2/E^{tot}_{e+e-} \lesssim 10^{-5}$, where M_B is the mass of the baryons engulfed by the expanding ultrarelativistic e^+e^- plasma with total energy E^{tot}_{e+e-}. The emission from short GRBs mainly consists in a first emission, the proper GRB (P-GRB), followed by a softer emission squeezed on the first one. The typical separation between the two components in the lightcurve is expected to be shorter than 1–10 ms and therefore there is no afterglow emission from these sources.

It is widely accepted that the most likely progenitors of short GRBs, based on their observational features, are binary neutron star mergers. The emission of gravitational waves signals from such binary systems are the most expected signals to be detect by the interferometers called Advanced LIGO[1]-VIRGO[2] and they have been planned for to be operational in a few years with a improved sensitivity approximately a factor of 10 better than the first generation of detectors.

The time-resolved spectral analysis of the Fermi-GBM and Konus-Wind satellites data of GRB 090227B has led to an estimate of its baryon load of $B = (4.13 \pm 0.05) \times 10^{-5}$ (Muccino et al. 2013). The parameters inferred for GRB 090227B (see below) have led indeed to the identification of the progenitor of the genuine short GRB in a neutron star binary: (1) the natal kicks velocities imparted to a neutron star binary at birth can be even larger than $200\,\mathrm{km\,s^{-1}}$ and therefore a binary system can runaway to the halo of its host galaxy, clearly pointing to a very low average number density of the circumburst medium (CBM); (2) the very large total energy, which we can indeed infer in view of the absence of beaming, and the very short time scale of emission point again to a neutron star binary; (3) as we shall show the very small value of the baryon load is strikingly consistent with two neutron stars having small crusts, in line with the recent neutron star theory (Rotondo et al. 2011; Rueda et al. 2011; Belvedere et al. 2012).

The aim of this work is to give a brief review of the recent results (Oliveira et al. 2014) of the gravitational waves emission expected from GRB 090227B and how it compares and contrasts with the electromagnetic emission. In doing so the structure parameters of the neutron star components are determined and a comparison of the classic description of the orbital decay with respect to the dynamics accounting for general relativity corrections is also presented.

[1] http://www.advancedligo.mit.edu.

[2] http://www.cascina.virgo.inft.it.

2 The Short GRB 090227B

We first recall that the canonical GRB within the Fireshell model has two components: an emission occurring at the transparency of the optically thick expanding e^+e^- baryon plasma (Ruffini et al. 2000), the Proper-GRB (P-GRB), followed by the extended afterglow, due to the interactions between the accelerated baryons and the CBM of average density $\langle n_{CBM} \rangle$. Such an extended afterglow comprises the prompt emission as well as the late phase of the afterglow (Bianco and Ruffini 2005a, b). The relative energy of these two components, for a given total energy of the plasma $E^{tot}_{e+e-} = E^{GRB}_{tot}$, where E^{GRB}_{tot} is the observed GRB energy, is uniquely a function of the above defined baryon load B.

According to the data analysis of GRB 090227B: $E^{GRB}_{tot} = 2.83 \times 10^{53}$, $B = 4.13 \times 10^{-5}$, the Lorentz factor at transparency is $\Gamma_{tr} = 1.44 \times 10^4$, the cosmological redshift is $z = 1.61$, the intrinsic duration of the GRB is $\Delta t = 0.35$, and $\langle n_{CBM} \rangle = 1.9 \times 10^{-5}$; we refer to Muccino et al. (2013) for further details. As we mentioned, an extremely low value of the baryon load, $B \lesssim 10^{-5}$ together with a low density of the CBM are imprints of a genuinely short GRB, in which no afterglow emission is observed. This is indeed the case of GRB 090227B.

The location of the binary in the very low interstellar density medium of galactic halos makes possible to probe the neutron star theory and equation of state (EoS) via the knowledge of the baryon load B inferred from the fitting of the GRB light curve and spectrum. For the modeling of the neutron star we follow (Belvedere et al. 2012) where configurations of equilibrium were computed satisfying the strong, weak, electromagnetic and gravitational interactions in a general relativistic framework and the condition of global charge neutrality, instead of the traditionally adopted ansatz of local charge neutrality, which have been shown to be incompatible with the equilibrium equations of motion (see Rotondo et al. 2011; Rueda et al. 2011, for further details).

For the nuclear EoS we use here the NL3, NL-SH, TM1 and TM2 models, which lead to critical masses and corresponding radii of globally neutral neutron stars: ($2.67\,M_\odot$, $R = 12.33$ km; $2.68\,M_\odot$, $R = 12.54$ km; $2.58\,M_\odot$, $R = 12.31$ km; and $2.82\,M_\odot$, $R = 13.28$ km), respectively (Belvedere et al. 2012).

The baryonic matter which the GRB interact with is in these systems provided by the material of the neutron star crusts ejected during the binary coalescence. Thus, a theoretical expectation of the baryon load B left in a binary neutron star merger is $B_{\rm th} = \eta M_{\rm crust} c^2 / E^{GRB}_{tot}$, where η is the fraction of the crustal mass ejected. In Fig. 1 we have plotted $B_{\rm th}$ for GRB 090227B, namely using $E^{GRB}_{tot} = 2.83 \times 10^{53}$ erg, as a function of the mass M of the globally and locally neutral neutron stars (Belvedere et al. 2012).

The agreement of the observed baryon load of GRB 090227B with the low mass of the crust obtained from the globally neutral neutron stars of Belvedere et al. (2012) is evident. It can be compared and contrasted with the ones obtained enforcing the local charge neutrality condition. For the specific binary neutron star system adopted here we obtain a theoretical prediction of the baryon load with $\eta = 1$, $B_{\rm th} \sim 4.5 \times 10^{-4}$, or a mass of the baryons $M_B = E^B_{\rm crust}/c^2 \sim 7.2 \times 10^{-5}\,M_\odot$,

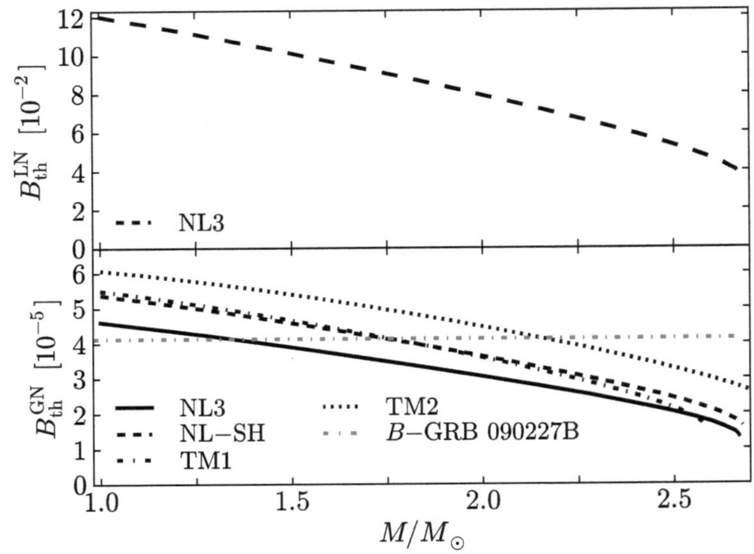

Fig. 1 Baryon load expected to be left by a binary neutron star merger, given by $B_{\rm th} = \eta M_{\rm crust}c^2/E_{tot}^{GRB}$ for $\eta = 1$, as a function of the total mass M of globally (*lower panel, solid black curve*, units 10^{-5}) and locally neutral (*upper panel, dashed black curve*, units 10^{-2}) neutron stars, for the case of GRB 090227B. We have indicated the observed baryon load of GRB 090227B, $B = 4.13 \times 10^{-5}$, with the *dotted-dashed gray horizontal* (Muccino et al. 2013)

to be confronted with the one obtained from the fitting procedure of GRB 090227B, $B \sim 4.13 \times 10^{-5}$, corresponding to $M_B = B \times E_{tot}^{GRB}/c^2 \sim 0.7 \times 10^{-5}\,{\rm M}_\odot$. A perfect agreement would require $\eta \approx 0.1$ for the NL3 nuclear model, while the other nuclear parameterizations require a slightly lower value of η. These theoretical predictions of the neutron star crust mass $M_{\rm crust}$ and the value of $E_{\rm crust}^B$ and $B_{\rm th}$ have been inferred for a crust with a density at its edge equal to the neutron drip density, $\rho_{\rm drip} \sim 4.3 \times 10^{11}$ g cm^{-3}. Neutron star crusts with densities lower than $\rho_{\rm drip}$ are predicted by the globally neutral neutron stars (Belvedere et al. 2012), and therefore there is still the possibility of having smaller values of the baryonic matter ejected in a binary process, and consequently to still shorter genuinely short GRBs.

3 Gravitational Wave Emission

We turn to the analysis of the binary dynamics. Classically, the loss of orbital binding energy by emission of gravitational waves from the binary in spiral phase is obtained for non-relativistic and point-like particles, and can be written as a function of the gravitational waves frequency f as Landau and Lifshitz (1980)

$$\frac{dE_b}{df} = -\frac{1}{3}(\pi G)^{2/3}\mathscr{M}^{5/3}f^{-1/3}, \tag{1}$$

where $\mathcal{M} = (M_1 M_2)^{3/5}/(M_1 + M_2)^{1/5}$ is the called chirp mass. From the above equation is already possible to estimate the characteristic amplitude of the gravitational waves, see below Eq. (6).

Now we briefly review how to obtain the gravitational waves energy spectrum, dE/df, through the effective one-body (EOB) dynamics (Buonanno and Damour 1999; Damour 2000; Damour and Nagar 2010). The EOB formalism maps the conservative dynamics of a binary system of non spinning objects onto the geodesic dynamics of one body of reduced mass $\mu = M_1 M_2/M$, with $M = M_1 + M_2$ the total binary mass. The effective metric is a modified Schwarzschild metric where the rescaled radial coordinate $r = c^2 r_{12}/(GM)$ has been introduced, with r_{12} the distance between the two stars. The radial potential is given by

$$A(u; \nu) = 1 - 2u + 2\nu u^3 + a_4 \nu u^4 + a_5 \nu u^5, \tag{2}$$

where $u = 1/r = GM/(c^2 r_{12})$, $\nu = M_1 M_2/(M_1 + M_2)^2$ is the symmetric mass ratio, with the values of the 3 and 4 post-Newtonian (PN)-level coefficients given by $a_4 = 94/3 - (41/32)\pi^2$ and $a_5(\nu) = a5^{c0} + \nu a5^{cl}$ see details in Bini and Damour (2013). We will denote to as P_n^m the Padè approximant of order (n, m), which when applied to $A(u; \nu)$ ensures the convergence of the solution near the merger point (Damour and Nagar 2009).

The EOB Hamiltonian is $H = Mc^2 \sqrt{1 + 2\nu(\hat{H}_{\rm eff} - 1)}$, and effective Hamiltonian is described by $\hat{H}_{\rm eff}^2 = A(u) + p_\phi^2 B(u)$, where $B(u) = u^2 A(u)$ and the angular momentum for the circular orbit is given by $p_\phi^2 = -A'(u)/[u^2 A(u)]'$, where prime stands for derivative with respect to u.

We need to write $\hat{H}_{\rm eff}$ as a function of the orbital angular velocity Ω, or orbital frequency f. For this, we need to write the u-parameter as a function of Ω, or f, which is obtained from the angular Hamilton equation of motion in the circular case

$$GM\Omega(u) = \frac{1}{u}\frac{\partial H}{\partial p_\phi} = \frac{MA(u) p_\phi(u) u^2}{H \hat{H}_{\rm eff}}. \tag{3}$$

The binding energy as a function of the orbital frequency is

$$E_b(\Omega) = H - Mc^2 = Mc^2[\sqrt{1 + 2\nu(\hat{H}_{\rm eff} - 1)} - 1], \tag{4}$$

and the gravitational energy spectrum is obtained through the derivative $dE_b/d\Omega$. The signal-to-noise ratio (SNR) making an rms average over all the possible source orientations, positions, and wave polarizations is given by,

$$\langle {\rm SNR}^2 \rangle = \int_{f_{\rm min}}^{f_{\rm max}} df_d \frac{h_c^2(f_d)}{5 f_d^2 S_h^2(f_d)}, \tag{5}$$

where $S_h(f)$ is the strain noise spectral density (units $1/\sqrt{\text{Hz}}$) in the interferometer and we have introduced the characteristic gravitational waves amplitude, h_c, defined using the Fourier transform of the gravitational waveform $h(t)$, $h_c(f) = f|\tilde{h}(f)|$, and it is given by

$$h_c^2(f) = \frac{2(1+z)^2}{\pi^2 d_L^2} \frac{dE_b}{df}[(1+z)f_d]. \tag{6}$$

with z the cosmological redshift, $f_d = f/(1+z)$ the gravitational wave frequency at the detector, $f = \Omega/\pi$ the frequency in the source frame, Ω is the orbital frequency, the minimal bandwidth frequency of the detector is f_{min}, and $f_{\text{max}} = f_c/(1+z)$ is the maximal bandwidth frequency, where $f_c = \Omega_c/\pi$ is the binary contact frequency and d_L is the luminosity distance.

Summarizing, we have shown that the observations of the genuinely short GRB 090227B lead to crucial information on the binary neutron star progenitor. The data obtained from the electromagnetic spectrum allows to probe crucial aspects of the correct theory of neutron stars and their equation of state.

The baryon load parameter B obtained from the analysis of GRB 090227B, leads to a most remarkable agreement of the baryonic matter expected to be ejected in a neutron star binary merger and validate the choice of the parameters of the binary components, $M_1 = M_2 = 1.34\,M_\odot$, and $R_1 = R_2 = 12.24$ km. This represents a test of the actual neutron star parameters described by the recent developed self-consistent theory of neutron stars (Belvedere et al. 2012) that takes into account the strong, weak, electromagnetic and gravitational interactions within general relativity, and satisfying the condition of global charge neutrality.

We estimate the detectability of GRB 090227B by the Advanced LIGO interferometer, by computing the SNR integrated up to the contact point of the binary components, for the theoretically inferred cosmological redshift, $z = 1.61$ (Muccino et al. 2013); see Fig. 2. We find that at such a redshift, the gravitational waves signal would produce a $\langle \text{SNR} \rangle \approx 0.32$, a value lower than the one needed for a positive detection, $\langle \text{SNR} \rangle = 8$. We turn to estimate the redshift at which Advanced LIGO would detect this GRB with a $\langle \text{SNR} \rangle = 8$ we obtained $z \approx 0.05$ or a distance to the source $d_L \approx 200$ Mpc. Unfortunately, in the last 40 years, no such a GRB has been observed.

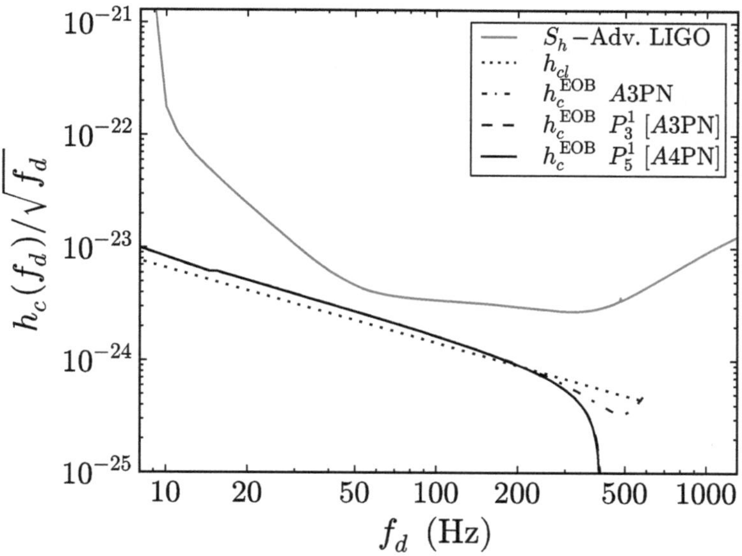

Fig. 2 Comparison of the characteristic gravitational waves amplitude per unit square frequency, $h_c(f_d)/\sqrt{f_d}$, see Eq. (6), with the noise density spectrum $S_h(f)$ of the Advanced LIGO interferometer. We use the binary neutron star parameters inferred for the short GRB 090227B, including the cosmological redshift $z = 1.61$. The comparison is made for both the dynamics given by the non-relativistic point-like particles approximation (*dotted black curve*) and the dynamics obtained from the EOB formalism. In the case of the EOB approach, the radial potential $A(u;\nu)$ was calculated using post-Newtonian approximation (PN). The *dotted-dashed black curve* is $A(u;\nu) = 3\,\text{PN}$, using the Padè approximant we calculated the $P_3^1[A(u;\nu) = 3\,\text{PN}]$ (*dashed black curve*) and the $P_5^1[A(u;\nu) = 4\,\text{PN}]$ (*solid black curve*). The noise spectral density of Advanced LIGO, $S_h(f)$, is represented by the *solid gray curve*

Acknowledgements F.G. Oliveira acknowledges the support given by the International Relativistic Astrophysics Erasmus Mundus Joint Doctorate Program under the Grant 2012-1710 from EACEA of the European Commission.

References

R. Belvedere, D. Pugliese, J.A. Rueda, R. Ruffini, S.S. Xue, Nucl. Phys. A **883**, 1 (2012)
C.L. Bianco, R. Ruffini, ApJ **620**, L23 (2005a)
C.L. Bianco, R. Ruffini, ApJ **633**, L13 (2005b)
D. Bini, T. Damour, Phys. Rev. D **87**, 121501 (2013)
A. Buonanno, T. Damour, Phys. Rev. D **59**, 084006 (1999)
T. Damour, Phys. Rev. D **62**, 064015 (2000)
T. Damour, A. Nagar, Phys. Rev. D **79**, 081503 (2009)
T. Damour, A. Nagar, Phys. Rev. D **81**, 084016 (2010)
S. Kobayashi, P. Meszaros, ApJ **589**, 861–870 (2003)
L.D. Landau, E.M. Lifshitz, *Statistical Physics. Part1* (Pergamon Press, Oxford, 1980)
M. Muccino, R. Ruffini, C.L. Bianco, L. Izzo, A.V. Penacchioni, ApJ **763**, 125 (2013)

F.G. Oliveira, J.A. Rueda, R. Ruffini, ApJ **787**, 150 (2014)
M. Rotondo, J.A. Rueda, R. Ruffini, S.S. Xue, Phys. Lett. B **701**, 667 (2011)
J.A. Rueda, R. Ruffini, S.S. Xue, Nucl. Phys. A **872**, 286 (2011)
R. Ruffini, J.D. Salmonson, J.R. Wilson, S.S. Xue, A&A **359**, 855 (2000)
R. Ruffini, C.L. Bianco, P. Chardonnet, F. Fraschetti, S.S. Xue, ApJ **555**, L113 (2001)
R. Ruffini, C.L. Bianco, P. Chardonnet, F. Fraschetti, S.S. Xue, ApJ **581**, L19 (2002)

Localizing Gravitational Wave Sources with Optical Telescopes and Combining Electromagnetic and Gravitational Wave Data

Shaon Ghosh and Gijs Nelemans

Abstract Neutron star binaries, which are among the most promising sources for the direct detection of gravitational waves (GW) by ground based detectors, are also potential electromagnetic (EM) emitters. Gravitational waves will provide a new window to observe these events and hopefully give us glimpses of new astrophysics. In this paper, we discuss how EM information of these events can considerably improve GW parameter estimation both in terms of accuracy and computational power requirement. And then in return how GW sky localization can help EM astronomers in follow-up studies of sources which did not yield any prompt emission. We discuss how both EM source information and GW source localization can be used in a framework of multi-messenger astronomy. We illustrate how the large error regions in GW sky localizations can be handled in conducting optical astronomy in the advance detector era. We show some preliminary results in the context of an array of optical telescopes called BlackGEM, dedicated for optical follow-up of GW triggers, that is being constructed in La Silla, Chile and is expected to operate concurrent to the advanced GW detectors.

1 Introduction

The second generation GW detectors are scheduled to come online from 2015 and should be operating at their design sensitivities by the end of 2019 (Aasi et al. 2013). A worldwide network of these advanced detectors will enable us in conducting GW astronomy for the first time in history. Compact binary coalescing (CBC) systems are the most promising candidates for detection of gravitational waves by these detectors as their GW frequency sweep through the sensitivity band towards merger. Binary neutron stars (BNS) and neutron star black hole binaries (NSBH) are also prospective progenitors of short duration gamma ray bursts (GRB). Current models of GRB mechanisms (Nakar 2007; Piran 1999) predicts that the prompt emissions are launched just before the merger of the compact binary. This presents

S. Ghosh (✉) • G. Nelemans
Radboud University Nijmegen and Nikhef, The Netherlands
e-mail: shaon@astro.ru.nl; nelemans@astro.ru.nl

C.F. Sopuerta (ed.), *Gravitational Wave Astrophysics*, Astrophysics and Space Science Proceedings, DOI 10.1007/978-3-319-10488-1_5

to us a unique opportunity for joint EM-GW astronomy. EM astronomy could complement GW astronomy in a number of ways. On one hand, gravitational wave sky-localization for the advanced ground based detectors could have large uncertainties, typically ~10–100 square degrees (Fairhurst 2009). Typical sky-localization of arc-second accuracy could be achieved in conventional EM astronomy. On the other hand, gravitational waves observation allows us to make an independent and direct measurement of masses and luminosity distances of these sources. Evidently, a tremendous amount of scientific merit lies in combining EM and GW observations. Implementation of EM information for detection of gravitational waves is already an active area of research. LIGO Scientific collaboration (LSC) and Virgo Scientific collaboration have been conducting search for gravitational waves from progenitors of short duration gamma ray bursts (sGRB) by following up on GCN (Gamma-ray Coordination Network) triggers (Abadie et al. 2010, 2012). In the context of GW parameter estimation, the use of EM source information gets an even richer dimension. A non-spinning compact binary system can be characterized by nine parameters, the luminosity distance d_L, the binary component masses m_1 and m_2, the sky position of the binary (α, δ), the inclination angle ι of the axis of the binary w.r.t the observer's line of sight, the polarization angle ψ of the binary orbit, the time of arrival of the signal t_a and the coalescence phase of the binary δ_0. EM information can fix (or constrain) a subset of these parameters reducing the dimensionality of the posterior probability distribution of Röver et al. (2007) and Shah et al. (2012) which directly translate into considerable reduction in computational power requirements for the estimation of the source parameters. In this paper, we first show in Sect. 2 how the EM information can improve the estimation of the parameters. We examine how different types of EM information can help in the estimation of inclination angle of binary and luminosity distance to the source. Then in Sect. 3 we explore how GW parameter estimations can in turn aid EM astronomy. Here we focus entirely on sky localization of gravitational wave sources. We discuss how to generate telescope pointing for optical telescope from error regions of the sky localization. In that context, we acknowledge the efforts towards construction of a dedicated GW follow-up facility.

2 Gravitational Wave Parameter Estimation in Presence of Electromagnetic Information

Most likely sources of electromagnetic information for gravitational wave parameter estimation could be in the form of prompt emissions like short duration gamma ray bursts, or from associated afterglows and kilonovae. A fully coherent parameter estimation incorporates a Bayesian Markov chain Monte Carlo technique which is computationally expensive. Thus it is extremely important that we choose a representative system. For this study, we injected inspiral gravitational wave signals from the TaylorT4 waveform family with order 3.5PN (Boyle et al. 2007) in initial

LIGO colored Gaussian noise. The source was chosen to be a neutron star—black hole binary (NSBH), with a $1.4M_{\odot}$ neutron star and a $10M_{\odot}$ black hole, inclined at angle of 10° w.r.t the observers line of sight, located at a distance of $\sim$30 Mpc from earth.

The choice of a NSBH system is motivated by the fact that a fair amount of research is ongoing on binary neutron star (BNS) systems while parameter estimation studies of NSBH systems are virtually non-existent. The choice of the inclination angle was motivated by the fact that even though there exists some uncertainties regarding the bounds on short GRB opening angles, recent studies (Fong et al. 2014) estimated the median opening angles of short GRBs to be $\approx 10^{\circ}$.

For the comparative analysis of how EM information can help GW parameter estimation we conducted three classes of studies. The first study constitutes the control of our experiment in the form of "blind" parameter estimation where we have no EM information available at our disposal. In the second class of analysis, we used the sky position of the source information, an astrophysical instance of which would be the case where a GRB was detected and located by the gamma ray observatory. And finally the third class of analysis involved the use of sky position and distance information. This corresponds to the case where a GRB and a subsequent afterglow associated with the same were discovered thus conveying to us the sky position and the distance information in the form of redshift. We found that not all parameters displayed improvement in estimation upon use of EM information. However, at the same time we note that certain parameters that are strongly correlated with another parameter, showed considerable improvement in its estimation upon using the correlated parameter EM information. As an example, we show the estimation of inclination angle in the presence of various EM information in Fig. 1. We note from this study that an sky position information does not improve inclination angle estimation. However, knowledge of distance greatly improves the precision and accuracy of the estimation as shown in Fig. 1 (top left). This is due to the well known strong correlation between the inclination angle and distance that we have shown in Fig. 1 (top right). Though we have seen in Fig. 1 (bottom left) that the knowledge of sky position itself do not provide any improvement in the estimation of distance, however one should note that if the knowledge of the sky position came from the detection of a GRB, then one might put constrains on the inclination angle of the source. This stems from the fact that for a GRB to be observed it would indicate that the source might be more favorably oriented to the observer. We show that as a result of constraining of inclination angle to 20°, the estimation of the distance in Fig. 1 (bottom right) improves compared to unconstrained distance estimation.

Studies of edge-on systems revealed some interesting features. Edge-on systems are different from face-on systems owing to the lack of the cross-polarization. Parameter estimation algorithm very quickly realizes its absence in the date and consequently estimates the inclination angle very precisely. This is observed in Fig. 2 (left), where we have compared the estimation of the inclination angle for an edge-on source with face-on oriented source. We conducted a similar study for sources with EM information in the form of sky position in Fig. 2 (right). While

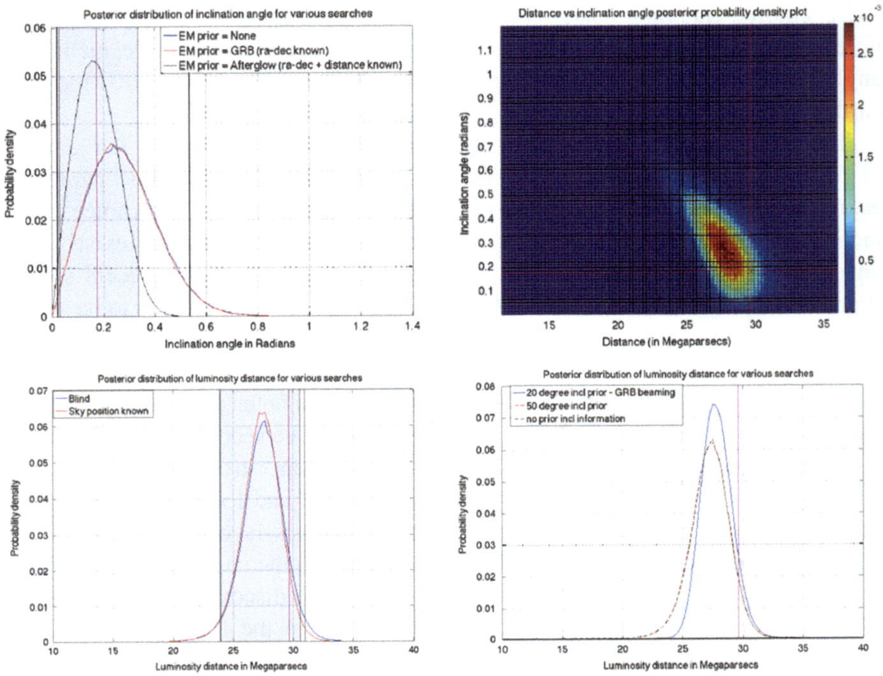

Fig. 1 *Top left*: Estimation of inclination angle for the three classes of studies, blind, sky position known and sky position and distance known. The *shaded region* denotes the 2σ spread of the distribution and the *vertical magenta line* denotes the actual injected inclination angle. Note that in the presence of distance information the inclination angle estimation gets better. *Top right*: The $2D$ posterior probability distribution plot for the inclination angle and distance shows the strong correlation between the two parameters. *Bottom left*: No measurable improvement in estimation of luminosity distance upon utilizing sky position information. *Bottom right*: Improvement in estimation of distance in the presence of inclination angle constraints (Color figure online)

sky position and inclination do not have strong correlation and therefore not much benefit is expected in the inclination angle estimation from sky position information, as we have seen in Fig. 1 (bottom left) red and blue plots, the situation is very different for edge-on systems. In Fig. 2 we compare the results of inclination angle estimation of edge-on and face-on systems in the absence (left) and presence (right) of sky position information. It is clear that the presence of sky position information strongly aids the inclination angle estimation for the edge-on configuration, unlike the previously shown face-one case. This can be explained as follows. The strain at a detector due to a particular polarization does not depend only on the inclination angle but also depends on the sky position.

$$h(t) = F_{+}(\theta,\phi,\psi)h_{+}(r,\iota,m_1,m_2,\delta_c,t_c) + F_{\times}(\theta,\phi,\psi)h_{\times}(r,\iota,m_1,m_2,\delta_0,t_c)\,, \tag{1}$$

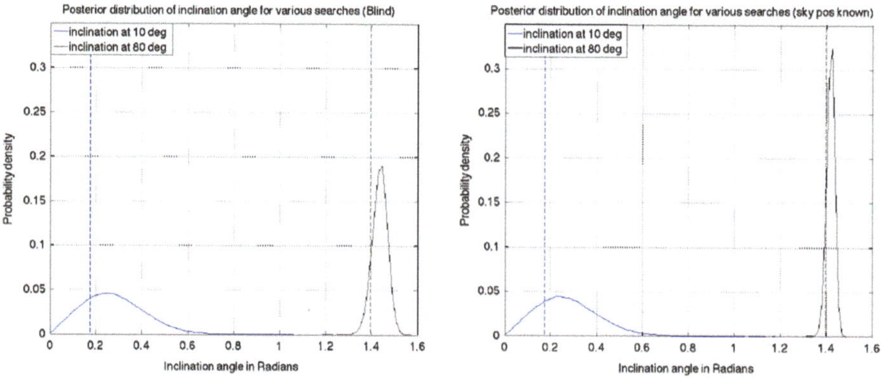

Fig. 2 *Left*: Inclination angle posterior probability distributions showing dependence on the inclination angle of the injected source. We observe that the edge-on source has the smallest spread. Note that higher PDFs are narrower. *Right*: Similar study conducted for systems for which the sky position is known from EM observations

where the $F_+(\theta, \phi, \psi)$ and $F_\times(\theta, \phi, \psi)$ are the antenna pattern functions and h_+ and $h_\times$ are the two gravitational wave polarization components. In absence of sky position information, we essentially have the freedom to change the antenna pattern functions during the estimation of the inclination angle and a wider posterior probability distribution could be accommodated due to this freedom. Thus, when the sky position information is used, we are likely to get narrower estimate of the inclination angle.

3 Gravitational Wave Sky Localization for Electromagnetic Follow-Up Studies

Past observations have not yielded any detection of short duration GRB within the typical advanced LIGO range $\sim$ 300 Mpc, for which we have redshift information (Nakar 2007). This could partly be explained by the lower event rate compared to their longer duration counterparts and strong beaming mechanism. Gravitational wave radiation on the other hand is expected to be more isotropic. Thus, gravitational wave source localization could act as a pointer for optical follow-up studies of short GRBs. Short GRB afterglows progressively gets more isotropic as we go from lower wavelength X-ray to higher wavelength optical band. An optical afterglow associated with a prompt emission is expected to be visible by ground based telescopes for typically $\sim$1 day. Kilonovae produced by r-process decay of neutron rich radioactive elements created due the disruption of the neutron star in a neutron star binary is expected to be highly isotropic. The emission from these events could predominantly be in the near-infrared regime (Tanvir et al. 2013; Barnes and Kasen 2013). These events might or might not accompany a short

GRB, thus have the potential of being an optical/near-infrared transient candidate associated with a class of compact binary mergers independent of whether or not they produce a prompt emission.

In order to meaningfully follow-up EM counterparts of gravitational wave triggers it is of paramount importance to identify these events very quickly (Cannon et al. 2010), and then localize them with very low latency. Typical time scales of a fully Bayesian Markov chain Monte Carlo (MCMC) study is of the order of a day, hence not useful for the purpose of optical followup. In order to address this issue, a rapid sky localization technique, known as BAYESTAR, has been developed that is capable of localizing gravitational wave events in the sky within minutes of receiving the trigger using techniques based on timing, phase and amplitude consistency (Singer et al. 2014). We generate gravitational wave error regions for these events which can then be followed-up by optical and infrared telescopes around the world. However, it is important to recognize that in the first year of advanced detectors only the LIGO detectors will be operating. A less sensitive Virgo detector will be coming online a year after. The error regions in these first 2 years are going to be elongated shaped and large with a median of 500 (200) square degrees with 90 % confidence level in first year (second year) of operation respectively (Singer et al. 2014). Under such circumstances, it is important to have telescopes that can cover such error regions within the optical transient timescales. We are currently constructing an array of telescopes in La Silla, Chile, called the BlackGEM that will initially (2016) have four telescopes, each with a field of view of 2.7 square degrees, that can potentially reach up to 23rd magnitude in 5 min of integration time. The full BlackGEM array of 15 telescopes is aimed to become operational in 2018. The multiple telescopes in the array can adjust their configurations to morph the field of view of the array to the shape of the error region. We created telescope pointing for BlackGEM using a number of sky-localizations received from the low latency timing pipeline to see how well BlackGEM is able to cover them. In our preliminary studies, we used ten injections that were localized using the low latency pipeline. Half of the injections were found as doubles and half as triples. We found that a four-telescope array will be pressed hard to cover error region for two detectors. Typical error region from LIGO only sky-localizations required >50 pointing from BlackGEM. With each pointing requiring 5 min of integration time, it will require many hours to cover the error regions completely. Moreover, we found that the error regions can span across both the hemispheres Fig. 3 indicating it may not always be feasible to cover the entire error region by one observatory. The situation gets better when Virgo comes online Fig. 4. With the error regions generally more localized and located within a single hemisphere, we found that we are in a position to cover the desired region within 6–20 BlackGEM pointing for sources lasting in time scales of hours. With more number of telescopes in the array, it will be realistically possible to cover most of the error regions from three detector localizations, and we expect that the full array of 15 telescopes should be able to cover a large fraction of two detector error regions.

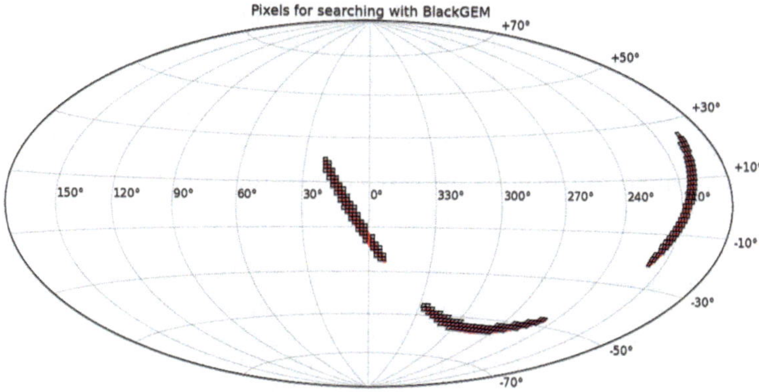

Fig. 3 BlackGEM pointing of 2σ error region for a double coincident detection in LIGO interferometer. Note that the long arcs span across both the hemisphere. One telescope might not be able to fully follow-up the entire error regions for all cases

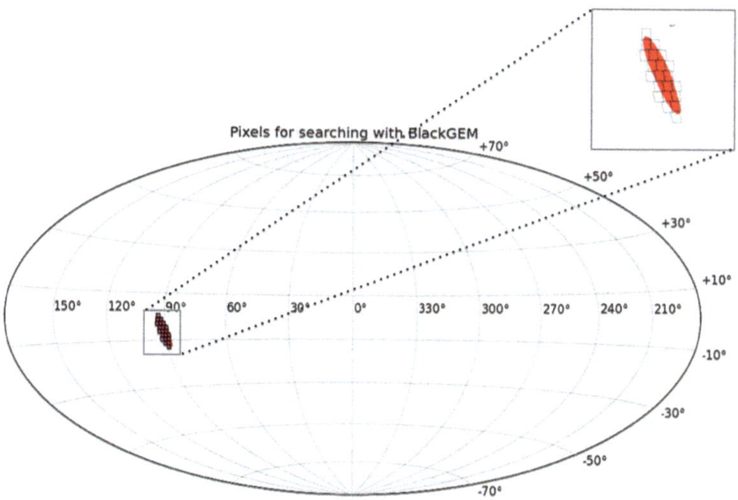

Fig. 4 BlackGEM pointing of 2σ error region for a triple coincident detection in LIGO-Virgo interferometers. The error region in this case can be more realistically followed-up with BlackGEM

4 Discussion

In this work, we explored various avenues of improving GW parameter estimation of NSBH systems using EM information from GRB and/or associated afterglows and Kilovovae. Our results show that EM information in the form of redshift from afterglows could be very useful for GW parameter estimation due to strong $d_L - \iota$ correlation. We also examined the improvement upon constraining inclination angle to 20° for known GRB triggers. We then explored the methods of conducting optical follow-up of GW triggers and presented some preliminary results of this study for the BlackGEM array of telescopes. We observed that it will be difficult to cover the full LIGO 2σ error regions in the 5 years when the Virgo detector is not operating. With the second year of operation of LIGO as the sensitivities improve, as Virgo comes online and as the full BlackGEM array is constructed, we would be seeing realistic possibilities of covering the GW error regions and possibly start making concurrent observations of EM counterparts to GW events.

Acknowledgements We would like to acknowledge Steven Bloemen and Paul Groot for providing us the sky pixel data of BlackGEM that was utilized to construct the pointing. SG is thankful to Marc van der Sluys for the discussions on the techniques of parameter estimation and MCMC. We would also like to thank Larry Price and Leo Singer for their assistance regarding the use of the low latency sky localization pipeline and constructive feedback regarding the work.

References

J. Aasi et al., *Prospects for Localization of Gravitational Wave Transients by the Advanced LIGO and Advanced Virgo Observatories*. arXiv: 1304.0670. (2013)

J. Abadie et al., Search for gravitational-wave inspiral signals associated with short gamma-ray bursts during LIGO's fifth and Virgo's first science run. Astrophys. J. **715**, 1453–1461 (2010)

J. Abadie et al., Search for gravitational waves associated with gamma-ray bursts during ligo science run 6 and virgo science runs 2 and 3. Astrophys. J. **760**(1), 12 (2012)

J. Barnes, D. Kasen, Effect of a high opacity on the light curves of radioactively powered transients from compact object mergers. Astrophys. J. **775**(1), 18 (2013)

M. Boyle et al., High-accuracy comparison of numerical relativity simulations with post-newtonian expansions. Phys. Rev. D **76**, 124038 (2007)

K. Cannon, A. Chapman, C. Hanna, D. Keppel, A.C. Searle, A.J. Weinstein, Singular value decomposition applied to compact binary coalescence gravitational-wave signals. Phys. Rev. D **82**, 044025 (2010)

S. Fairhurst, Triangulation of gravitational wave sources with a network of detectors. New J. Phys. **11**(12), 123006 (2009)

W. Fong, E. Berger, B.D. Metzger, R. Margutti, R. Chornock, G. Migliori, R.J. Foley, B.A. Zauderer, R. Lunnan, T. Laskar, S.J. Desch, K.J. Meech, S. Sonnett, C. Dickey, A. Hedlund, P. Harding, Short grbă130603b: Discovery of a jet break in the optical and radio afterglows, and a mysterious late-time x-ray excess. Astrophys. J. **780**(2), 118 (2014)

E. Nakar, Short-hard gamma-ray bursts. Phys. Rep. **442**, 166–236 (2007)

T. Piran, Gamma-ray bursts and the fireball model. Phys. Rep. **314**, 575–667 (1999)

C. Röver, R. Meyer, N. Christensen, Coherent bayesian inference on compact binary inspirals using a network of interferometric gravitational wave detectors. Phys. Rev. D **75**, 062004 (2007)

S. Shah, M.V.D. Sluys, G. Nelemans, Using electromagnetic observations to aid gravitational-wave parameter estimation of compact binaries observed with lisa. Astron. Astrophys. **544**, A153 (2012)

L. Singer et al., The First Two Years of Electromagnetic Follow-up with Advanced Ligo and Virgo. arXiv: 1404.5623. (2014)

N.R. Tanvir, A.J. Levan, A.S. Fruchter, J. Hjorth, R.A. Hounsell, K. Wiersema, R.L. Tunnicliffe,A 'kilonova' associated with the short-duration gamma-ray burst grb 130603b. Nature **500**, 547 (2013)

Advanced Models of Black Hole–Neutron Star Binaries and Their Astrophysical Impact

Zachariah B. Etienne, Vasileios Paschalidis, and Stuart L. Shapiro

Abstract Fully general relativistic simulations of black hole–neutron star (BHNS) binary inspiral and merger indicate that the NS simply plunges into the BH over much of the likely binary parameter space, without leaving a remnant disk. Plunging mergers are unlikely to generate useful information about the NS equation of state (EOS) from the gravitational waves (GWs) alone. However, when the initial BH possesses a moderate to high spin, aligned with the orbital angular momentum, or when the NS possesses a low compaction, the NS may tidally disrupt outside the BH's innermost stable circular orbit radius, generating a massive accretion disk and a long tidal tail. When observed by Advanced LIGO/VIRGO, the GWs from this scenario may in fact constrain the NS EOS. After disruption, some of the neutron-rich tidal tail may be unbound, favoring formation of r-process elements. The subsequent decay may be observable in the electromagnetic (EM) spectrum. Meanwhile, the remnant BH accretion disk may provide the energy reservoir for a short gamma-ray burst. Finally, even in cases for which no EOS information may be gleaned, the BHNS binary is expected to act as a unipolar inductor during inspiral, potentially releasing an EM signature that distinguishes certain binary parameters. Taken together, these phenomena make BHNS binary mergers rich and exciting systems for theoretical modeling, and we outline the latest results from the most advanced, fully general relativistic simulations.

Z.B. Etienne (✉)
Department of Mathematics, West Virginia University, 320 Armstrong Hall, P.O. Box 6310, Morgantown, WV 26506-6310, USA
e-mail: zachetie@gmail.com

V. Paschalidis
Department of Physics, University of Illinois, 1110 W. Green St., Urbana, IL 61801-3080, USA
e-mail: vpaschal@illinois.edu

S.L. Shapiro
Department of Physics, University of Illinois, 1110 W. Green St., Urbana, IL 61801-3080, USA

Department of Astronomy, University of Illinois, 1110 W. Green St., Urbana, IL 61801-3080, USA
e-mail: slshapir@illinois.edu

C.F. Sopuerta (ed.), *Gravitational Wave Astrophysics*, Astrophysics and Space Science Proceedings, DOI 10.1007/978-3-319-10488-1_6

1 Introduction

Black hole–neutron star (BHNS) binaries are one of the most promising candidates for the first incident detection of gravitational waves (GWs) by Advanced LIGO/VIRGO. However, the insights gained from GW and possible coincident electromagnetic (EM) observations will depend sensitively on our theoretical understanding of these systems, and our best theoretical understanding lies with our most advanced models.

This proceeding will therefore focus on theoretical predictions of EM and GW signatures from BHNS binary late inspiral and merger, as generated by *the most sophisticated, fully general relativistic BHNS binary simulations* to date. These numerical relativity (NR)-based models are very difficult to generate, as they combine the challenges associated with BHs (singularities) moving on numerical grids with the subtleties of relativistic hydrodynamics and magnetohydrodynamics (MHD) for the NS, including strong shocks and ultrarelativistic, highly-magnetized flows. In addition, the most interesting regions of BHNS parameter space involve highly spinning BHs, which generally require very high numerical resolutions to model. As a result, BHNS modeling has traditionally lagged behind binary neutron star (NSNS) simulations. Many scientists have been involved in the modeling of BHNS binary systems, from early, Newtonian simulations (see e.g. Kluźniak and Lee 1998; Ruffert and Janka 1999; Lee and Kluźniak 1999a, b; Janka et al. 1999; Lee 2000; Rosswog et al. 2004; Ishii and Shibata 2004 and references therein), to simulations using a Paczyński-Wiita potential (see e.g. Rosswog 2005; Setiawan et al. 2006; Ruffert and Janka 2010 and references therein), to the state-of-the-art, fully general relativistic (GR) BHNS simulations of today (see e.g. Etienne et al. 2008, 2009; Matthew et al. 2010; Chawla et al. 2010; Foucart et al. 2011; Kyutoku et al. 2011; Lackey et al. 2012; Foucart et al. 2013b; Etienne et al. 2012, b, a; Lovelace et al. 2013 and references therein; see also Shibata and Taniguchi 2011 for a review).

It is well known that electromagnetic (EM) observations of these mergers will be critically important to sky localization of the sources, particularly in the next few years when the first incident detections of GWs are made. In addition to improving sky localization, coincident EM signatures are *critically important* to BHNS GW observations in particular, because over much of the likely BHNS parameter space, the NS simply plunges into the BH, without forming a remnant disk. In this case, BHNS and binary black hole (BHBH) gravitational waveforms with the same masses can be nearly *indistinguishable*. Though discerning between the two binary types may be possible from the chirp mass alone, i.e., there is no known formation mechanism for a $1.4M_{\odot}$ BH (see, e.g., Bulik and Belczyński 2003 for chirp mass histograms for expected binary types), it may be that large error bars in the first GW observations will confuse binary parameters sufficiently to make definitive determination of the type of binary impossible from GWs alone. In this case, an EM counterpart during inspiral or merger might be *necessary* in distinguishing binary types. For example, unlike BHBHs, inspiraling BHNS binaries may generically behave as unipolar inductors, driving currents that will be dissipated electromagnetically. Advanced modeling of such EM counterparts will be discussed at length in Sect. 4.1 below.

The most interesting part of parameter space will likely be when the NS compactions are low, or the aligned BH spin is high. In these cases, the BH tidally disrupts the NS well before merger, leading to the possibility that information about the NS equation of state might be encoded in the GWs themselves. We review current literature addressing this topic in Sect. 3. As for EM counterparts, this part of BHNS parameter space is also expected to yield EM signatures beginning with the unipolar inductor mechanism during inspiral (see Sect. 4.1). Then in the violence of disruption, the NS becomes severely tidally distorted, forming a large tidal tail. Typically, some of the matter within this neutron-rich tail is unbound and r-process elements are likely to form. The subsequent decay of these elements may generate a kilonova. The most advanced NR-based models of this phenomenon are described in Sect. 4.3. Well after NS disruption, a quasi-stable, hot, magnetized accretion disk generally forms around the BH with sufficient mass to provide an energy reservoir for a short gamma-ray burst (sGRB). This possibility has been explored by self-consistent, advanced GR magnetohydrodynamic (GRMHD) modeling, which is reviewed in Sect. 4.2.

But before analyzing results from the most advanced theoretical models of BHNS binary mergers, let us first focus on the broader question: *Based on astrophysical considerations, what parts of BHNS binary parameter space are most likely to be occupied?*

2 BHNS Binary Parameter Space

As a point of comparison, binary black hole calculations set the standard for GW modeling accuracy. This is because BHBH systems are far simpler to model than binaries with matter, as BHBH waveform modeling does not require simulations of GR hydrodynamics or GRMHD. Further, spanning the parameter space of possible BHBH waveforms is also much easier than other compact binaries, as the parameter space is much smaller. Given that GWs from BHBHs in the LIGO/VIRGO band will generally circularize the orbit (Peters and Mathews 1963), gravitational waveforms may be uniquely determined by specifying the BH spins and masses.

For BHNS binaries, population synthesis models and basic astrophysical insights limit the likely parameter space. For example, unlike BHs, NS masses will not span orders of magnitude. Thus it seems unlikely we will observe a BH:NS mass ratio less than or close to unity. In fact, population synthesis models seem to favor BH:NS mass ratios between 5:1 and 7:1 (Belczynski et al. 2008, 2010).

So according to these rather crude, preliminary population synthesis models , with a standard NS mass of around $1.4M_{\odot}$, the most likely BH companion mass is around $10M_{\odot}$. Aside from the mass ratio, which appears to be highly constrained by astrophysical considerations (as compared to BHBHs, at least), there are two other highly relevant parameters for theoretical models. The first is the BH spin vector. The BH spin magnitude in GR is constrained only by the maximal Kerr limit. The second is the NS EOS, which is still uncertain, so theoretical models must vary over different possibilities for the EOS, restricting them to those yielding maximum NS masses $\approx 2.0 M_{\odot}$ or higher (cf. PSR J1614-2230 (Demorest et al. 2010) and PSR J0348+0432 (Antoniadis et al. 2013)). Finally, tidal locking will not occur, as the tidal locking timescale is expected to be much longer than the GW infall timescale (Kochanek 1992; Bildsten and Cutler 1992). Thus the NS is not expected to corotate and is likely to be closer to irrotational.

With these parameter space constraints, modeling BHNS binaries is somewhat less daunting. However the NS EOS is not a simple scalar like the mass ratio or a three-vector like the BH spin. The NS EOS is itself characterized by many parameters, making it exceedingly difficult to generate BHNS simulations that adequately sample all plausible NS EOSs. The next section reviews the current state of our understanding regarding how well GW observations from BHNS binaries alone might constrain the NS EOS.

3 Constraining the NS EOS from BHNS Binary GWs

Constraining the NS EOS through GW observations would provide a huge advance in our understanding of matter at temperatures and densities well outside of the range we are able to study in the lab. So just how likely will we be able to constrain the NS EOS with observations of GWs from BHNS mergers alone? To address this question, we will now take a hard look at some BHNS binary merger simulations (Foucart et al. 2013a) produced by the SXS (Simulating eXtreme Spacetimes) collaboration.

3.1 What Can Be Learned About the NS Equation of State from a Plunging Merger into the BH?

Figure 1 shows a BHNS system as the NS (in blue) *plunges* into the BH. This case has a BH:NS mass ratio of 6:1, which is within the most probable range according to population synthesis models, and the BH is initially nonspinning. Figure 2 shows that the gravitational waveforms from this BHNS merger are practically indistinguishable from those of a BHBH with precisely the same total mass, mass ratio, and spins. So for this *realistic* case, we learn *nothing* about the NS EOS from the GWs alone. But is there some region of BHNS parameter space for which we might learn more about the NS EOS from GWs?

Based on these results, it seems reasonable to focus on cases in which NS does not behave so much as a point mass during merger, instead choosing the part of

Fig. 1 Snapshot of a 6:1 BH:NS mass ratio binary during merger. The BH was initially nonspinning. *Blue* denotes the NS and *grey* the BH apparent horizon. Reprinted with author permission, from Foucart et al. (2013a) (Color figure online)

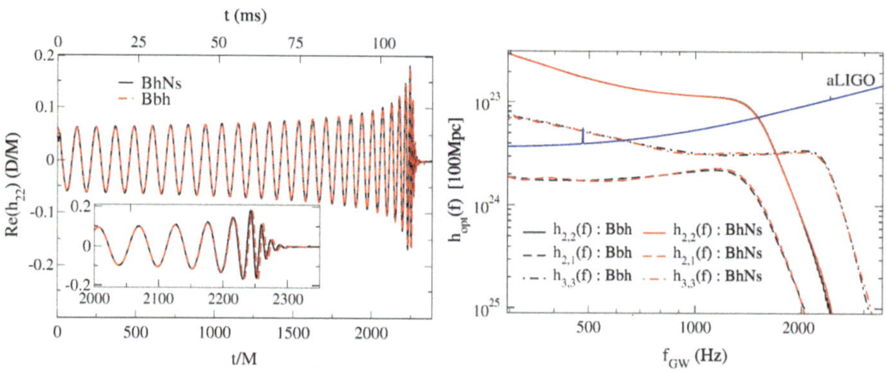

Fig. 2 *Left panel*: Dominant 2,2 mode of GW strain from inspiral, through merger and ringdown, comparing BHNS and BHBH gravitational waveforms in which both systems have the same initial total mass, mass ratio, and BH spins. *Right panel*: Power spectra for the three most dominant GW modes of the same cases, as compared to the Advanced LIGO noise curve. Reprinted with author permission, from Foucart et al. (2013a)

parameter space in which the NS *tidally disrupts* prior to merger, generating a significantly different waveform from that of a BHBH system. But what parameters would potentially push NS tidal disruption far from the BH? We address this question next.

3.2 *What Parameters Are Important for Tidal Disruption?*

From a Newtonian perspective, tidal disruption will occur just inside an orbital separation d at which the gravitational force $F_{\rm NS}$ on a test particle at the NS surface is balanced by the BH tidal force $F_{\rm BH,tidal}$. Expressed more precisely, in geometrized units ($G = c = 1$),

$$F_{\rm NS} \sim F_{\rm BH,tidal} \tag{1}$$

$$\rightarrow \frac{M_{\rm NS}}{R_{\rm NS}^2} \sim \frac{M_{\rm BH}}{d^3} R_{\rm NS}. \tag{2}$$

Defining the NS compaction $C = M_{\rm NS}/R_{\rm NS}$ and BH:NS mass ratio $q = M_{\rm BH}/M_{\rm NS}$, Eq. (2) can be expressed in terms of the dimensionless quantity $d/M_{\rm BH}$:

$$\frac{d}{M_{\rm BH}} \sim q^{-2/3} C^{-1}. \tag{3}$$

The relevant lengthscale for comparing d is the innermost stable circular orbit (ISCO) radius of the BH, $r_{\rm ISCO}$. It seems reasonable to expect that if the NS disrupts farther from the BH ISCO, the NS merger will not mimick that of a point source, leading to a BHNS gravitational waveform that is less like a BHBH waveform. That is to say, the stronger the inequality $d > r_{\rm ISCO}$ is satisfied, the more distinguishable BHNS and BHBH waveforms may be, giving us some chance of extracting NS EOS information from BHNS gravitational waveforms.

Therefore, according to Eq. (3), as the BH:NS mass ratio or the NS compaction decrease, we would expect the GWs to carry more information about the NS EOS. In addition, test particle dynamics about a Kerr BH imply that $r_{\rm ISCO}$ decreases as the BH spin aligned with the orbital angular momentum increases. Hence, keeping d fixed and increasing the aligned spin of the BH also yields a stronger inequality, which in turn may result in less of a plunge-like merger, a more massive remnant disk and a BHNS gravitational waveform that might give us insight into the NS EOS. But how do these Newtonian arguments hold up to NR simulations?

3.3 Tidal Disruption in BHNS Mergers: Results from Fully General Relativistic Simulations

Figure 3 presents a measure of NS disruption by the remnant disk mass. This plot is based on a number of fully GR simulations of BHNS binary mergers from quasicircular orbits in which the initial NS compaction is fixed at a conservative value (0.155), while the mass ratio and aligned BH spin are varied. Consistent with our analytic estimates, when the compaction is kept fixed, as the BH aligned spin decreases or the NS mass ratio increases, the result tends toward a NS plunging into the BH without disrupting and forming a disk. The corresponding gravitational waveform will be more like a BHBH in this region, making it exceedingly difficult or impossible to glean information about the NS EOS from the merger signal.

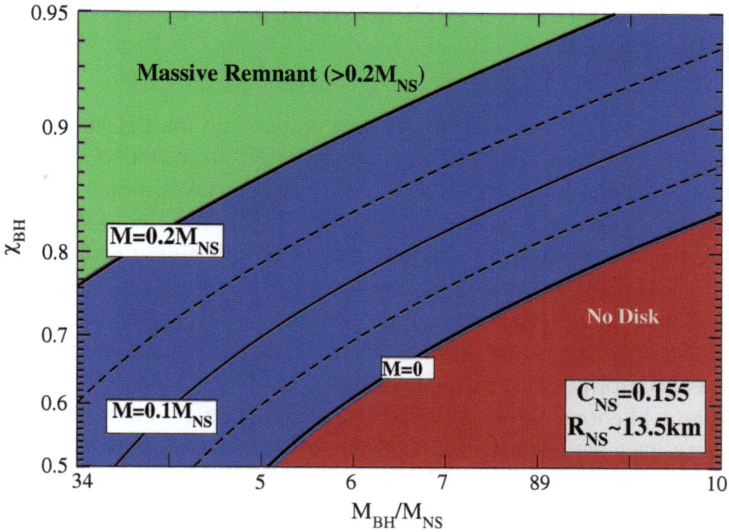

Fig. 3 Contours of remnant disk rest mass following NS disruption by a BH on the M_{BH}/M_{NS}–χ_{BH} plane, where $\chi_{BH} = J_{BH}/M_{BH}^2$ is the BH dimensionless spin parameter. *Red* implies no disk formation (the NS plunges into the BH intact, $M_{disk} = 0$), *blue* implies a $0 < M_{disk} \leq 0.2M_{NS}$, and *green* implies $M_{disk} \geq 0.2M_{NS}$. Reprinted with author permission, from Foucart (2012) (Color figure online)

Notice also that for BH:NS mass ratios ∼7:1,[1] only cases with a highly spinning, aligned BH result in significant NS tidal disruption and a remnant disk, consistent with the plunging merger observed in the 6:1 mass ratio, nonspinning case of Sect. 3.1. In such low-aligned-spin cases, GWs from BHNS mergers may be indistinguishable from BHBH gravitational waveforms with the same masses and mass ratios. So if all BH spins are equally likely for this mass ratio range and NS compaction, GWs from *most* of BHNS parameter space are unlikely to yield information about the NS EOS. This being said, it may well be the case that binary parameters are not all equally likely, and that strong, aligned BH spins may actually be preferred. Let us focus on this particularly interesting region of parameter space.

[1]The BH:NS mass ratio is typically defined as the ratio of the BH Christodoulou mass to the ADM mass of a NS in isolation with the same rest-mass.

3.4 *Can We Extract NS EOS Information from BHNS Merger GWs Alone?*

For the cases in which the NS does not just plunge into the BH intact, but instead tidally disrupts, we may be able to distinguish BHNS from BHBH waveforms, and at least in principle constrain the NS EOS from the GWs alone. This region of parameter space has been most thoroughly studied by the Kyoto/Princeton/UWM collaboration, who performed a staggering 134 BHNS binary evolutions, varying mass ratio, BH spin, and NS EOS (Lackey et al. 2014). Using these waveforms, they found that the NS EOS-parameter best measured from GWs is the NS's tidal Love number Λ, which is a measure of tidal deformability such that larger Λ implies a more easily deformed NS, and BHs possess $\Lambda = 0$.

Building a phenomenological BHNS waveform model on these NR waveforms and varying over parameters, they find that Λ is constrained only to around 10–100 % at 1-σ error by Advanced LIGO. One reason for this is simple: when the NS disrupts, the disruption event itself encodes the most information about the NS EOS into the GWs, but disruption happens at frequencies at the high-frequency edge of the Advanced LIGO noise curve, as shown in Fig. 4. Another reason for the poor constraint on Λ from GW observations is the fact that Λ correlates with other

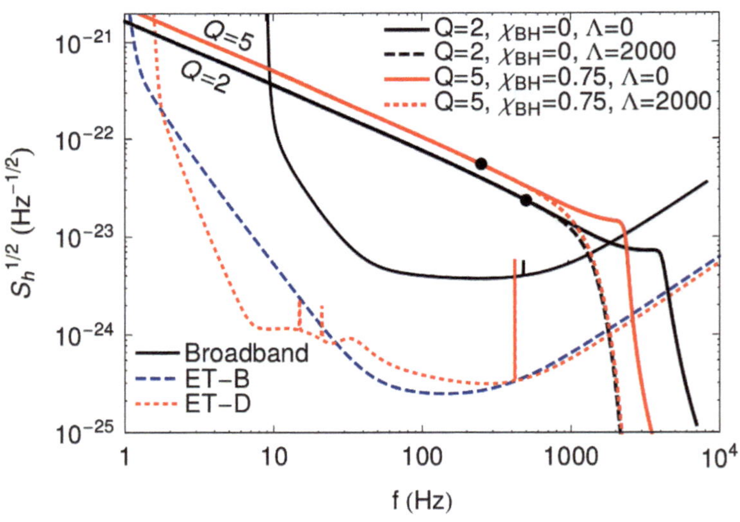

Fig. 4 Power spectra of four phenomenological BHNS waveforms, where Q is the BH:NS mass ratio and Λ the tidal Love number. $\Lambda = 0$ (*thick solid red* and *black lines*) denotes a waveform power spectrum consistent with a BHBH of the same total mass, spins, and mass ratio, and $\Lambda = 2,000$ (*thick dashed red* and *black lines*) represents a BHNS binary merger waveform with a NS that possesses a very high tidal deformability. *Black dots* indicate the waveform frequency of $Mf = 0.01$. The broadband Advanced LIGO (*thin black*), Einstein Telescope-B (*thin blue dashed*), and Einstein Telescope-D (*thin red dotted*) noise curves are plotted for comparison. Reprinted with author permission, from Lackey et al. (2014) (Color figure online)

binary parameters near merger and GW ringdown. As a result, they find that Λ is constrained best by Advanced LIGO by GWs observed prior to merger.

The discussion so far has focused on current-generation GW observatories. As seen in Fig. 4, the proposed *next*-generation ground-based interferometer, the Einstein Telescope (Punturo et al. 2010), should be able to reduce these errors in Λ significantly; In Lackey et al. (2014) it was found that ET estimates of Λ from BHNS binary tidal disruptions will have 1-σ error bars around 1–10 %.

In summary, prospects are not great for constraining NS EOS from BHNS GW observations alone with Advanced LIGO-era GW interferometers. But what about EM counterparts?

4 EM Counterparts to BHNS GW Signals

Recall from the previous section that any disruption signal in the GWs will likely fall on the border of Advanced LIGO detectability. Now suppose Advanced LIGO/VIRGO detects a GW, and the chirp mass falls near the range expected for BHNS binaries. Without a clear disruption signal, how could we convincingly distinguish a BHNS binary from a BHBH with the same chirp mass? One answer is observation of an EM counterpart to the inspiral or merger. In Sect. 4.1, we review simulations of the BHNS unipolar inductor mechanism, which may generate a detectable EM counterpart to the *inspiral*, regardless of binary parameters.

Then, for the cases in which the NS tidally disrupts prior to merger, we review the exciting EM counterparts that may accompany the GW signal. Sections 4.2 and 4.3 review post-merger EM signatures in the form of sGRBs and kilonovae, respectively.

4.1 Electromagnetic Counterparts to BHNS Inspiral: Magnetospheric Phenomena

During inspiral the BHNS binary may act as a unipolar inductor (Goldreich and Lynden-Bell 1969; Lai 2012), with the BH sweeping across the NS dipolar magnetic fields, inducing an EMF, which drives currents that will be dissipated. Near merger, the dissipation rate was first estimated by McWilliams and Levin (2011) to be of order $10^{46}(B_{\rm NS,p}/10^{13}{\rm G})^2({\rm M_{BH}}/10{\rm M_\odot})^{-4}$ ergs/s at the light ring radius, where $B_{\rm NS,p}$ is the magnetic field strength at the NS pole, and $\sim 10^{42}$ ergs/s under the same assumptions, except at a separation $d \simeq 6.6R_{\rm NS}$.

The Illinois group performed the first fully GR force-free evolutions of the BHNS unipolar inductor scenario (Paschalidis et al. 2013), adopting quasiequilibrium BHNS systems in quasicircular orbits with a fixed mass ratio $q = 3$ and orbital separation of $d \simeq 6.6R_{\rm NS}$, varying the BH dimensionless spin parameter $\chi_{\rm BH}$

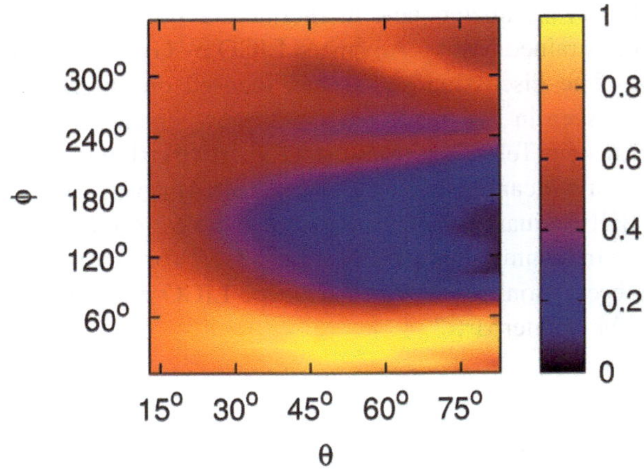

Fig. 5 Angular distribution of Poynting flux, as measured on a sphere far from the center of mass of a widely-separated BHNS binary system with BH dimensionless spin of 0.75 and mass ratio $q = 3$. From Paschalidis et al. (2013)

between $-0.5, 0$ and 0.75 aligned with the orbital angular momentum. In all our simulations we find that after a short transient period during which our initial magnetic fields settle into quasistationary configuration, the McWilliams & Levin's order-of-magnitude-estimates for the dissipation rate are in good agreement with the outgoing EM luminosity computed from our simulations ($\sim 10^{42}$ ergs/s). We also find that the outgoing EM flux peaks within a broad beam of $\sim 40°$ in the azimuthal direction, demonstrating a distinctive lighthouse effect, as shown in Fig. 5. If such a signature can be observed, it could trigger targeted GW searches.

Though EM signals preceding merger are exciting, particularly in the context of GW observations, the merger itself may trigger an extremely luminous EM outburst that, unlike the BHNS unipolar inductor scenario, has a rich observational history: a short gamma-ray burst (sGRB).

4.2 Electromagnetic Counterparts to BHNS Merger: Short Gamma-Ray Burst

There are a number of models for sGRB engines (see e.g. Piran 2004; Woosley and Bloom 2006 for reviews). One idea of how an sGRB might be formed is the magnetic launching mechanism: a magnetized NS gets ripped apart over the timescale of tens of milliseconds, resulting in a black hole–magnetized disk system. At this point, the Blandford-Znajeck and/or Blandford-Payne mechanisms act, and the BH and disk act to wind up and collimate magnetic fields above the BH pole, increasing the magnetic-to-gas pressure ratio until the magnetic fields overcome the inertia and ambient pressure, launching a relativistic outflow.

Of course, since sGRBs last around a second or so, this scenario depends on the matter in the disk feeding these jets for around a second. The Illinois group was among the first to simulate this scenario with GRMHD simulations, in an attempt to

Fig. 6 Remnant disk at the end of a 3:1 BH:NS mass ratio simulation in which the BH has an initial dimensionless spin parameter of 0.75, aligned with the orbital angular momentum. *Yellow lines* denote magnetic field lines, with density corresponding roughly to local magnetic energy density. From Etienne et al. (2012b) (Color figure online)

determine its feasibility (Etienne et al. 2012, b). What follows is a rather humbling tale, that teaches us just how much farther we have to go to adequately model the formation of an sGRB engine.

We launched a 3:1 BH:NS mass ratio simulation in which the BH has an initial dimensionless spin parameter of 0.75, aligned with the orbital angular momentum. We initially seed the *interior* of the NS with purely poloidal B fields set such that the magnetic to gas pressure is half a percent. So what happens when we perform this simulation? Are collimated jets launched with highly relativistic outflows, characteristic of an sGRB engine?

Figure 6 shows the remnant disk at the end of this magnetized BHNS binary merger simulation. Notice the strong, *toroidal* magnetic fields, wound up by the MHD fluid flow, and no evidence of magnetic field collimation or outflows along the BH polar regions. But there is a potentially critically important magnetic phenomenon that is not being modeled here.

After tidal disruption and disk formation, we expect an MHD instability known as the magnetorotational instability, or MRI, to amplify our originally weak poloidal fields exponentially, creating a turbulent mixture of toroidal+poloidal magnetic fields. The resulting MHD turbulence acts as an effective viscosity, driving accretion onto the BH. This may be the missing link in launching jets and producing an sGRB engine from BHNS binary mergers. But MRI is difficult to model numerically. Keeping all else fixed, the characteristic MRI wavelength $\lambda_{\rm MRI}$ of the fastest growing mode is proportional to the poloidal magnetic field strength in the remnant disk. Notice the apparent lack of such turbulence in Fig. 6. As it turns out, in this simulation we would need numerical resolutions an order of magnitude higher to see any effects from MRI, which translates to simulations about 10,000 times more expensive. So the question is: *How might we increase λ_{MRI}?*

As it turned out, the simulation shown in Fig. 6 was completely symmetric about the orbital plane, and imposing this symmetry enabled us to cut the computational cost in half. However, imposing this symmetry implies that fluid cannot move across the symmetry plane. Our idea was, if we no longer impose this symmetry and choose an *asymmetric* magnetic field, we might end up with stronger poloidal fields in our remnant disk, making it easier to numerically resolve MRI. So we addressed our hypothesis: Will asymmetric magnetic fields initially in the NS yield stronger poloidal fields in the remnant disk?

The answer is *Yes.* As a rule of thumb, the characteristic MRI wavelength λ_{MRI} must be resolved by at least ten gridpoints to see any effects from MRI in our GRMHD models. That is to say, $\lambda_{\mathrm{MRI}}/\Delta x \geq 10$ to capture MRI, where Δx is our gridspacing. We found that as the initial tilt angle of the NS magnetic fields increased from 0° to 90°, $\lambda_{\mathrm{MRI}}/\Delta x$ increased from $\approx$1 to nearly 8—almost an order of magnitude. And although eight points is not quite enough to adequately resolve the characteristic MRI wavelength and see the effects of this critically important instability, we should start seeing the effects of MRI with only 25 % higher resolution, with a 90° tilt angle on the NS magnetic moment, with respect to the orbital angular momentum vector.

But what about lower degrees of tilt? In many compact binary systems, there are regions of parameter space that are exceedingly difficult to model using numerical relativity codes, requiring resolutions that are unfeasibly high. What we have found is that, given the high resolution requirements for modeling MRI, exploration of this magnetic launching mechanism in global GRMHD simulations of BHNS binaries as sGRB progenitors in the near future may be restricted to magnetic fields that are strongly tilted with respect to the orbital angular momentum.

Finally, we must ask: Under the right conditions, might BHNS mergers form an sGRB engine? To answer this, we seeded the remnant disk from an unmagnetized BHNS simulation with purely poloidal fields that are dynamically unimportant initially, but strong enough to resolve MRI. Magnetic turbulence occurs in the disk, driving accretion and supporting Poynting-dominated jet outflows sufficient to power an sGRB.

4.3 Electromagnetic Counterparts to BHNS Merger: Kilonovae

GRBs are certainly not the only possible EM counterpart to the merger and post-merger of BHNS binary systems. For example, in certain cases involving high, aligned-spin BHs, parts of the neutron-rich tidal tail are unbound, and can be the sites of r-process element production. The subsequent decay of these elements may power an observable, nearly isotropic EM signature and may help explain the observed abundances of heavy elements in the galaxy (Li and Paczyński 1998; Kulkarni 2005; Metzger et al. 2010).

The Kyoto-Tokyo-Milwaukee group was the first to study r-process luminosity from the unbound material ejected from a number of BHNS and NSNS simulations (Hotokezaka et al. 2013; Tanaka et al. 2014). They model the released heat and resulting EM spectra from r-process-element decays approximately, using a radiation transfer code, finding that this decay could result in a peak luminosity of $\sim 5 \times 10^{41}$ ergs/s. Also, unlike a NSNS merger, in which ejecta are uniformly distributed from the remnant, the ejecta from a BHNS binary merger is confined to a narrow tidal tail. As a result, their calculations find the BHNS EM signal to be bluer, potentially paving a way to distinguish NSNS and BHNS mergers from their EM signals alone.

Conclusions

Advanced NR-based modeling of BHNS binary mergers has shown that over much of the "likely" BHNS binary parameter space, the NS simply plunges into BH, making GWs from BHNS and BHBH inspirals practically indistinguishable. However, this degeneracy might be broken if EM counterparts are observed. For example, during inspiral, magnetospheric effects in the BHNS binary may extract energy from the system, yielding $\sim 10^{42}(B_{\rm NS,p}/10^{13}{\rm G})^2({\rm M_{NS}}/1.4{\rm M_\odot})^2$ ergs/s average luminosities near merger, with a distinctive broad-beaming, "lighthouse" effect that may help distinguish BHNS from BHBH systems when coincident GWs are observed. Also, BHNS mergers are prime candidates for the generation of GRB engines, but adequately modeling of this extreme EM phenomenon through NR simulations will require higher resolutions to properly model the magnetic fields, as well as advanced EOSs and accurate neutrino transport.

From a GW perspective, the part of parameter space in which the NS is expected to be tidally disrupted by the BH and form a disk is far more interesting, as it may allow for crude estimates of the tidal Love number, constraining the NS EOS. As for EM signatures, in addition to the unipolar inductor EM signal during inspiral, tidal disruption far from the BH may result in a BH accretion disk sufficiently massive to act as an energy reservoir for an sGRB, and the decay of r-process elements formed in the ejected material from a BHNS merger may produce its own observable EM signature. This decay may help distinguish NSNS from BHNS mergers, as BHNS merger signals are likely to be bluer.

Decades have passed since the first, Newtonian-based simulations of BHNS binary mergers, and only now do numerical relativity codes exist that have some chance of capturing all of the relevant physics. However, for the

(continued)

most interesting parts of parameter space (i.e., when the NS tidally disrupts, forming a massive BH accretion disk), no simulations have captured *all* the relevant physics with sufficiently detailed algorithms and resolutions. So now is a critical time to prepare these theoretical models, as the first incident detection of GWs from BHNS binaries may be only a few years away.

References

J. Antoniadis, P.C.C. Freire, N. Wex, T.M. Tauris, R.S. Lynch et al., A massive pulsar in a compact relativistic binary. Science **340**, 6131 (2013)

K. Belczynski, R.E. Taam, E. Rantsiou, and M. van der Sluys. Black hole spin evolution: implications for short-hard gamma-ray bursts and gravitational wave detection. Astrophys. J., **682**, 474–486, July 2008.

K. Belczynski, M. Dominik, T. Bulik, R. O'Shaughnessy, C. Fryer, D.E. Holz, The effect of metallicity on the detection prospects for gravitational waves. ApJ **715**, L138–L141 (2010)

L. Bildsten, C. Cutler, Tidal interactions of inspiraling compact binaries. ApJ **400**, 175–180 (1992)

T. Bulik, K. Belczyński, Constraints on the binary evolution from chirp mass measurements. ApJ **589**, L37–L40, (2003)

S. Chawla, M. Anderson, M. Besselman, L. Lehner, S.L. Liebling et al., Mergers of magnetized neutron stars with spinning black holes: disruption, accretion and fallback. Phys. Rev. Lett. **105**, 111101 (2010)

P.B. Demorest, T. Pennucci, S.M. Ransom, M.S.E. Roberts, J.W.T. Hessels, A two-solar-mass neutron star measured using Shapiro delay. Nature **467**, 1081–1083 (2010)

M.D. Duez, F. Foucart, L.E. Kidder, C.D. Ott, S.A. Teukolsky, Equation of state effects in black hole-neutron star mergers. Class. Quantum Grav. **27**, 114106 (2010)

Z.B. Etienne, J.A. Faber, Y.T. Liu, S.L. Shapiro, K. Taniguchi, T.W. Baumgarte, Fully general relativistic simulations of black hole-neutron star mergers. Phys. Rev. D **77**, 084002 (2008)

Z.B. Etienne, Y.T. Liu, S.L. Shapiro, T.W. Baumgarte, General relativistic simulations of black-hole-neutron-star mergers: effects of black-hole spin. Phys. Rev. D **79**(4), 044024 (2009)

Z.B. Etienne, Y.T. Liu, V. Paschalidis, S.L. Shapiro, General relativistic simulations of black-hole-neutron-star mergers: effects of magnetic fields. Phys. Rev. D **85**(6), 064029 (2012)

Z.B. Etienne, V. Paschalidis, Y.T. Liu, S.L. Shapiro, Relativistic magnetohydrodynamics in dynamical spacetimes: improved electromagnetic gauge condition for adaptive mesh refinement grids. Phys. Rev. D **85**(2), 024013 (2012)

Z.B. Etienne, V. Paschalidis, S.L. Shapiro. General-relativistic simulations of black-hole-neutron-star mergers: effects of tilted magnetic fields. Phys. Rev. D **86**(8), 084026 (2012)

F. Foucart, Black-hole-neutron-star mergers: disk mass predictions. Phys. Rev. D **86**(12), 124007 (2012)

F. Foucart, M.D. Duez, L.E. Kidder, S.A. Teukolsky, Black hole-neutron star mergers: effects of the orientation of the black hole spin. Phys. Rev. D **83**, 024005 (2011)

F. Foucart, L. Buchman, M.D. Duez, M. Grudich, L.E. Kidder, I. MacDonald, A. Mroue, H.P. Pfeiffer, M.A. Scheel, B. Szilagyi, First direct comparison of nondisrupting neutron star-black hole and binary black hole merger simulations. Phys. Rev. D **88**(6), 064017 (2013)

F. Foucart, M.B. Deaton, M.D. Duez, L.E. Kidder, I. MacDonald, C.D. Ott, H.P. Pfeiffer, M.A. Scheel, B. Szilagyi, S.A. Teukolsky, Black-hole-neutron-star mergers at realistic mass ratios: equation of state and spin orientation effects. Phys. Rev. D **87**(8), 084006 (2013)

P. Goldreich, D. Lynden-Bell, Io, a jovian unipolar inductor. ApJ **156**, 59–78 (1969)

K. Hotokezaka, K. Kyutoku, M. Tanaka, K. Kiuchi, Y. Sekiguchi et al., Progenitor models of the electromagnetic transient associated with the short gamma ray burst 130603B. Astrophys. J. **778**, L16 (2013)

M. Ishii, M. Shibata, Numerical study of the tidal disruption of neutron stars moving around a black hole—compressible jeans and roche problems. Prog. Theor. Phys. **112**, 399–413 (2004)

H.-T. Janka, T. Eberl, M. Ruffert, C.L. Fryer, Black hole-neutron star mergers as central engines of gamma-ray bursts. ApJ **527**, L39–L42 (1999)

W. Kluźniak, W.H. Lee, Simulations of binary coalescence of a neutron star and a black hole. ApJ **494**, L53–L55 (1998)

C.S. Kochanek, Coalescing binary neutron stars. ApJ **398**, 234–247 (1992)

S.R. Kulkarni, Modeling supernova-like explosions associated with gamma-ray bursts with short durations. Preprint (arXiv:astro-ph/0510256) (2005).

K. Kyutoku, H. Okawa, M. Shibata, K. Taniguchi, Gravitational waves from spinning black hole-neutron star binaries: dependence on black hole spins and on neutron star equations of state. Phys. Rev. D **84**, 064018 (2011)

B.D. Lackey, K. Kyutoku, M. Shibata, P.R. Brady, J.L. Friedman, Extracting equation of state parameters from black hole-neutron star mergers. I. Nonspinning black holes. Phys. Rev. D **85**, 044061 (2012)

B.D. Lackey, K. Kyutoku, M. Shibata, P.R. Brady, J.L. Friedman, Extracting equation of state parameters from black hole-neutron star mergers: aligned-spin black holes and a preliminary waveform model. Phys. Rev. D **89**, 043009 (2014)

D. Lai, DC circuit powered by orbital motion: magnetic interactions in compact object binaries and exoplanetary systems. ApJ **757**, L3 (2012)

W.H. Lee, Newtonian hydrodynamics of the coalescence of black holes with neutron stars - III. Irrotational binaries with a stiff equation of state. MNRAS **318**, 606–624 (2000)

W.H. Lee, W. Kluźniak, Newtonian hydrodynamics of the coalescence of black holes with neutron stars - II. Tidally locked binaries with a soft equation of state. MNRAS **308**, 780–794 (1999)

W.H. Lee, W. Kluźniak, Newtonian hydrodynamics of the coalescence of black holes with neutron stars. i. Tidally locked binaries with a stiff equation of state. ApJ **526**, 178–199 (1999)

L.-X. Li, Paczyński, ApJ **507**, 59 (1998)

G. Lovelace, M.D. Duez, F. Foucart, L.E. Kidder, H.P. Pfeiffer et al., Massive disc formation in the tidal disruption of a neutron star by a nearly extremal black hole. Class. Quantum Grav. **30**, 135004 (2013)

S.T. McWilliams, J. Levin, Electromagnetic extraction of energy from black-hole-neutron-star binaries. ApJ **742**, 90 (2011)

B.D. Metzger, G. Martínez-Pinedo, S. Darbha, E. Quataert, A. Arcones, D. Kasen, R. Thomas, P. Nugent, I.V. Panov, N.T. Zinner, Electromagnetic counterparts of compact object mergers powered by the radioactive decay of r-process nuclei. MNRAS **406**, 2650–2662 (2010)

V. Paschalidis, Z.B. Etienne, S.L. Shapiro, General relativistic simulations of binary black hole-neutron stars: precursor electromagnetic signals. Phys. Rev. D **88**(2), 021504 (2013)

P.C. Peters, J. Mathews, Gravitational radiation from point masses in a keplerian orbit. Phys. Rev. **131**, 435–440 (1963)

T. Piran, The physics of gamma-ray bursts. Rev. Mod. Phys. **76**, 1143–1210 (2004)

M. Punturo et al., The einstein telescope: a third-generation gravitational wave observatory. Class. Quantum Grav. **27**(19),194002 (2010)

S. Rosswog, Mergers of neutron star-black hole binaries with small mass ratios: nucleosynthesis, gamma-ray bursts, and electromagnetic transients. ApJ **634**, 1202–1213 (2005)

S. Rosswog, R. Speith, G.A. Wynn, Accretion dynamics in neutron star-black hole binaries. MNRAS **351**, 1121–1133 (2004)

M. Ruffert H. Janka, Simulations of coalescing neutron star and black hole binaries. Prog. Theor. Phys. Suppl. **136**, 287–299 (1999)

M. Ruffert, H.-T. Janka, Polytropic neutron star - black hole merger simulations with a Paczyński-Wiita potential. A&A **514**, A66 (2010)

S. Setiawan, M. Ruffert, H.-T. Janka, Three-dimensional simulations of non-stationary accretion by remnant black holes of compact object mergers. A&A **458**, 553–567 (2006)

M. Shibata, K. Taniguchi. Coalescence of Black Hole-Neutron Star Binaries. Living Rev. Relativity, **14**(6), August 2011.

M. Tanaka, K. Hotokezaka, K. Kyutoku, S. Wanajo, K. Kiuchi et al., Radioactively powered emission from black hole-neutron star mergers. Astrophys. J. **780**, 31 (2014)

S.E. Woosley, J.S. Bloom, The supernova gamma-ray burst connection. Ann. Rev. Astron. Astrophys. **44**, 507–556 (2006)

Coincidence Searches of Gravitational Waves and Short Gamma-Ray Bursts

Andrea Maselli and Valeria Ferrari

Abstract Black-hole neutron-star coalescing binaries have been invoked as one of the most suitable scenario to explain the emission of short gamma-ray bursts. Indeed, if the black-hole which forms after the merger, is surrounded by a massive disk, neutrino annihilation processes may produce high-energy and collimated electromagnetic radiation. In this paper, we devise a new procedure, to be used in the search for gravitational waves from black-hole-neutron-star binaries, to assign a probability that a detected gravitational signal is associated to the formation of an accreting disk, massive enough to power gamma-ray bursts. This method is based on two recently proposed semi-analytic fits, one reproducing the mass of the remnant disk surrounding the black hole as a function of some binary parameters, the second relating the neutron star compactness, with its tidal deformability. Our approach can be used in low-latency data analysis to restrict the parameter space searching for gravitational signals associated with short gamma-ray bursts, and to gain information on the dynamics of the coalescing system and on the neutron star equation of state.

1 Introduction

Coalescing binary systems formed by neutron stars (NSs) and/or black holes (BHs) represent one of the most promising sources of gravitational waves (GWs) to be detected by interferometric detectors of second (AdvLIGO/Virgo) and third (the Einstein Telescope, ET) generation (http://www.ligo.caltech.edu, http://www.ego-gw.it; http://www.et-gw.eu). Moreover, these events have recently been proposed as a candidate for the central engine of short gamma-ray bursts (SGRB), provided the stellar-mass BH which forms after merging is surrounded by a hot and sufficiently massive accreting disk (see for instance Lee and Ramirez-Ruiz 2007 and references therein). Since the electromagnetic emission is produced at large distance from the central engine, it does not give strong information on the source. In addition, the

A. Maselli (✉) • V. Ferrari
Dipartimento di Fisica, Università di Roma "La Sapienza" & Sezione INFN Roma1, P.A. Moro 5, 00185 Rome, Italy
e-mail: andrea.maselli@roma1.infn.it; valeria.ferrari@roma1.infn.it

C.F. Sopuerta (ed.), *Gravitational Wave Astrophysics*, Astrophysics and Space Science Proceedings, DOI 10.1007/978-3-319-10488-1_7

emission is beamed, and consequently these events may not be detected if one is looking in the wrong direction. Conversely, the gravitational wave (GW) emission is not beamed, and exhibits a characteristic waveform (the chirp), which should allow a non-ambiguous identification of the source. GRBs are characterized by a prompt emission, which lasts a few seconds, and an afterglow, whose duration ranges from hours to days. Thus, gravitational wave detection may be used to trigger the afterglow search of GRBs that have not been detected by the on-axis prompt observation and to validate the "jet model" of SGRB. Or, alternatively, the observation of a SGRB may be used as a trigger to search for a coincident GW signal. Indeed, this kind of search has already been done in the data of LIGO and Virgo (Abadie et al. 2012a, b).

However, not all coalescences of compact bodies produce a black hole with an accreting disk sufficiently massive to power a SGRB: it is therefore crucial to devise a strategy to extract those having the largest probability to produce a SGRB. This is one of the purposes of this work. The LIGO-Virgo Collaboration has recently developed a plausible observing schedule, according to which within this decade the advanced detectors, operating under appropriate conditions, will be able to determine the sky location of a source within 5 and 20 deg^2 (Aasi et al. 2013). Given the cost of spanning this quite large region of sky to search for a coincident SGRB with electromagnetic detectors, indications on whether a detected signal is likely to be associated with a SGRB are valuable information.

The procedure we propose has several applications. It can be used in the data analysis of future detectors (1) to gain information on the range of parameters that are more useful to span in the low-latency search for GWs emitted by BH-NS sources (Abadie et al. 2012c), (2) for an externally triggered search for GW coalescence signals following GRB observations (Abadie et al. 2012a, b), and (3) when the binary parameters are measured with sufficient accuracy and in a sufficiently short time to allow for an electromagnetic follow-up to search for off-axis GRB afterglows.

2 Selecting Candidates for Gamma-Ray Bursts Emission

In the last years a large number of numerical studies of BH-NS coalescence, have allowed to derive two interesting fits. The first (Foucart 2012) gives the mass of the accretion disk, $M_{\rm rem}$, as a function of the NS compactness $\mathscr{C} = M_{\rm NS}/R_{\rm NS}$, where $M_{\rm NS}$ and $R_{\rm NS}$ are the NS gravitational mass and its radius, the dimensionless BH spin, $\chi_{\rm BH} \in [-1, 1]$, and the mass ratio $q = M_{\rm BH}/M_{\rm NS}$:

$$\frac{M_{\rm rem}}{M_{\rm NS}^b} = K_1(3q)^{1/3}(1 - 2\mathscr{C}) - K_2 q\, \mathscr{C}\, R_{\rm ISCO}. \qquad (1)$$

Here $M_{\rm NS}^{\rm b}$ is the NS baryonic mass which, following Giacomazzo et al. (2013), we assume to be 10 % larger than the NS gravitational mass; $R_{\rm ISCO}$ is the radius of the innermost, stable circular orbit for a Kerr black hole (Bardeem et al. 1972). The two coefficients $K_1 = 0.288 \pm 0.011$ and $K_2 = 0.1248 \pm 0.007$ have been derived (Foucart 2012) through a least-square fit of the results of fully relativistic numerical simulations (Kyutoku et al. 2011; Etienne et al. 2009; Foucart et al. 2012, 2011).

$M_{\rm rem}$ is a key parameter in our study. Indeed, neutrino-antineutrino annihilation processes extract energy from the disk (Piran 2005), and several studies have shown that this process could supply the energy required to ignite a short gamma-ray burst, if $M_{\rm rem} \in (0.01 \div 0.05) M_{\rm NS}$ (Stone et al. 2013). In the following we shall assume as a threshold for SGRB formation $M_{\rm rem} = 0.01\ M_{\rm NS}$.

The second fit (Maselli et al. 2013) is a universal relation between the NS compactness $\mathscr{C}$ and the tidal deformability $\lambda_2 = -Q_{ij}/C_{ij}$, where Q_{ij} is the NS star traceless quadrupole tensor, and C_{ij} is the tidal tensor,

$$\mathscr{C} = 0.371 - 3.9 \times 10^{-2} \ln \bar{\lambda} + 1.056 \times 10^{-3} (\ln \bar{\lambda})^2 \quad , \quad \bar{\lambda} = \lambda_2 / M_{\rm NS}^5. \tag{2}$$

Hereafter, we shall denote by $\mathscr{C}_\lambda$ the NS compactness obtained from this fit.

Let us now assume that the gravitational wave signal emitted in a BH-NS coalescence is detected; a suitable data analysis allows us to find the values of the mass-ratio $q = M_{\rm BH}/M_{\rm NS}$, of the chirp mass $\mathscr{M} = (M_{\rm NS} M_{\rm BH})^{3/5}/(M_{\rm NS} + M_{\rm BH})^{1/5}$, and of the black hole spin $\chi_{\rm BH}$, with the corresponding errors. Knowing $q \pm \sigma_q$ and $\chi_{\rm BH} \pm \sigma_{\chi_{\rm BH}}$, using the fit (1) we can trace the plot of Fig. 1 in the $q - \mathscr{C}$ plane, for an assigned disk mass threshold, say $M_{\rm rem} = 0.01 M_{\rm NS}$. This plot allows us to identify the parameter region where a SGRB may occur, i.e., the region $M_{\rm rem} \gtrsim 0.01 M_{\rm NS}$ (below the fit curve in the figure), and the forbidden region above

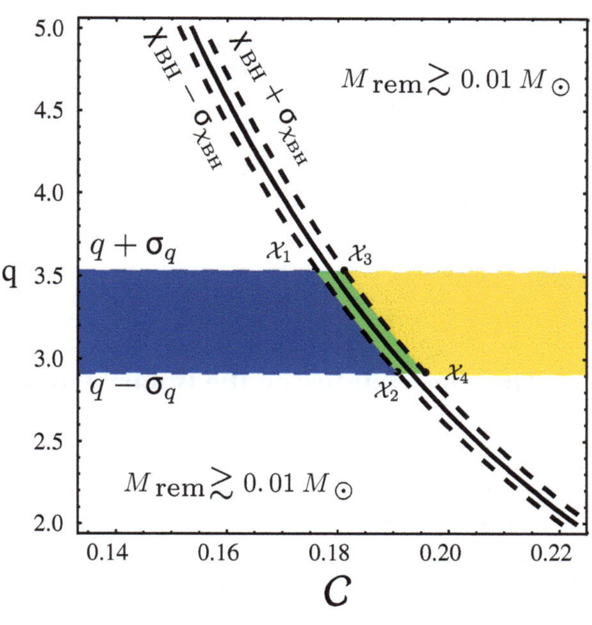

Fig. 1 Contour plot of the fit (1) in the q-$\mathscr{C}$ plane, for $M_{\rm NS} = 1.2 M_\odot$, $\chi_{\rm BH} = 0.5$ and $M_{\rm rem} = 0.01 M_\odot$. The fit separates the region allowed for SGRB ignition (below the fit curve) from the forbidden region (above the fit). Given the measured values of $q \pm \sigma_q$ and $\chi_{\rm BH} \pm \sigma_{\chi_{\rm BH}}$, a detected signal can correspond to a NS with compactness $\mathscr{C}$ which falls in one of the regions bounded by the *dashed curves*. Since $\mathscr{C}$ also comes with an error $\sigma_{\mathscr{C}}$, in order to infer if it can be associated with a SGRB, we need to evaluate the probability $P(\mathscr{C} \leq \mathscr{C}_4)$ and $P(\mathscr{C} \leq \mathscr{C}_1)$ (see text)

the fit ($M_{\rm rem} \lesssim 0.01 M_{\rm NS}$). In addition, we identify four points $\mathscr{X}_1, \dots \mathscr{X}_4$, which are the intersection between the contour lines for $\chi_{\rm BH} \pm \sigma_{\chi_{\rm BH}}$ and the horizontal lines $q \pm \sigma_q$. Let us indicate as $\mathscr{C}_1, \dots, \mathscr{C}_4$ the corresponding values of the neutron star compactness. Since the fit (1) is monotonically decreasing, $\mathscr{C}_1 < \mathscr{C}_2 < \mathscr{C}_3 < \mathscr{C}_4$. At this stage we still cannot say whether the detected binary falls in the region allowed for the formation of a SGRB or not. In order to get this information, we need to evaluate $\mathscr{C}$. As discussed in Del Pozzo et al. (2013), Damour et al. (2012), Read et al. (2013), Pannarale et al. (2011), Maselli et al. (2013), Read et al. (2009), Advanced LIGO/Virgo are expected to measure the gravitational wave phase with an accuracy sufficient to estimate the NS tidal deformability λ_2. Thus, using the fit (2), the neutron star compactness $\mathscr{C}_\lambda$ and the corresponding uncertainty $\sigma_{\mathscr{C}_\lambda}$ can be derived (see Maselli et al. 2013 for details on how to compute the compactness error).

Knowing the parameters and their uncertainties, the probability that a SGRB is associated to the detected coalescence can now be evaluated. We assume that $(q, \mathscr{C}_\lambda, \chi_{\rm BH})$ are described by a multivariate Gaussian distribution,

$$\mathscr{P}(q, \mathscr{C}_\lambda, \chi_{\rm BH}) = \frac{1}{(2\pi)^{3/2}|\Sigma|^{1/2}} \exp\left[-\frac{1}{2}\Delta^{\rm T}\Sigma^{-1}\Delta\right] , \tag{3}$$

where $\Delta = (\boldsymbol{x} - \boldsymbol{\mu})$, $\boldsymbol{\mu} = (q, \mathscr{C}_\lambda, \chi_{\rm BH})$, and Σ is the covariance matrix. Then, we define the maximum and minimum probability that the binary coalescence produces an accretion disk with mass over the threshold, $\bar{M}_{\rm rem}$, as

$$\mathrm{P}_{\rm MAX}(M_{\rm rem} \gtrsim \bar{M}_{\rm rem}) \equiv \mathrm{P}(\mathscr{C}_\lambda \leq \mathscr{C}_4) \ , \quad \mathrm{P}_{\rm MIN}(M_{\rm rem} \gtrsim \bar{M}_{\rm rem}) \equiv \mathrm{P}(\mathscr{C}_\lambda \leq \mathscr{C}_1) \ , \tag{4}$$

where $\mathrm{P}(\mathscr{C}_\lambda \leq \mathscr{C}_i)$ is the cumulative distribution of Eq. (3), which gives the probability that the measured compactness $\mathscr{C}_\lambda$, estimated through the fit (2), is smaller than an assigned value $\mathscr{C}_i$.

As an illustrative example, we now evaluate the probability that a given BH-NS coalescing binary produces a SGRB, assuming a set of equations of state for the NS matter and evaluating the uncertainties on the relevant parameters using a Fisher matrix approach.

3 The Uncertainties on the Binary Parameters

The accuracy with which future interferometers will measure a set of binary parameters $\boldsymbol{\theta}$ is estimated by comparing the gravity-wave data stream with a set of theoretical templates. For strong enough signals, $\boldsymbol{\theta}$ are expected to have a Gaussian distribution centered around the true values, with covariance matrix $\mathrm{Cov}^{ab} = (\Gamma^{-1})^{ab}$, where Γ^{ab} is the Fisher information matrix which contains the partial derivatives of the template with respect to the binary parameters (Poisson and Will 1995).

To model the waveform we use the TaylorF2 approximant in the frequency domain, assuming the stationary phase approximation $h(f) = \mathscr{A}(f)e^{i\psi(f)}$ (Damour et al. 2000). The post-Newtonian expansion of the phase includes spin-orbit and tidal corrections. It can be written as $\psi(f) = \psi_{\rm PP} + \psi_{\rm T}$, i.e. a sum of a point-particle term (see Blanchet et al. (2004a, b) for the complete expression) and a tidal contribution (Vines et al. 2011; Damour et al. 2012). The latter is given by

$$\psi_{\rm T} = -\frac{117\Lambda}{8}\frac{(1+q)^2}{qm^5}x^{5/2}\left[1+\frac{3115}{1248}x - \pi x^{3/2} + \left(\frac{23073805}{3302208}+\frac{120}{81}\right)x^2 - \frac{4283}{1092}\pi x^{5/2}\right], \tag{5}$$

where $x = (m\pi f)^{5/3}$, $m = M_{\rm BH} + M_{\rm NS}$ is the total mass of the system, and Λ is the averaged tidal deformability, which for BH-NS binaries reads (Flanagan and Hinderer 2008): $\Lambda = \lambda_2\frac{(1+12q)}{26}$.

We consider non-rotating NSs, as this is believed to be a reliable approximation of real astrophysical systems (Bildsten and Cutler 1992; Kochanek 1992). Therefore, our template is fully specified by six parameters[1] $\boldsymbol{\theta} = (t_c, \phi_c, \ln\mathscr{M}, q, \Lambda, \beta)$ where t_c, ϕ_c are the time and phase at the coalescence and β is the 2 PN spin-orbit contribution in $\psi_{\rm PP}$. We choose the BH spin aligned with the orbital angular momentum. Moreover, since $\chi_{\rm BH} \leq |1|$, $\beta \lesssim 9.4$; therefore we consider the following prior probability distribution on β: $p^{(0)}(\beta) \propto \exp\left[-\frac{1}{2}(\beta/9.4)^2\right]$.

In our analysis we consider both second and third generation detectors. For AdvLIGO/Virgo we use the `ZERO_DET_high_P` noise spectral density of AdvLIGO (Shoemaker xxxx), in the frequency ranges [20 Hz, $f_{\rm ISCO}$]; for the Einstein Telescope we use the analytic fit of the sensitivity curve provided in Sathyaprakash and Schultz (2009), in the range [10 Hz, $f_{\rm ISCO}$]. $f_{\rm ISCO}$ is the frequency of the Kerr ISCO including corrections due to NS self-force (Favata 2011).

We model the NS structure by means of piecewise polytropes (Read et al. 2009). Indeed we consider four EoS, labeled as `2H`,`H`, `HB` and `B`, which denote very stiff, stiff, moderately stiff and soft nuclear matter, respectively. The stellar parameters for $M_{\rm NS} = (1.2, 1.35)M_\odot$, are shown in Table 1.

[1]The signal amplitude $\ln\mathscr{A}$ is uncorrelated with the other variables, so we perform derivatives only with respect to the remaining parameters.

Table 1 For each EoS we show the NS mass, the compactness $\mathscr{C} = M_{NS}/R_{NS}$, and the tidal deformability λ_2

EoS	$M_{NS}(M_\odot)$	$\mathscr{C}$	λ_2 (km^5)	$M_{NS}(M_\odot)$	$\mathscr{C}$	λ_2 (km^5)
2H	1.2	0.117	75,991	1.35	0.131	72,536
H	1.2	0.145	21,232	1.35	0.163	18,964
HB	1.2	0.153	15,090	1.35	0.172	13,161
B	1.2	0.162	10,627	1.35	0.182	8,974

Table 2 We show the probability range [P_{MIN},P_{MAX}] that the coalescence of a BH-NS binary produces a disk mass larger than $\bar{M}_{rem} = 0.01 M_{NS}$ for AdLIGO/Virgo (AdV), for binaries with $q = 3$ and $q = 7$, NS masses $(1.2,1.35) M_\odot$, and BH spin $\chi_{BH} = (0.2, 0.5, 0.9)$

AdV		$q = 3$			$q = 7$		
$M_{NS} = 1.2 M_\odot$		χ_{BH}			χ_{BH}		
EOS	$\mathscr{C}_\lambda$	0.2	0.5	0.9	0.2	0.5	0.9
2H	0.118	1	1	1	0.4	[0.8–0.9]	1
H	0.147	[0.6–0.9]	1	1	0.4	0.4	[0.8–0.9]
HB	0.155	[0.5–0.7]	[0.9–1]	1	0.4	0.4	[0.7–0.8]
B	0.164	[0.4–0.6]	[0.7–0.8]	1	0.4	0.4	[0.6–0.7]
$M_{NS} = 1.35 M_\odot$							
2H	0.132	1	1	1	0.3	[0.4–0.5]	1
H	0.164	[0.4–0.6]	[0.8–0.9]	1	0.4	0.4	0.7
HB	0.173	[0.4–0.5]	[0.6–0.8]	1	0.4	0.4	0.6
B	0.184	[0.4–0.5]	[0.5–0.6]	1	0.4	0.4	0.5

Sources are assumed to be at $d = 100$ Mpc. The star compactness $\mathscr{C}_\lambda$ is estimated throughout the universal relation (2)

4 Numerical Results

Following the strategy previously outlined, we compute the minimum and maximum probabilities (4) that the coalescence of a BH-NS system produces a remnant disk with mass above a threshold $\bar{M}_{rem}$, for the NS models listed in Table 1 and different values of the mass ratio q. The results are given in Tables 2 and 3, for $q = 3$ and $q = 7$, black hole spin $\chi_{BH} = (0.2, 0.5, 0.9)$, $M_{NS} = (1.2, 1.35)\ M_\odot$, and disk mass thresholds $\bar{M}_{rem} = 0.01 M_{NS}$.

For AdvLIGO/Virgo we put the source at a distance of 100 Mpc. For ET the binary is at 1 Gpc. In this case the signal must be suitably redshifted (Cutler and Flanagan 1994; Maselli et al. 2013), and we have assumed that z is known with a fiducial error of the order of 10 % (Messenger and Read 2012).

The first clear result is that as the BH spin approaches the highest value we consider, $\chi_{BH} = 0.9$, and for low mass ratio $q = 3$, the probability that a BH-NS coalescence produces a disk with mass above the threshold is insensitive to the NS internal composition, and it approaches unity for all considered configurations. These would be good candidates for GRB production. For the highest mass ratio we

Table 3 Same of Table 2, but for the Einstein Telescope (ET). In this case we assume prototype BH-NS binaries at $d = 1$ Gpc

ET		$q = 3$			$q = 7$		
$M_{NS} = 1.2 M_\odot$		χ_{BH}			χ_{BH}		
EOS	$\mathscr{C}_\lambda$	0.2	0.5	0.9	0.2	0.5	0.9
2H	0.118	1	1	1	[0.3–0.4]	[0.7–0.8]	1
H	0.147	[0.9–1]	1	1	0.3	[0.3–0.4]	1
HB	0.155	[0.7–0.8]	1	1	0.3	0.3	0.9
B	0.164	[0.5–0.6]	1	1	0.3	0.3	[0.7,0.8]
$M_{NS} = 1.35 M_\odot$							
2H	0.132	1	1	1	0.2	[0.4,0.5]	1
H	0.164	[0.4–0.6]	1	1	0.3	0.3	0.8
HB	0.173	[0.3–0.4]	0.8	1	[0.3–0.4]	[0.3–0.4]	0.6
B	0.184	[0.3–0.4]	0.6	1	0.4	0.4	[0.5–0.6]

consider, $q = 7$, the probability to form a sufficiently massive disk depends on the NS mass and EoS, and on the detector. In particular, it decreases as the EoS softens, and as the NS mass increases. This is a general trend, observed also for smaller values of χ_{BH}. However, when $\chi_{BH} = 0.9$ the probability that the coalescence is associated to a SGRB is always $\gtrsim 50\,\%$.

Let us now consider the results for $\chi_{BH} = 0.2$. If the NS mass is 1.2 $M_\odot$ the probability that a detected GW signal from a BH-NS coalescence is associated to the formation of a black hole with a disk of mass above threshold is $\gtrsim 50\,\%$ for both AdvLIGO/Virgo and ET, provided $q = 3$. For larger NS mass, this remains true only if the NS equation of state is stiff (2H or H). High values of q are disfavored.

When the black hole spin has an intermediate value, say $\chi_{BH} = 0.5$, Table 2 shows that, the NS compactness plays a key role in the identification of good candidates for GRB production, for both detectors. Again large values of the mass ratio yield small probabilities.

The range of compactness shown in Tables 2 and 3 includes neutron stars with radius ranging within $\sim[10-15]$ km. From the table it is also clear that if we choose a compactness smaller than the minimum value, the probability of generating a SGBR increases, and the inverse is true if we consider compactness larger than our maximum.

Conclusions

The method developed in this paper can be used in several different ways. In the future, gravitational wave detectors are expected to reach a sensitivity sufficient to extract the parameters on which our analysis is based, i.e., chirp mass, mass ratio, source distance, spin and tidal deformability. We can

(continued)

also expect that the steady improvement of the efficiency of computational facilities experienced in recent years will continue, reducing the time needed to obtain these parameters from a detected signal. Moreover, the higher sensitivity will allow us to detect sources in a much larger volume space, thus increasing the detection rates. In this perspective, the method we envisage in this paper will be useful to trigger the electromagnetic follow-up of a GW detection, searching for the afterglow emission of a SGRBs.

Until then, the method we propose can be used in the data analysis of advanced detectors as follows:

- Tables 2 and 3 indicate the systems that are more likely to produce accretion disks sufficiently massive to generate a SGRB. The table can be enriched including more NS equations of state or more binary parameters; however, it already contains a clear information on which is the range of parameters to be used in the GW data analysis, if the goal is to search for BH-NS signals which may be associated to a GRB. For instance, Tables 2 and 3 suggests that searching for mass ratio smaller than, or equal to, 3–4, and values of the black-hole angular momentum larger than 0.5–0.6 would allow us to save time and computational resources in low-latency search. In addition, it would allow us to gain sensitivity in externally triggered searches performed in time coincidence with short GRBs observed by gamma-ray satellites.
- If a SGRB is observed sufficiently close to us in the electromagnetic waveband, the parameters of the GW signal detected in coincidence would allow us to set a threshold on the mass of the accretion disk. If the GW signal comes, say, from a system with a BH with spin $\chi_{BH} = 0.5$, mass ratio $q = 7$, and neutron star mass $M_{NS} = 1.2 M_{\odot}$, from Table 2, equations of state softer than the EoS 2H would be disfavored. Thus, we would be able to shed light on the dynamics of the binary system, on its parameters and on the internal structure of its components. We would enter into the realm of gravitational wave astronomy.

Finally, it is worth stressing that as soon as the fit (1) is extended to NS-NS coalescing binaries, this information will be easily implemented in our approach. With the rate of NS-NS coalescence higher than that of BH-NS, our approach will acquire more significance, and will be a very useful tool to study these systems.

References

J. Aasi et al., arXiv:1304.0670 (2013)

J. Abadie et al. (LIGO Scientific Collaboration), Astrophys. J. **755**, 2 (2012)

J. Abadie et al. (LIGO Scientific Collaboration), Astrophys. J. **760**, 12 (2012)
J. Abadie et al. (LIGO Scientific Collaboration), Astron. Astrophys. **541** A155 (2012)
J.M. Bardeem, W.H. Press, S.A. Teukolsky, Astrophys. J. **178**, 347 (1972)
L. Bildsten, C. Cutler, Astrophys. J **400**, 175 (1992)
L. Blanchet, G. Faye, B.R. Iyer, B. Jouget, Phys. Rev. D **65**, 061501 (2004)
L. Blanchet, T. Damour, G. E. Farese, B.R. Iyer, Phys. Rev. Lett. **93**, 091011 (2004)
C. Cutler, E.E. Flanagan, Phys. Rev. D **49**, 2658 (1994)
T. Damour, B. Iyer, B. Sathyaprakash, Phys. Rev. D **62**, 084036 (2000)
T. Damour, A. Nagar, L. Villain, Phys. Rev. D **85**, 123007 (2012)
W. Del Pozzo et al., Phys. Rev. Lett. **111**, 071101 (2013)
Z.B. Etienne, Y.T. Liu, S.L. Shapiro, T.W. Baumgarte, Phys. Rev. D **79**, 044024 (2009)
M. Favata, Phys. Rev. D **83**, 024028 (2011)
E.E. Flanagan, T. Hinderer, Phys. Rev. D **77**, 021502 (2008)
F. Foucart, Phys. Rev. D **86**, 124007 (2012)
F. Foucart, M.D. Duez, L.E. Kidder, S.A. Teukolsky, Phys. Rev. D **83**, 024005 (2011)
F. Foucart et al., Phys. Rev. D **85**, 044015 (2012)
B. Giacomazzo, R. Perna, L. Rezzolla, E. Troja D. Lazzati, Astrophys. J. Lett., **762**, L18 (2013)
C.S. Kochanek, Astrophys. J. **398**, 234 (1992)
K. Kyutoku, H. Okawa, M. Shibata, K. Taniguchi, Phys. Rev. D **84**, 064018 (2011)
W.H. Lee, E. Ramirez-Ruiz, New J. Phys. **9**, 17 (2007)
A. Maselli, V. Cardoso, V. Ferrari, L. Gualtieri, P. Pani, Phys. Rev. D **88**, 023007 (2013)
A. Maselli, L. Gualtieri, V. Ferrari, Phys. Rev. D **88**, 104040 (2013)
C. Messenger, J. Read, Phys. Rev. Lett. **108** 09110 (2012)
F. Pannarale, L. Rezzolla, F. Ohme, J.S. Read, Phys. Rev. D **84**, 104017 (2011)
T. Piran, Rev. Mod. Phys. **76**, 1143 (2005)
E. Poisson, C.M. Will, Phys. Rev. D **52**, 2 (1995)
J. Read et al., Phys. Rev. D **79**, 124033 (2009)
J. Read et al., Phys. Rev. D **88** 88, 044042 (2013)
B.S. Sathyaprakash, B.F. Schultz, Living Rev. Relativity **12**, 2 (2009)
D. Shoemaker, https://dcc.ligo.org/cgi-bin/DocDB/ShowDocument?docid=2974 (2010)
N. Stone, A. Loeb, E. Berger, Phys. Rev. D **87** 084053 (2013)
J. Vines, E.E. Flanagan, T. Hinderer, Phys. Rev. D **83**, 084051 (2011)

Gravitational Waves from Rapidly Rotating Neutron Stars

Brynmor Haskell, Nils Andersson, Caroline D'Angelo, Nathalie Degenaar, Kostas Glampedakis, Wynn C.G. Ho, Paul D. Lasky, Andrew Melatos, Manuel Oppenoorth, Alessandro Patruno, and Maxim Priymak

Abstract Rapidly rotating neutron stars in Low Mass X-ray Binaries have been proposed as an interesting source of gravitational waves. In this chapter we present estimates of the gravitational wave emission for various scenarios, given the (electromagnetically) observed characteristics of these systems. First of all we focus on the r-mode instability and show that a "minimal" neutron star model (which does not incorporate exotica in the core, dynamically important magnetic fields or superfluid degrees of freedom), is not consistent with observations. We then present estimates of both thermally induced and magnetically sustained mountains in the crust. In general magnetic mountains are likely to be detectable only if the buried magnetic field of the star is of the order of $B \approx 10^{12}$ G. In the thermal mountain case we find that gravitational wave emission from persistent systems may be detected by ground based interferometers. Finally we re-asses the idea that gravitational wave emission may be balancing the accretion torque in these systems, and show that in most cases the disc/magnetosphere interaction can account for the observed spin periods.

B. Haskell (✉) • P.D. Lasky • A. Melatos • M. Priymak
School of Physics, The University of Melbourne, Parkville, VIC 3010, Australia
e-mail: bhaskell@unimelb.edu.au

N. Andersson • W.C.G. Ho
Mathematical Sciences and STAG Research Centre, University of Southampton, Southampton SO17 1BJ, UK

N. Degenaar
Department of Astronomy, University of Michigan, Ann Arbor, MI 48109, USA

K. Glampedakis
Departamento de Fisica, Universidad de Murcia, 30100 Murcia, Spain

M. Oppenoorth
Department of Earth Sciences, Utrecht University, Budapestlaan 4, 3584 CD Utrecht, The Netherlands

C. D'Angelo • A. Patruno
Leiden Observatory, Leiden University, Postbus 9513, 2300 RA Leiden, The Netherlands

ASTRON, the Netherlands Institute for Radio Astronomy, Postbus 2, 7990 AA Dwingeloo, The Netherlands

C.F. Sopuerta (ed.), *Gravitational Wave Astrophysics*, Astrophysics and Space Science Proceedings, DOI 10.1007/978-3-319-10488-1_8

1 Introduction

Neutron Stars (NSs) are one of the most fascinating fundamental physics laboratories in the Universe. With masses comparable to that of the sun compressed in a 10 km radius, these objects have internal densities that can easily exceed the nuclear saturation density, $\rho_0 \approx 2.4\times10^{14}$ g/cm^3, allowing us to probe a regime of the strong interaction that is not accessible with terrestrial experiments. In fact, although the internal temperatures of NS can be around $T \approx 10^8$ K, at such high densities the thermal energy of the constituents is negligible compared to their Fermi energy. Neutron Stars are thus essentially cold objects. While colliders, such as GSI at Darmstadt to the LHC at CERN, allow us to probe the high temperature regime of the QCD phase diagram (generally at low densities) and study phases such as quark gluon plasmas (Brambilla et al. 2014), NSs give us the opportunity to probe the high density, low temperature regime of QCD. At asymptotically high densities one expects quarks to pair in the so-called Colour-Flavour-Locked (CFL) phase (Alford et al. 2008). At realistic NS densities, however, there is still significant uncertainty on what the ground state of matter will be, and only astrophysical observations can shed light on this fundamental problem.

In order to interpret astrophysical data it is necessary to model the interior dynamics of NSs in detail. This is a formidable task, as several complex physical processes are at work in these systems. The outer layers of the star form a crystalline crust, that can support shearing and effectively insulate the hot interior and lead to the observable electromagnetic emission (Gudmundsson et al. 1983). At higher densities neutrons begin to drip out of the nuclei and are expected to be superfluid. At even higher densities one has a transition to a core fluid of superfluid neutrons, superconducting protons (most likely in a type II superconducting state) and electrons. Finally, at densities higher than saturation density, the composition of the star is unknown and may include exotic particles such as hyperons or deconfined quarks. Further complications arise from the fact that in many cases one has to deal with rapid rotation (up to considerable fractions of the Keplerian breakup frequency) and with dynamically important magnetic fields (of up to $B \approx 10^{15}$ G in magnetars). The extreme compactness of NSs further complicates the problem, as the effects of General Relativity become significant, and must be taken into account.

In order to understand the different physical mechanisms at work it is thus necessary to take a multi messenger approach and combine all observational signatures of NSs. NSs are observed in several electromagnetic bands, from the radio to gamma rays, but are also likely to be interesting sources of GWs. This opens a new and exciting window, as electromagnetic radiation originates mainly from the outer layers of the star while GWs interact weakly with matter and carry a strong imprint of the physics at work in the high density interior. There are several GW emission mechanisms that involve NSs and could lead to emission at the level detectable with current and next generation ground based GW detectors, such as Advanced LIGO, Advanced Virgo Kagra or the Einstein Telescope (ET). The most promising sources are clearly NS-NS binaries, which are the prime target for

Advanced LIGO and could carry the imprint of the equation of state of dense matter (Del Pozzo et al. 2013). In the following, however, we shall focus on a different class of sources, continuous sources. In particular we shall discuss several mechanisms that may lead to continuous GW emission from a rapidly rotating NS in a Low Mass X-ray Binary (LMXB) and assess the detectability of such a signal.

2 Gravitational Wave Emission Mechanisms

There are several mechanisms that can lead to GW emission from a rapidly rotating NS. All of them are based on the idea that a non-axisymmetric perturbation will be dragged around by rotation and lead to assess GW emission. The types of perturbation can roughly be divided into two categories: either "mountains", i.e. deformations that are static (at least on dynamical timescales) in the frame of the star, or hydrodynamical modes of oscillation being excited in the star. The most natural location for a NS mountain is the crust, as the finite shear modulus of the crystalline crust offers the possibility of supporting a deformation (Bildsten 1998). Recent estimates of the breaking strain of the crust have shown that high pressure and gravity lead to a remarkably strong material (Horowitz and Kadau 2009), that could sustain mountains large enough to be detected ground based interferometers (Haskell et al. 2006; Johnson-McDaniel and Owen 2013).

Strong magnetic fields can also confine material and lead to deformations that could be potentially quite large in magnetars (Ciolfi and Rezzolla 2013; Haskell et al. 2008). The situation is even more interesting in accreting systems, in which, although the magnetic field is globally much weaker than in magnetars, the accretion process can lead to material spreading equatorially and compressing the field, making it locally strong enough to sustain a sizeable mountain (Melatos and Payne 2005; Payne and Melatos 2004; Priymak et al. 2011).

Modes of oscillation of the star can also grow to large amplitude and lead to gravitational radiation. The prime candidate for this kind of mechanism in LMXBs is the r-mode. This is a toroidal mode of oscillation for which the restoring force is the Coriolis force. To leading order in the slow-rotation approximation it is purely toroidal and the Eulerian velocity perturbation $\delta\mathbf{v}$ takes the form:

$$\delta\mathbf{v} = \alpha \left(\frac{r}{R}\right)^l R\Omega \mathbf{Y}^B_{lm} e^{i\omega t} \tag{1}$$

where $\mathbf{Y}^B_{lm} = [l(l+1)]^{-1/2} r\nabla \times (r\nabla Y_{lm})$ is the magnetic-type vector spherical harmonic (with Y_{lm} the standard spherical harmonics), R is the stellar radius and α the dimensionless amplitude of the mode (Owen et al. 1998), Ω the rotation frequency of the star and ω the frequency of the mode.

It is of particular interest because not only is its frequency in the right range to be detected by ground based GW detectors (if the star is rotating at millisecond periods), but it is also generically unstable to GW emission (Andersson 1998;

Friedman and Morsink 1998). As we shall see in the following, this means that the mode can grow to large amplitudes, provided viscosity does not damp it on a shorter timescale than GW emission can drive it unstable.

3 Low Mass X-Ray Binaries

Before discussing the above mechanisms in detail, let us examine why NSs in LMXBs are interesting from a GW perspective. In an LMXB a compact object (a NS in the case that interests us) is accreting material from a low mass star ($M \lesssim 1 M_{\odot}$) which fills its Roche lobe. Matter leaves the secondary star and forms an accretion disc around the NS, eventually interacting with the magnetic field of the star and being accreted. Angular momentum is transfer to the NS, spinning it up and allowing for old, slow pulsars to be recycled to millisecond spin periods. This is believed to be the primary formation channel for Millisecond Radio Pulsars (MSRPs) (Alpar et al. 1982).

In this scenario, and provided that the magnetic fields of these systems are weak, one would expect the NS to spin up to its Keplerian break up frequency. The exact value of such frequency is equation of state dependent but, quite generally, is expected to be above $\approx$1.5 kHz. This expectation is, however, not borne out by observations of both LMXBs and MSRPs. In both cases the frequency distribution appears to have a cutoff around 700 Hz, well below the breakup frequency (Chakrabarty et al. 2003; Patruno 2010). An additional mechanism thus needs to be at work to remove angular momentum from the system and halt the accretion induced spin up of the NS. A natural candidate would be the interaction of the stellar magnetosphere with the accretion disc. If the magnetic field is strong enough (above $\approx 10^8$ G) it can disrupt the accretion disc above the stellar surface. Matter is accreted at the truncation radius and transfers its angular momentum to the star. This can, however, only happen as long as matter at the truncation radius is rotating faster than the star. Once the NS spin exceeds this limit, accretion is centrifugally inhibited, and matter can either be expelled from the system (Illarionov and Sunyaev 1975) or accrete unstably (Spruit and Taam 1993).

The first possibility was examined by White and Zhang (1997), who considered the implications of assuming that the observed spin period of the LMXB is the spin equilibrium period, as set by the torque balance mechanism described above. Their conclusion was, based on the models and data available, that this is unlikely, as it would require both stronger magnetic fields that observed in MSRPs (i.e. it would require fields in the range $B \approx 10^9$–10^{10} G or above) and an unexpected correlation between the magnetic field strength and the mass accretion rate. We shall discuss recent reassessments of this analysis later on and discuss how it may actually be an explanation for the observed spin distribution. Nevertheless the original analysis by White and Zhang (1997) led to renewed interest in GW emission as a mechanism to remove angular momentum from rapidly rotating neutron stars (Papaloizou and

Pringle 1978; Bildsten 1998) and to detailed analysis of the physical mechanisms (described in Sect. 2) that could lead to it.

4 The r-Mode Instability

Let us begin our analysis from the r-mode instability. As already mentioned the mode can only grow to large amplitudes if GW emission can drive it on a shorter timescale than viscosity can damp it. The competition between different mechanisms depends on several parameters, mainly the mass and equation of state of the star, its temperature and spin frequency. Given an equation of state (and thus a composition) for the star we can fix the mass and define an instability "window" in the temperature vs frequency plane. In Fig. 1 we show the instability window for a "minimal" neutron star model, i.e. a model in which we assume no exotica in the core, no dynamically important magnetic field or superfluid degrees of freedom, and take a simple $n = 1$ polytrope as an equation of state. We consider a typical 1.4 $M_{\odot}$ NS with a 10 km radius and show the curve on which the driving and damping

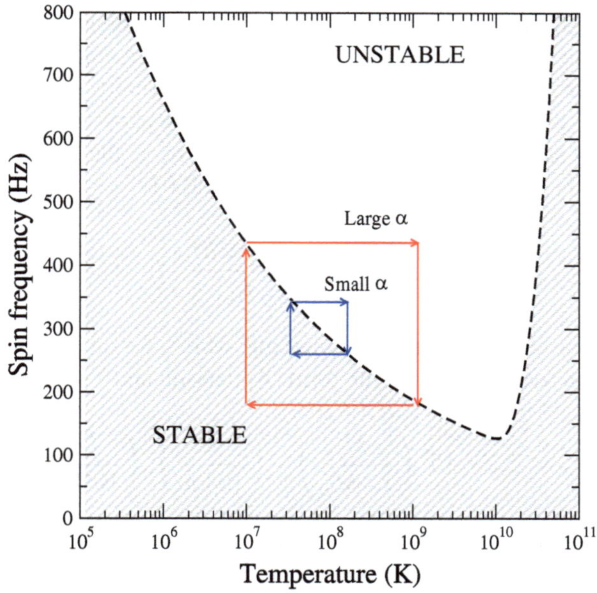

Fig. 1 The r-mode instability window for a 1.4 $M_{\odot}$, $R = 10$ km NS, described by an $n = 1$ polytrope. We assume the "minimal" model described in the text, with no exotica in the core. At low temperatures the main sources of damping are shear viscosity and dissipation at the crust-core boundary. At high temperatures bulk viscosity gives the main contribution. We also schematically illustrate the cycle that a system would follow in the temperature-frequency plane, both for small and large amplitude α of the mode

timescale are equal, i.e. the solutions of:

$$\frac{1}{\tau_{gw}} = \sum_i \frac{1}{\tau_{Vi}} \tag{2}$$

where τ_{gw} is the timescale on which GWs drive the mode unstable, which for an $n = 1$ polytrope is given by (Andersson and Kokkotas 2001)

$$\tau_{gw} = -47 \left(\frac{M}{1.4M_\odot}\right)^{-1} \left(\frac{R}{10\text{km}}\right)^{-4} \left(\frac{P}{1\text{ms}}\right)^{6} \text{s}, \tag{3}$$

while τ_{Vi} is the viscous damping timescale for process i acting in the star. At high temperature bulk viscosity provides the main damping mechanism, while at low temperatures the main contribution is from shear viscosity, due to standard scattering processes (neutron-neutron in non-superfluid matter or electron-electron in superfluid matter (Andersson et al. 2005a)), or from viscosity at the crust-core interface.

In such a scenario an accreting neutron star, with a typical *core* temperature of around 10^8 K would spin up into the unstable region due to accretion, leading to the onset of the instability. The r-mode would then rapidly grow to large amplitude, leading to fast heating. Eventually the thermal runaway is halted by neutrino emission and the star spins down due to GW emission, re-enters the stable region and cools, starting the cycle again, as depicted schematically in Fig. 1. The amount of heating, i.e. how far into the instability window a system can go, depends critically on the saturation amplitude for the mode, α, as the energy dissipated by viscosity takes the form (Andersson and Kokkotas 2001)

$$\frac{dE}{dt} = 1.31 \frac{\alpha^2 \nu^2 M R^2}{\tau_{sv}} \tag{4}$$

with τ_{sv} the shear viscosity damping timescale, which, for an $n = 1$ polytrope and electron-electron scattering, takes the form (Andersson and Kokkotas 2001):

$$\tau_{sv} = 2.2 \times 10^5 \left(\frac{M}{1.4M_\odot}\right)^{-1} \left(\frac{R}{10\text{ km}}\right)^{5} \left(\frac{T}{10^8\text{K}}\right)^{2} s \tag{5}$$

If the mode can grow to large amplitudes ($\alpha \approx 1$) the system will enter well into the unstable region, but the duty cycle will be very short, less than $\approx 1\,\%$ (Levin 1999; Spruit 1999). If, on the other hand, the mode saturates at a relatively low amplitude ($\alpha \approx 10^{-5}$), as calculations of non-linear couplings to other modes suggest (Bondarescu et al. 2007), the duty cycle is much longer but the system will never depart significantly from the instability curve (Heyl 2002). In either scenario it is highly unlikely to observe a system in the unstable region.

An analysis by Haskell et al. (2012a), confirmed by Mahmoodifar and Strohmayer (2013), has however revealed that if one populates the instability

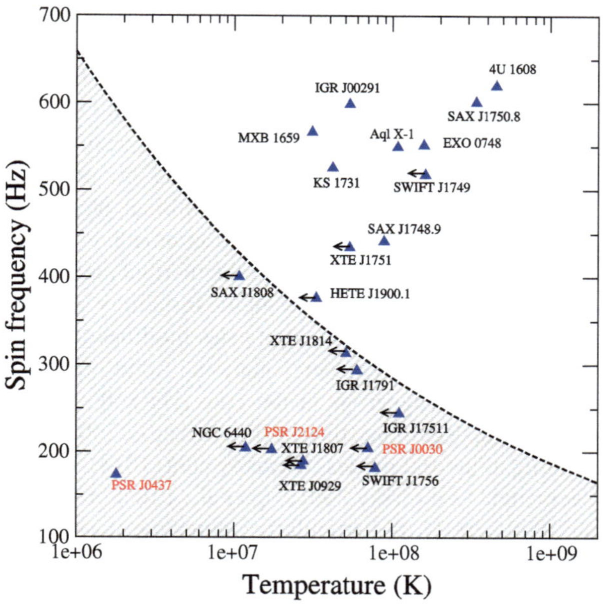

Fig. 2 The "minimal" instability window described in the text, compared to the spin frequency and core temperatures inferred in Haskell et al. (2012a) for the known LMXBS. Clearly there are many systems in the instability window, which is inconsistent with the predicted window. To reconcile theory and observations it is thus necessary to either include additional sources of damping in our model. Another possibility is that the r-mode saturation amplitude is small enough that it is indeed unstable, but does not impact on the thermal and spin evolution of this systems

window with data from systems where there is both an estimate of the spin period of the NS and its internal temperature (obtained from fits to the surface temperature), many systems would sit inside the instability window, as shown in Fig. 2. This issue has also been examined on a theoretical basis by Ho et al. (2011). The conclusion is robust despite the uncertainty introduced by the unknown mass of the star, and the need to model the atmosphere to map the surface temperature to the core (see Haskell et al. 2012a for a detailed description of the different assumptions). It is thus clear that the "minimal" NS model described above is not consistent with observations. One can also assess the viability of the spin equilibrium scenario by calculating the internal temperature that a star would have if the spin-up torque due to accretion is balanced by an r-mode at the observed spin period. Given the observed accretion luminosity L_{acc}, the heat dissipated at equilibrium is (Brown and Ushomirsky 2000)

$$L_{heat} = 0.064 \left(\frac{\nu}{300\text{Hz}} \right) L_{acc}. \tag{6}$$

The core temperature can be obtained by assuming that the heat in (6) is carried away by neutrinos. In Fig. 2 we show the inferred core temperatures obtained by balancing heating with neutrino emission processes, calculated accounting for modified Urca processes and Copper pair formation, as described in Ho et al. (2011). It can be seen that in most cases the stars are too cold to allow for an r-mode to balance the spin-up torque, except for the hotter, faster systems. This is important for GW target selection, as the energy emitted in GWs increases steeply with frequency, making these systems the best targets for next-generation detectors.

Our understanding of the r-mode instability can be made consistent with observations in two ways: either we include additional physics in our models, allowing for additional sources of damping, or we assume that the r-mode saturation amplitude is so small that it has no impact on the thermal and frequency evolution of the star, and that a system can thus "live" in the instability window.

The first possibility, i.e. that additional physics modifies the r-mode instability window, has been considered by several authors. Additional viscosity may be provided, among others, by hyperons in the core (Haskell and Andersson 2010; Nayyar and Owen 2006), deconfined quarks (Haskell et al. 2007, 2012b; Andersson et al. 2010), coupling to inertial modes (Gusakov et al. 2014), torsional oscillations of the crust (Levin and Ushomirsky 2001; Ho et al. 2011) or by magnetic braking (Rezzolla et al. 2000, 2001). A particularly interesting possibility is that strong superfluid mutual friction, due to superfluid vortices cutting through magnetic flux tubes in the superconducting interior of the NS, could lead to increased damping. In Fig. 3 we show the effect on the instability window of assuming a superfluid drag parameter $\mathscr{R} \approx 0.01$, of the order expected if vortices are continuously cutting through flux tubes. The dimensionless parameter $\mathscr{R}$ represents the strength of the mutual friction between the superfluid neutrons and the protons, and couples the two components on a timescale $\tau \approx (1+\mathscr{R}^2)/2\Omega$, with Ω the rotation frequency of the

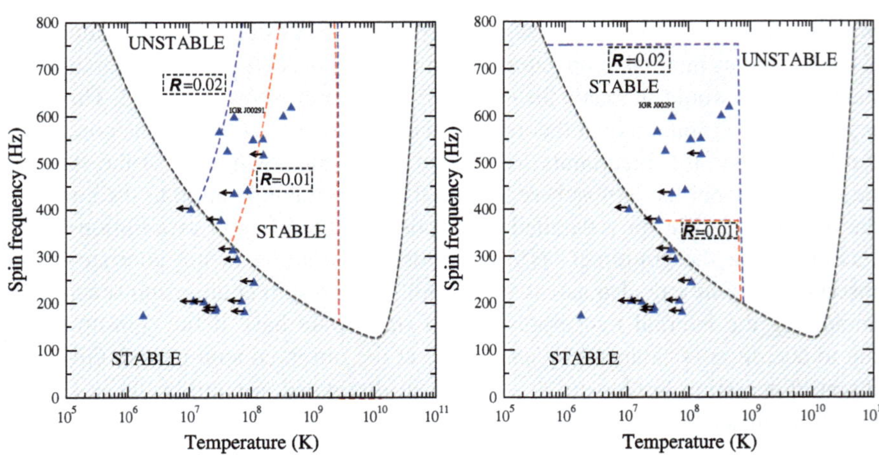

Fig. 3 The instability window for strong mutual friction due to vortex/flux tube cutting in the core. We show the results for two different models for the superfluid pairing gaps, the "strong" (*left*) and "weak" (*right*) models described in Haskell et al. (2009). In both cases the window can be reconciled with observations for superfluid drag parameters of $\mathscr{R} \approx 0.01$, which is in the possible range for the vortex flux tube cutting mechanism

star. Note that for standard mutual friction, due to electron scattering off magnetised vortex cores, the drag parameter is much smaller, $\mathscr{R} \approx 10^{-4}$, and the effect on the instability window is negligible (Haskell et al. 2009; Passamonti et al. 2009). In the estimates in Fig. 3, we have, however, assumed that vortices are free to cut through flux tubes. This may not be the case, as the energy cost associated with cutting effectively "pins" vortices to flux tubes until a sufficient lag builds up between the neutron an proton fluid, and classical hydrodynamical lift forces, Magnus forces, can push vortices out. This provides a non-linear saturation mechanism for the r-mode, as the mode can only grow to the point where the velocity perturbation is large for the Magnus force to push vortices through flux-tubes (more specifically it is the counter-moving component of the velocity perturbation that grows, but as it grows at the same rate as the total velocity perturbation, this complication can be avoided in the following discussion. See Haskell et al. 2009 for a detailed analysis). At this point the process is strongly dissipative and rapidly damps the mode, thus setting a saturation amplitude α_s, which takes the form (Haskell et al. 2014):

$$\alpha_s \approx 10^{-6}\left(\frac{\lambda_0}{0.1}\right)^{-1}\left(\frac{\nu}{500\text{Hz}}\right)^{-3}\left(\frac{B}{10^8\text{G}}\right)^{1/2}, \tag{7}$$

where ν is the spin frequency of the star, and λ_0 is the ratio between the amplitude of the counter moving component of the mode to the amplitude of the oscillation in the total velocity, as described in Haskell et al. (2014).

A similar effect could be at work in the deep core, if there is a transition to quark matter. In this case a large enough velocity perturbation could lead to strong bulk viscosity due to the different reactions on the two sides of the interface, which saturates the mode (Alford et al. 2014). Another possibility is that, if viscosity is weak at the crust-core interface [due e.g. to the presence of so called "pasta" phases (Lorenz et al. 1993)], non linear couplings saturate the mode at very low amplitudes (Bondarescu and Wasserman 2013). In all these cases the saturation amplitude α could be low enough to allow systems to be r-mode unstable, without any observable signature. In this scenario old systems such as LMXBs are unlikely to lead to strong GW emission due to unstable r-modes, with young NSs being a much more promising GW source (Alford and Schwenzer 2014a, b).

5 Mountains on Neutron Stars

Let us now move on to discuss "mountains". As already mentioned the crust of the NS can sustain shearing and a sizeable mountain. The leading order contribution to GW emission will come from the mass quadrupole Q_{22}, and theoretical calculations of the yield point of the crust suggest that it could sustain quadrupoles of up to $Q_{22} \approx 10^{39}$–10^{40} g cm^2 (Haskell et al. 2006; Johnson-McDaniel and Owen 2013), depending on the mass of the star. This is more than enough to allow for torque balance in LMXBs, which requires (Ushomirsky et al. 2000):

$$Q_{eq} = 3.5 \times 10^{37} \left(\frac{M}{1.4M_\odot}\right)^{1/4} \left(\frac{R}{10^6\text{cm}}\right)^{1/4} \left(\frac{\dot{M}}{10^{-9}M_\odot/\text{yr}}\right)^{1/2} \times \left(\frac{300\text{Hz}}{\nu}\right)^{5/2} \text{g cm}^2, \tag{8}$$

where $\dot{M}$ is the average mass accretion rate. It is, however, necessary to understand which physical mechanisms will be at work in a real system, and whether they would allow for a "maximal" mountain to build up in an LMXB. To address this problem we consider the two main mechanisms that have been suggested: thermal mountains and magnetic mountains.

5.1 *Thermal Mountains*

In the thermal case the mountain arises as the crust is heated by reactions that occur as accreted material is submerged deep into the crust. As it reaches higher densities several pycno-nuclear reactions occur, which heat the star locally by an amount (Ushomirsky and Rutledge 2001):

$$\delta T \approx 10^3 C_k^{-1} \left(\frac{p_d}{10^{30}\text{erg cm}^{-3}}\right)^{-1} \left(\frac{Q_n}{\text{MeV}}\right) \left(\frac{\Delta M}{10^{22}\text{g}}\right) \text{K}, \tag{9}$$

where C_k is the heat capacity per baryon in units of the Boltzman constant, p_d is the pressure at which the reaction occur and Q_n the deposited heat per unit baryon. ΔM is the amount of mass that is accreted. If part of this heating is asymmetric, and quadrupolar in particular, this can lead to a mass quadrupole (Ushomirsky et al. 2000)

$$Q_{22} \approx 1.3 \times 10^{35} \left(\frac{R}{10^6\text{cm}}\right)^4 \left(\frac{Q}{30\text{MeV}}\right)^3 \left(\frac{\delta T_q}{10^5\text{K}}\right) \text{g cm}^2, \tag{10}$$

where δT_q is the quadrupolar component of the temperature variation due to the reactions and Q the threshold energy for different reactions. Combining Eqs. (9) and (10), we can thus estimate how large a quadrupole can be built up during an outburst of a certain system, given the observed accretion rate and outburst duration. In Fig. 4 we show the results of Haskell et al. (in preparation) for known LMXB transients, that swing from accretion outburst to periods of quiescence. We plot the expected gravitational wave amplitude:

$$h = \frac{256}{5} \left(\frac{\pi^5}{3}\right)^{1/2} \frac{G Q_{22} \nu^2}{dc^4}, \tag{11}$$

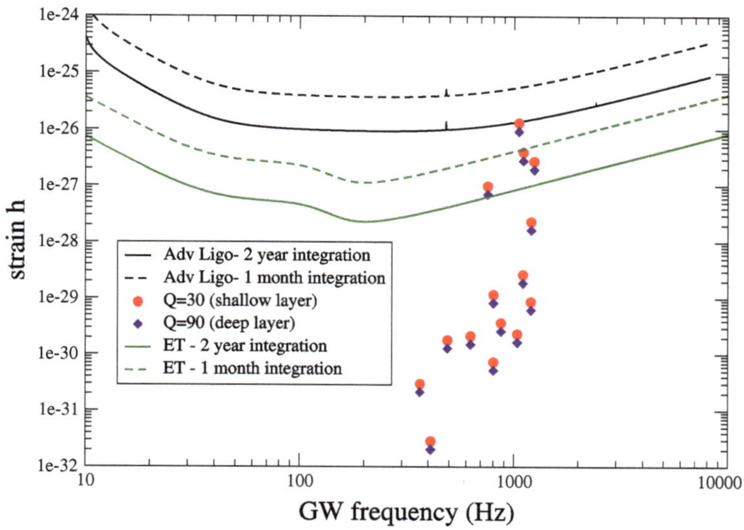

Fig. 4 The LIGO and ET sensitivity curves compared to the GW amplitude estimated by Haskell et al. (in preparation) for transient LMXBs. We can see that assuming that the main contribution is from reactions in a deep or shallow layer makes little difference (i.e. with larger or smaller thresholds Q for the reactions). We show results for both 1 month and 2 years integrations with both detectors, although the maximum time one can track the signal for will be set by the outburst duration, and in most cases will be closer to a month

with G the gravitational constant, c the speed of light and d the distance to the source. Given that the thermal timescale of the crust is generally quite short (of order a few years for the very deep crust) compared to the quiescence timescale, we assume that in this case the deformation is washed away in-between outbursts, and has to be rebuilt each time. Comparing our results to LIGO and ET sensitivities show that detecting this kind of emission will be very challenging, as also discussed in Watts et al. (2008). The situation is more promising for persistent systems, which undergo long periods of accretion, as illustrated in Fig. 5. Here we have assumed that it is possible to build a "maximal" mountain in the crust. Even accounting for uncertainties in mass and equation of state, it is clear that these are the most promising targets for Advanced LIGO and ET. Note that it may be possible for compositional asymmetries to persist in accreting systems even in quiescence (Ushomirsky et al. 2000), leading to larger quadrupoles that those estimated by Haskell et al. (in preparation). Finally it is also possible for the core to sustain shearing, if it contains a condensate of quarks in the CFL phase (Haskell et al. 2007; Owen 2005), although in this case the deformations would be much larger and in some cases can start to be constrained by current LIGO upper limits (Aasi et al. 2014).

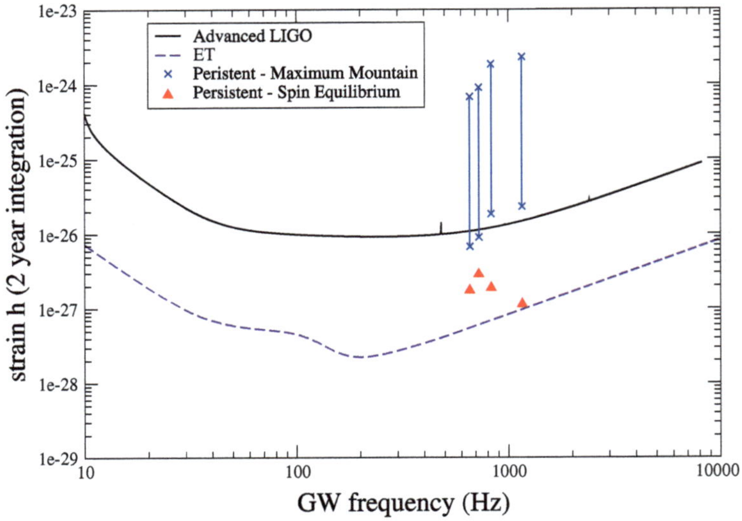

Fig. 5 The LIGO and ET sensitivity curves compared to the GW amplitude estimated by Haskell et al. (in preparation), for persistent LMXBs. We assume the maximum deformation that the crust can sustain and the error bars account for uncertainties in mass and equation of state. For comparison we also show the GW amplitude that would be needed for torque balance. The deformation needed for spin equilibrium, in fat, smaller than the maximal mountain, so if persistent LMXBs are emitting GWs at the maximal level they will be spinning down

5.2 Magnetic Mountains

Accretion does not only lead to thermal perturbations in the crust, but also perturbs the magnetic field structure. After matter is accreted at the magnetic poles it spreads towards the equator, compressing the field and leading to an overall suppression of the large-scale dipolar structure, but also to local enhancements that can support a sizeable mountain (Melatos and Payne 2005; Payne and Melatos 2004; Vigelius and Melatos 2009b). Given an amount of accreted mass M_a, the mass quadrupole is given by Shibazaki et al. (1989) and Priymak et al. (2011):

$$Q_{22} = 10^{45} A \left(\frac{M_a}{M_\odot}\right) \left(1 + \frac{M_a}{M_c}\right)^{-1} \tag{12}$$

where $A \approx 1$, M_c is the critical mass at which the process saturates, and the exact value of both quantities is equation of state dependent. In Figs. 6 and 7 we show the results obtained by Haskell et al. (in preparation) for the model E equation of state of Priymak et al. (2011). The critical mass depends on the assumed background field of the systems and we consider two possibilities, a background field of $B = 10^{12}$ G and one of $B = 10^{10}$ G. The latter may be a more realistic limit, given that Grad-Shafranov simulations show that the external dipole can be quenched by

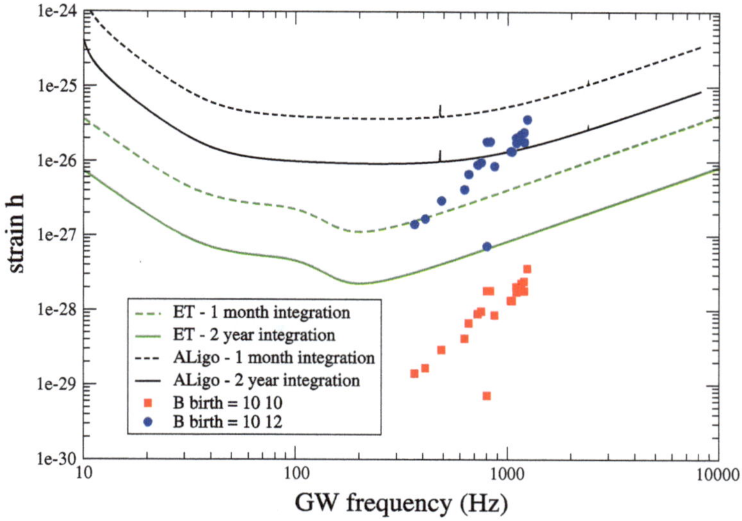

Fig. 6 Prediction by Haskell et al. (in preparation) for the GW emission from known LMXBs, given a magnetic mountain with a background magnetic field of $B = 10^{10}$ G or $B = 10^{12}$ G. We consider the case in which the mountain is stable in-between outbursts and can thus be built gradually over the life time of the system. Comparing to the sensitivity of Advanced LIGO and ET we can see that only the somewhat extreme case of a background (buried) magnetic field of $B = 10^{12}$ G would lead to a detectable signal

approximately an order of magnitude (Priymak et al. 2011), and the inferred dipolar fields of LMXBs and MSRPs are generally in the $B \approx 10^8$–10^9 G range.

Dynamical MHD simulations generally confirm stability of the mountain on timescales of $\tau \geq 10^8$ years (Vigelius and Melatos 2009a) and thus allow to construct the mountain over several outbursts, as assumed in Fig. 6. Nevertheless, given that MHD simulations may fail to resolve certain instabilities due to the finite grid size, we also present the case in which the mountain is dissipated between outbursts in Fig. 7. In both cases the detection prospects are quite pessimistic, with a detection likely only for the somewhat more extreme hypothesis of a buried 10^{12} G magnetic field.

If a GW signal were to be detected, it would thus have to be from a strong field system and one would expect to see cyclotron features in the electromagnetic emission of the NS, as compared to simple thermal asymmetries in the thermal mountain case. This would offer the possibility, given a GW detection of a continuous signal, to distinguish between the two kinds of NS mountains (Haskell et al. in preparation; Priymak et al. 2014). Note, however, that no cyclotron lines have been detected to date from the known LMXBs containing rapidly rotating NSs, and that the magnetic fields that are inferred for these systems are generally much weaker, of the order of $B \approx 10^8$ G.

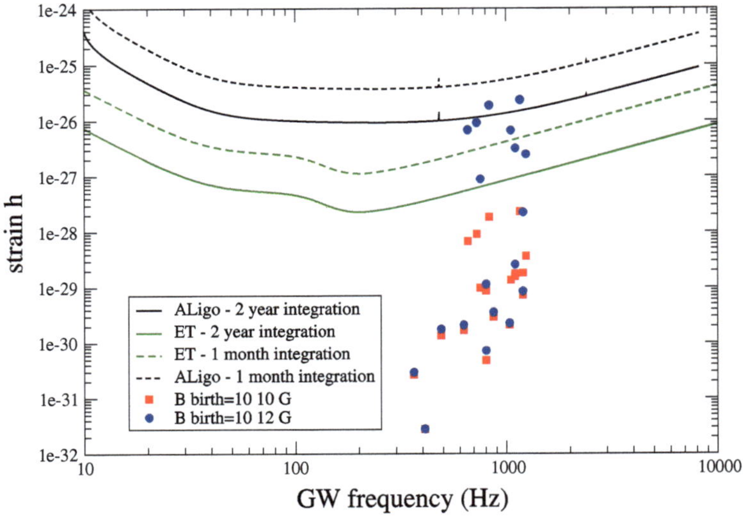

Fig. 7 Prediction by Haskell et al. (in preparation) for the GW emission from known LMXBs, given a magnetic mountain with a background magnetic field of $B = 10^{10}$ G or $B = 10^{12}$ G. We consider the case in which the mountain is unstable in-between outbursts and is thus dissipated in quiescence. Comparing to the sensitivity of Advanced LIGO and ET we can see, as in the previous case, that only the somewhat extreme case of a background (buried) magnetic field of $B = 10^{12}$ G would lead to a detectable signal

6 Torque Balance Revisited

Let us reconsider in more detail the idea that a GW braking torque is needed in LMXBs to balance the accretion spin-up torque. White and Zhang (1997) argue that the observed cutoff in the frequency distribution of LMXBs and MSRPs could only be explained by the disc/magnetosphere interaction by invoking an unexpected correlation between the magnetic field strength B and mass-accretion rate $\dot{M}$, and assuming that in the more luminous LMXBs the magnetic field strength is greater than what is typically observed in MSRPS (i.e. fields in the range $B \approx 10^9$–10^{10} G would be required). This led to GWs being suggested as a mechanism to remove angular momentum and brake the NS.

However the estimates presented above show that in many systems it would be quite challenging to build a large enough quadrupole with known mechanisms. Furthermore Haskell and Patruno (2011) recently considered in detail two systems: SAX J1808.4-3658 and XTE J1814-338. Both these systems are interesting, as the timing solution during an outburst suggests that the frequency is constant (Hartman et al. 2008, 2009), which is at odds with the theoretical expectation that they should be spinning up due to accretion. This opens up the intriguing possibility that this may be a direct consequence of GWs removing angular momentum from these two systems. Haskell and Patruno (2011) considered various GW mechanisms in detail but found that none of them could lead to a large enough quadrupole to balance the spin up torque. In all cases the outbursts are not very luminous and quite short, and SAX J1808.4-3658 has an inferred external dipolar magnetic field of $B \approx 10^8$ G. It is thus unlikely that a mountain, either thermal of magnetic, could provide the necessary quadrupole. Furthermore both systems are too cold to allow for the presence of a large mode of oscillation, such as an r-mode.

However, if one considers a slightly more sophisticated disc model than that used by White and Zhang (1997), such as those in Illarionov and Sunyaev (1975), Andersson et al. (2005b), Spruit and Taam (1993), and D'Angelo and Spruit (2010) which also account for magnetic torques and radiation pressure, it can be seen that the average accretion rate during the outbursts is actually quite close to the value required for torque balance. In this case, i.e. close to torque balance, the accretion torque is much weaker and can thus account for the absence of a significant increase in spin during the outbursts of these systems. This conclusion is further strengthened by the observation of 1 Hz Quasi-Periodic Oscillations (QPOs) during reflaring activity at the end of the outburst, which are likely to signal the onset of a propeller phase (Patruno et al. 2009, 2012; Patruno and D'Angelo 2013), which has also been observed in other systems (Ferrigno et al. 2013).

White and Zhang (1997) also included in their analysis a number of systems for which the spin frequency had been inferred from the separation between kilohertz QPOs. However it has since been shown that this might not be good proxy for the spin frequency of the star (Mendez and Belloni 2007; Yin et al. 2007). A recent analysis of the disc/magnetosphere spin balance scenario has found no correlation between B and $\dot{M}$ for the current sample of LMXBs (Patruno et al. 2012).

In Fig. 8 we also show the parameter space that is consistent with current accretion disc models, for a system at equilibrium at $\nu = 730$ Hz. It is clear that

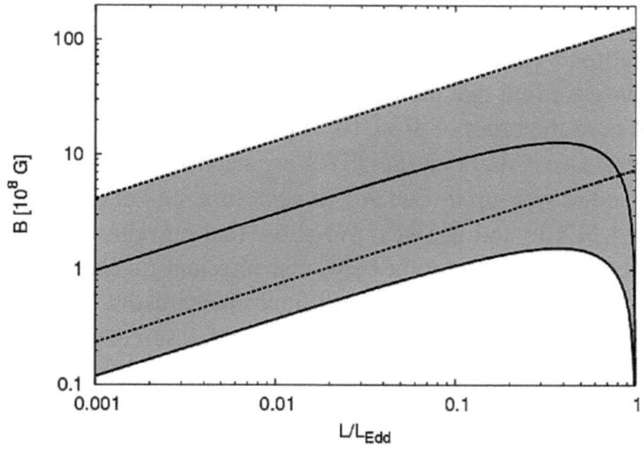

Fig. 8 The *shaded region* represents the region of parameter space allowed by current disc models, in the magnetic field vs luminosity (scaled to the Eddington luminosity) plane, for a hypothetical 730 Hz accreting neutron star at spin equilibrium, as described in Patruno et al. (2012). As we can see the uncertainties are large enough that the disc/magnetosphere interaction can lead to spin equilibrium at all luminosities for a magnetic field in the range $B \approx 10^8$ G

for all accretion rates one can account for the current spin period with a magnetic field in the range of $B \approx 10^8$ G, which is consistent with the values that are inferred from the spin down of MSRPs and of accreting pulsars that have been timed during multiple outbursts. In conclusion our analysis shows that GW emission is not needed to explain the observed spin distribution of LMXBs and MSRPs, but these systems may be emitting GWs at lower level that can still be detected by next-generation detectors, such as the Einstein Telescope.

7 Summary

In this chapter we review several GW emission mechanism that may be at work in LMXBs. First of all we consider the r-mode instability in rapidly rotating NSs. Following the analysis of Haskell et al. (2012a) we show that a "minimal" NS model, that does not include exotica in the core or dynamically significant magnetic fields and superfluid degrees of freedom, is not consistent with the inferred spins and temperatures of NSs in LMXBs. It is thus necessary to include additional physics in our model to account either for additional viscosity that stabilises the mode, or for a very small saturation amplitude that allows the system to be r-mode unstable without any observational impact on its spin and thermal evolutions. Furthermore most systems are too cold for the torque balance scenario, except for the faster, hotter systems, that may be good targets for next-generation GW detectors such as Advanced LIGO or ET.

We also consider the possibility of a thermal or magnetic mountain being built up during an accretion outburst. We find that persistent systems are a promising GW source, as they would allow to build up a large mountain that could be detected by Advanced LIGO or ET, while for transient systems the mountain is dissipated in quiescence, leading to much lower level emission (Haskell et al. in preparation). In the magnetic case we find that the GW signal would only be detectable if the buried magnetic field is of the order of $B \approx 10^{12}$ G.

Finally we re-assess the idea that GWs are needed to provide a braking torque that can balance the spin-up torque due to accretion, and explain the observed spin distribution of LMXBs and MSRPs. We show that current data is consistent with the disc/magnetosphere interaction being the physical mechanism responsible for the observed distribution (although additional mechanisms, such as spin glitches, may be at work in individual systems, see e.g. Andersson et al. 2014). Furthermore current disc models allow for systems to be at equilibrium with a magnetic field of $B \approx 10^8$ G. There is thus no need to invoke GWs as a necessary mechanism to provide spin equilibrium in LMXBs, but we have shown that many systems may still be emitting GWs, at a possibly lower level than that required for torque balance. They may thus be interesting sources for next-generation detectors such as the Einstein Telescope.

Acknowledgements BH acknowledges the support of the Australian Research Council via a Discovery Early Career Award (DECRA) fellowship. A.P. acknowledges support from the Netherlands Organization for Scientific Research (NWO) Vidi fellowship.

References

J. Aasi, J. Abadie, B.P. Abbott et al., ApJ **785**, 119 (2014)
M. Alford, K. Schwenzer, eprint: arXiv:1403.7500 (2014)
M. Alford, K. Schwenzer, ApJ **781**, 26 (2014)
M. Alford, A. Schmitt, K. Rajagopal, T. Schäfer, Rev. Mod. Phys. **80**, 1455 (2008)
M. Alford, S. Han, K. Schwenzer, eprint: arXiv:1404.5279 (2014)
M.A. Alpar, A.F. Cheng, M. Ruderman, J. Shaham, Nature **300**, 728 (1982)
N. Andersson, ApJ **502**, 708 (1998)
N. Andersson, K. Kokkotas IJMPD **10**, 381 (2001)
N. Andersson, G.L. Comer, K. Glampedakis, Nucl. Phys. A **763**, 212 (2005)
N. Andersson, K. Glampedakis, B. Haskell, A.L. Watts, MNRAS **361**, 1153 (2005)
N. Andersson, B. Haskell, G.L. Comer, Phys. Rev. D **82**, 023007 (2010)
N. Andersson, D.I. Jones, W.C.G. Ho, eprint: arXiv:1403.0860 (2014)
L. Bildsten, ApJ **501**, L89 (1998)
R. Bondarescu, I. Wasserman, ApJ **778**, 9 (2013)
R. Bondarescu, S.A. Teukolsky, I. Wasserman, Phys. Rev. D **76**, 064019 (2007)
N. Brambilla, S. Eidelman, P. Foka et al., eprint: arXiv:1404.3723 (2014)
E.F. Brown, G. Ushomirsky, ApJ **536**, 915 (2000)
D. Chakrabarty, E.H. Morgan, M.P. Muno et al., Nature **424**, 42 (2003)
R. Ciolfi, L. Rezzolla, MNRAS **435**, L43 (2013)
C. D'Angelo, H.C. Spruit, MNRAS **406**, 1208 (2010)
W. Del Pozzo, T.G.F. Li, M. Agathos, C. Van Den Brock, S. Vitale, PRL **111**, 071101 (2013)
C. Ferrigno, E. Bozzo, A. Papitto et al., eprint: arXiv:1310.7784 (2013)
J.L. Friedman, S.M. Morsink, ApJ **502**, 714 (1998)
E.H. Gudmundsson, C.J. Pethick, R.I. Epstein, ApJ **272**, 286 (1983)
M.E. Gusakov, A.I. Chugunov, E.M. Kantor, PRL **112**, 151101 (2014)
J.M. Hartman, A. Patruno, D. Chakrabarty et al., ApJ **675**, 1468 (2008)
J.M. Hartman, A. Patruno, D. Chakrabarty et al., ApJ **702**, 1673 (2009)
B. Haskell, N. Andersson, MNRAS **408**, 1897 (2010)
B. Haskell, A. Patruno A., ApJ **738**, L14 (2011)
B. Haskell, D.I. Jones, N. Andersson, MNRAS **397**, 1464 (2006)
B. Haskell, N. Andersson, D.I. Jones, L. Samuelsson, PRL **99**, 231101 (2007)
B. Haskell, L. Samuelsson, K. Glampedakis, N. Andersson, MNRAS **385**, 531 (2008)
B. Haskell, N. Andersson, A. Passamonti, MNRAS **397**, 1464 (2009)
B. Haskell, N. Andersson, G.L. Comer, Phys. Rev. D **86**, 063002 (2012)
B. Haskell, N. Degenaar, W.C.G. Ho, MNRAS **424**, 93 (2012)
B. Haskell, K. Glampedakis, N. Andersson, MNRAS **441**, 1662 (2014)
B. Haskell, A. Melatos, M. Priymak, P.D. Lasky, A. Patruno, M. Oppenoorth (in preparation)
J. Heyl, ApJ **574**, 57 (2002)
W.C.G. Ho, N. Andersson, B. Haskell, PRL **107**, 101101 (2011)
C.J. Horowitz, K. Kadau, PRL **103**, 191102 (2009)
A.F. Illarionov, R.A. Sunyaev, A&A **39**, 185 (1975)
N.K. Johnson-McDaniel, B.J. Owen, Phys. Rev. D **88**, 044004 (2013)
Y. Levin ApJ **517**, 328 (1999)
Y. Levin, G. Ushomirsky, MNRAS **324**, 917 (2001)
C.P. Lorenz, D.G. Ravenhall, C.J. Pethick, Phys. Rev. Lett. **70**, 379 (1993)

S. Mahmoodifar, T. Strohmayer, ApJ **773**, 140 (2013)
A. Melatos, D.J.B. Payne, ApJ **623**, 1044 (2005)
M. Mendez, T. Belloni, MNRAS **381**, 790 (2007)
M. Nayyar, B.J. Owen, Phys. Rev. D **73**, 084001 (2006)
B.J. Owen, PRL **95**, 211101 (2005)
B.J. Owen, L. Lindblom, C. Cutler, B.F. Schutz, A. Vecchio, N. Andersson, Phys. Rev. D **58**, 084020 (1998)
J. Papaloizou, J.E. Pringle, MNRAS **184**, 501 (1978)
A. Passamonti, B. Haskell, N. Andersson, MNRAS **396**, 951 (2009)
A. Patruno, ApJ **722**, 909 (2010)
A. Patruno, C. D'Angelo, ApJ **771**, 94 (2013)
A. Patruno, A.L. Watts, M. Klein Wolt, R. Wijnands, M. van der Klis, ApJ **707**, 1296 (2009)
A. Patruno, B. Haskell, C. D'Angelo, ApJ **746**, 9 (2012)
D.J.B. Payne, A. Melatos, MNRAS **351**, 569 (2004)
M. Priymak, A. Melatos, D.J.B. Payne, MNRAS **417**, 2696 (2011)
M. Priymak, A. Melatos, P.D. Lasky, accepted for publication in MNRAS, arXiv:1409.3327 (2014)
L. Rezzolla, F.K. Lamb, S.L. Shapiro, ApJ **531**, 139 (2000)
L. Rezzolla, F.K. Lamb, D. Markovic, S.L. Shapiro, Phys. Rev. D **64**, 104014 (2001)
N. Shibazaki, T. Murakami, J. Shaham, K. Nomoto, Nature **342**, 656 (1989)
H.C. Spruit, Astronomy & Astrophysics **341**, L1 (1999)
H.C. Spruit, R.E. Taam, ApJ **402**, 593 (1993)
G. Ushomirsky, R.E. Rutledge, MNRAS **325**, 1157 (2001)
G. Ushomirsky, C. Cutler, L. Bildsten, MNRAS **319**, 902 (2000)
M. Vigelius, A. Melatos, MNRAS **395**, 1985 (2009)
M. Vigelius, A. Melatos, ApJ **395**, 197 (2009)
A.L. Watts, B. Krishnan, L. Bildsten, B. Schutz, MNRAS **389**, 839 (2008)
N. White, W. Zhang, ApJ **490**, L87 (1997)
H.X. Yin, C.M. Zhang, Y.H. Zhao et al., A&A **471**, 381 (2007)

Supermassive Black Hole Binaries: The Search Continues

Tamara Bogdanović

Abstract Gravitationally bound supermassive black hole binaries (SBHBs) are thought to be a natural product of galactic mergers and growth of the large scale structure in the universe. They however remain observationally elusive, thus raising a question about characteristic observational signatures associated with these systems. In this conference proceeding I discuss current theoretical understanding and latest advances and prospects in observational searches for SBHBs.

1 Theoretical Background: Formation and Evolution of SBHBs

Over the past two decades it became apparent that the evolution of galaxies and their supermassive black holes (SBHs) is intricately connected (Ferrarese and Merritt 2000; Gebhardt et al. 2000; Tremaine et al. 2002). It has also been known for a while that galaxies evolve through mergers (White 1978; Schweizer 1986; Barnes and Hernquist 1992, 1996; Volonteri et al. 2003; Di Matteo et al. 2008), thus raising a question: *when galaxies merge, what happens to their SBHs?* In the aftermath of a generic merger the two parent galaxies form a new stellar bulge and their SBHs find themselves at a mutual separation of ~kpc. At this stage, widely separated SBHs are gravitationally bound to the surrounding gas and stars but not to one another. In this proceeding I will refer to the dual SBHs in this phase of evolution as *pairs* and to the gravitationally bound SBHBs as *binaries*.

Theoretical studies have established that evolution of SBH pairs from ~kpc to smaller scales is determined by gravitational interactions of individual black holes with their environment (see the groundbreaking work by Begelman et al. (1980) for description and Mayer 2013 for a most recent review). These include interaction of the SBHs with their own wakes of stars and gas, also known as the dynamical friction (Chandrasekhar 1943; Ostriker 1999; Milosavljević and Merritt

T. Bogdanović (✉)
Center for Relativistic Astrophysics, School of Physics, Georgia Institute of Technology, Atlanta, GA 30332-0430, USA
e-mail: tamarab@gatech.edu

C.F. Sopuerta (ed.), *Gravitational Wave Astrophysics*, Astrophysics and Space Science Proceedings, DOI 10.1007/978-3-319-10488-1_9

2001; Escala et al. 2004), as well as the scattering of the SBHs by massive gas clouds and spiral arms produced by the local and global dynamical instabilities during the merger (Fiacconi et al. 2013). During these interactions the SBHs exchange orbital energy and angular momentum with the ambient medium and can in principle grow though accretion (Escala et al. 2004, 2005; Dotti et al. 2006, 2007, 2009b; Callegari et al. 2011; Khan et al. 2012a; Chapon et al. 2013). These factors determine the SBH dynamics and whether they evolve to smaller separations to form a gravitationally bound binary. For example, Callegari et al. (2009, 2011) find that SBH pairs with mass ratios $q < 0.1$ are unlikely to form binaries within a Hubble time at any redshift. On the other hand SBH pairs with initially unequal masses can evolve to be more equal-mass, through preferential accretion onto a smaller SBH. It is therefore likely that SBH pairs with $q \gtrsim 0.1$ form a parent population of bound binaries at smaller separations.

Gravitationally bound binary forms at the point when the amount of gas and stars enclosed within its orbit becomes comparable to the mass of the SBHB. For a wide range of host properties and SBH masses this happens at orbital separations $\lesssim 10\,\mathrm{pc}$ (Mayer et al. 2007; Dotti et al. 2007; Khan et al. 2012a). The subsequent rate of binary orbital evolution sensitively depends on the nature of gravitational interactions that it experiences and is still an area of active research often abbreviated as *the last parsec problem*. The name refers to a possible slow-down and stalling in the orbital evolution of the parsec-scale SBHBs driven by the inefficient interactions with stars (Milosavljević and Merritt 2001) and gas (Escala et al. 2005). If present, a consequence of this effect would be that a significant fraction of SBHBs in the universe should reside at orbital separations of $\sim$1 pc. Several recent theoretical studies that focus on the evolution of *binaries in predominantly stellar backgrounds* however report that evolution of binaries to much smaller scales continues unhindered (Berczik et al. 2006; Preto et al. 2011; Khan et al. 2011, 2012b, 2013), although the agreement about the leading physical mechanism responsible for the evolution is still not universal (Vasiliev et al. 2014).

SBH binaries in predominantly gaseous environments have also been a topic of a number of theoretical studies (Armitage and Natarajan 2005; MacFadyen and Milosavljević 2008; Cuadra et al. 2009; Hayasaki 2009; Roedig et al. 2012; Shi et al. 2012; Noble et al. 2012; Kocsis et al. 2012a, b; D'Orazio et al. 2013; Farris et al. 2014). They find that binary torques can truncate the sufficiently cold circumbinary disks and create an inner low density cavity by evacuating the gas from the central portion of the disk (see Lin and Papaloizou 1979 and references above). As the binary orbit decays, the inner rim of the disk follows it inward until the timescale for orbital decay by gravitational radiation becomes shorter than the viscous timescale[1] of the disk (Armitage and Natarajan 2005). At that point, the rapid loss of orbital energy and angular momentum through gravitational radiation cause the binary to detach from the circumbinary disk and to accelerate towards coalescence. This final phase of binary evolution has been captured in a

[1]The time scale on which the angular momentum is transported outwards through the disk.

series of investigations based on fully relativistic particle (van Meter et al. 2010), electrodynamic and (magneto)hydrodynamic simulations of coalescing binaries (Palenzuela et al. 2009; Bode et al. 2010; Farris et al. 2010; Palenzuela et al. 2010a, b; Mösta et al. 2010; Moesta et al. 2012; Bode et al. 2012; Farris et al. 2011; Alic et al. 2012; Giacomazzo et al. 2012; Gold et al. 2014).

Through its dependance on the viscous time scale, the orbital evolution of a gravitationally bound SBHB in the circumbinary disk sensitively depends on the thermodynamic properties of the disk. These are uncertain, as they are still prohibitively computationally expensive to model from the first principles and are unconstrained by observations. More specifically, the thermodynamics of the disk is determined by the binary dynamics but also by the presence of magnetic field and radiative heating and cooling. While the role of magnetic field is beginning to be explored in some simulations, a fully consistent calculation of radiative heating and cooling is still beyond the computational reach.

Regardless of whether gravitationally bound SBH binaries evolve in mostly stellar or gas rich environments, the exchange of angular momentum with the ambient medium is likely to result in eccentric SBHB orbits (Armitage and Natarajan 2005; Cuadra et al. 2009; Roedig et al. 2011; Sesana et al. 2011; Chapon et al. 2013). An interesting implication of this finding is that eccentric binaries that evolve all the way to the gravitational wave (GW) coalescence may leave a clear imprint of the orbital eccentricity in their emitted waveforms, detectable by the future space-based GW observatories, such as eLISA (Key and Cornish 2011; Amaro-Seoane et al. 2013). Another property of astrophysical SBHs that will be possible to measure to high precision and high redshifts from eLISA observations are the magnitudes and orientations of the SBH spins prior to the coalescence (Amaro-Seoane et al. 2013). This is an exciting prospect as these would complement the existing spin estimates for about two dozen single SBHs, based on measurements of the relativistically broadened X-ray FeKα emission line profiles (see Reynolds 2013 for a recent review of this method).

Our understanding of spin magnitudes and orientations in *binary* SBHs on the other hand relies mostly on theoretical considerations. Interest in this topic was triggered by the prediction of numerical relativity that coalescence of SBHs with certain spin configurations can lead to the ejection of a newly formed SBH from its host galaxy. This effect arises due to the asymmetry in emission of GWs in the final stages of a SBH merger and can lead to a GW kick of up to $\sim$5,000 km s^{-1} (Campanelli et al. 2007; Lousto and Zlochower 2011). In majority of the binary configurations however, the GW kick velocity was found to be lower and is minimized whenever the SBH spin axes are aligned with the binary orbital axis.

Several subsequent theoretical studies found that accretion and gravitational torques can act to align the spin axes of SBHs evolving in gas reach environments and in such way minimize the GW recoil as long as the SBHs are orbiting within a rotationally supported, moderately geometrically thik disk (Bogdanović et al. 2007; Dotti et al. 2010, 2013; Sorathia et al. 2013; Miller and Krolik 2013) (see however Lodato and Gerosa 2013 for a different view). The mutual SBH spin alignment is on the other hand not expected to happen in gas poor environments, geometrically thick, turbulent and magnetically dominated disks (Fragile and Anninos 2005; Fragile et al. 2007; McKinney et al. 2013) hence, allowing a possibility that runaway SBHs and empty nest galaxies may exist. This realization stimulated lots of research activity on the astrophysical implications of the GW kick (Bonning et al. 2007; Volonteri 2007; Schnittman 2007; Sesana 2007; Loeb 2007; Gualandris and Merritt 2008; Komossa et al. 2008; Shields and Bonning 2008; Kornreich and Lovelace 2008; Berti and Volonteri 2008; Schnittman and Krolik 2008; Holley-Bockelmann et al. 2008; Komossa and Merritt 2008; Blecha and Loeb 2008; Volonteri and Madau 2008; O'Neill et al. 2009; Devecchi et al. 2009; Merritt et al. 2009; Volonteri et al. 2010; Robinson et al. 2010; Lovelace and Kornreich 2010; Corrales et al. 2010; Guedes et al. 2011; Blecha et al. 2011; Nixon et al. 2011; Nixon 2012; Lousto et al. 2012; Ponce et al. 2012; Komossa 2012; Stone and Loeb 2012; Shields and Bonning 2013; King and Nixon 2013; Gerosa and Sesana 2014). I will not dwell further on recoiling SBHs in this proceeding except to note that some signatures of recoiling SBHs can coincide with those of the subarsec scale SBHBs (see Sects. 2.1 and 2.3), effectively allowing to accomplish two searches with one observational strategy.

2 Observational Evidence for SBHBs

The key characteristic of gravitationally bound SBHBs is that they are observationally elusive and expected to be intrinsically rare. While the frequency of binaries is uncertain and dependent on their unknown rate of evolution on small scales (see previous section), theorists estimate that a fraction $< 10^{-3}$ of active galactic nuclei (AGNs) at redshift $z < 0.7$ may host SBHBs (Volonteri et al. 2009). This result has two important implications: (a) any observational search for SBHBs must involve a large sample of AGNs, making the archival data from large surveys of AGNs an attractive starting point and (b) observational technique used in a search must be able to distinguish signatures of binaries from those of AGNs powered by single SBHs. In the following sections I describe the application of imaging, photometric, and spectroscopic techniques, starting with the more direct methods and progressing towards less direct. I focus on the techniques that have been commonly used in SBHB searches and direct the reader to Dotti et al. (2012) and Schnittman (2013) for recent reviews of a broad range of SBHB signatures proposed in the literature.

2.1 Direct Imaging of Double Nuclei

A clearest manifestation of a parsec scale SBHB is an image of a binary AGN which forms a gravitationally bound system (as opposed to an accidental projection on the sky). A practical obstacle in the detection of such objects arises from their small angular separation on the sky: for example, a parsec-scale binary at a moderate redshift of $z \approx 0.2$ subtends an angle of only $\sim$3 mas on the sky (neglecting the projection effects). Such scales are below the angular resolution

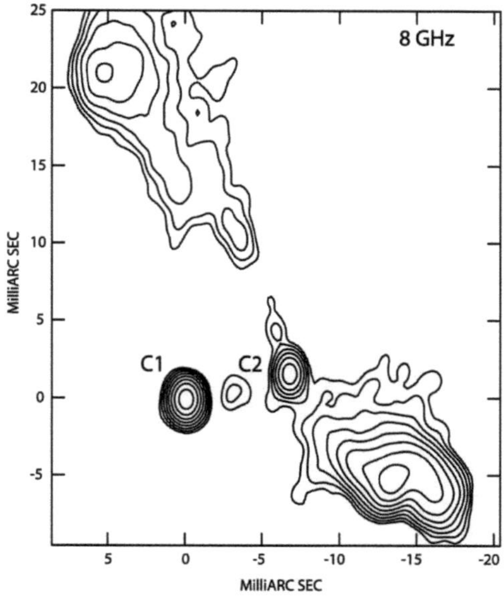

Fig. 1 VLBA image of the SBHB candidate in the radio galaxy 0402+379 observed at the frequency of 8 GHz. Two compact radio cores, C1 and C2, are separated by 7.3 pc on the sky and are thought to harbor SBHs. Figure adapted from Rodriguez et al. (2006)

of most astronomical instruments, except the very long baseline interferometers (VLBI) used at radio wavelengths.

A most convincing candidate for a SBHB in this class of objects was discovered in the radio galaxy 0402+379 by the Very Long Baseline Array (VLBA; Fig. 1) (Maness et al. 2004; Rodriguez et al. 2006, 2009; Morganti et al. 2009). This object shows two compact radio cores at the projected separation of 7.3 pc on the sky. Both cores are characterized by the flat radio spectra and have been identified as possible AGN based on this signature (Maness et al. 2004; Rodriguez et al. 2006). While the SBHB candidate in the galaxy 0402+379 was discovered serendipitously, it demonstrated the power of radio interferometry in imaging of the small separation SBHs.

In a subsequent investigation, Burke-Spolaor (2011) searched for binaries in the archival VLBI data. The search targeted spatially resolved, double radio-emitting nuclei with a wide range of orbital separations ($\sim$3 pc–5 kpc) among 3,114 radio-luminous AGNs in the redshift range $0 < z \leq 4.715$. Another investigation in addition to SBHBs searched for the recoiling SBHs spatially offset from the centers of their host galaxies (Condon et al. 2011). The latter study is based on the VLBA 8 GHz observations of 834 nearby radio-luminous AGN[2] with typical distances of $\sim$200 Mpc. Neither search unearthed new instances of the double-radio nuclei, leaving several possible interpretations: (a) there is a true paucity of SBHBs, (b) SBHBs may be present but have low radio brightness, or (c) only one component of the binary is radio-bright and it may or may not show a detectable spatial offset

[2]Radio sources brighter than 100 mJy were selected based on the NVSS catalogue at 1.4 GHz.

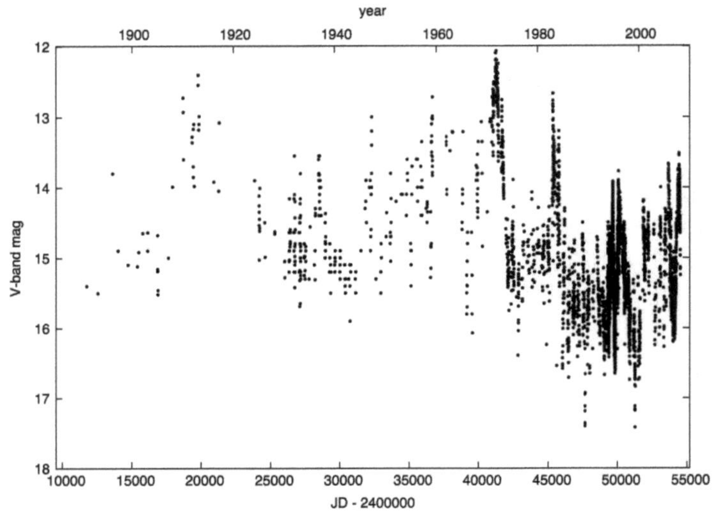

Fig. 2 Historical light curve of OJ 287 in V-magnitude recorded over more than 100 years. Quasi-periodic outbursts with a period of ∼12 years in the light curve of this object have been interpreted as a signature of the SBHB orbital motion. Figure from Valtonen et al. (2008)

relative to the center of the host galaxy. This points to difficulties in using the radio imaging as the primary technique to select the gravitationally bound SBHB candidates or their progenitors given their unknown radio properties. See however Burke-Spolaor et al. (2014) for a discussion of the optimal design of radio searches targeting the dual and binary SBHs.

2.2 *Photometric Measurements of Quasi-Periodic Variability*

The second most convincing line of evidence for the presence of a SBHB is a sustained periodic or nearly periodic variability on a time scale associated with the orbital period of the binary. This technique favors binaries with relatively short orbital periods, $P \lesssim 10$ years, for which multiple cycles can be recorded in observations (for e.g., Fan et al. 1998; Rieger and Mannheim 2000; De Paolis et al. 2002; Liu et al. 2014). A well known example of a SBHB candidate in this category is a blazar OJ 287 which exhibits outburst activity in its optical light curve with a period close to 12 years (see Fig. 2), interpreted as a signature of the orbital motion (Valtonen et al. 2008).

It is worth noting however that OJ 287 is unique among photometrically selected binary candidates because the first recorded data points in its light curve extend into the nineteenth century. Along similar lines, OJ 287 received an unprecedented level of observational coverage in modern times (from 1970s onwards), yielding a light curve with high frequency sampling. While indications of quasi-periodicity have been claimed in a handful of other objects, they are generally less pronounced and recorded over much shorter time span than in the case of OJ 287, thus preventing a strong case for SBHBs from being made in these sources. This observationally intensive technique may nevertheless play a more important role in the near future, as a number of high-cadence synoptic sky surveys come online.

2.3 Spectroscopic Measurements of Offset Broad Emission-Lines

The least direct of the three methods relies on the spectroscopic detection of the Doppler-shift that arises as a consequence of the binary orbital motion. It is based on a well established technique for detection of single- and double-line spectroscopic binary stars. In the double-line systems each offset line corresponds to one member of the binary, whereas in the single-line systems only one member is visible. In both classes of spectroscopic binaries, the lines are expected to oscillate about their rest frame wavelength on the time scale corresponding to the orbital period.

SBHs in the circumbinary disk phase (described in Sect. 1) can accrete by capturing gas from the inner rim of this disk. In the context of this model, the spectral lines are assumed to be associated with the gas accretion disks that are gravitationally bound to the individual SBHs (Bogdanović et al. 2008). Given the velocities of the bound gas, the emission line profiles from the SBH mini-disks are expected to be Doppler-broadened, similar to the emission lines originating in the broad line regions (BLRs) of AGNs. Moreover, several theoretical studies have shown that for binary mass ratios $q < 1$ accretion occurs preferentially onto the lower mass object (Artymowicz and Lubow 1996; Gould and Rix 2000; Hayasaki et al. 2007), rendering it potentially more luminous than the primary and indicating that some fraction of the SBHBs may appear as the single-line spectroscopic binaries.

This realization motivated several searches for SBHBs based on the criterion that the culprit sources exhibit broad optical lines offset with respect to the rest frame of the host galaxy (Bogdanović et al. 2009a; Dotti et al. 2009a; Boroson and Lauer 2009; Tang and Grindlay 2009; Decarli et al. 2010; Barrows et al. 2011; Tsalmantza et al. 2011; Tsai et al. 2013). Because this effect is also expected to arise in the case of a recoiling SBH receding from its host galaxy, the same approach has been used to flag candidates of that type (Komossa et al. 2008; Shields et al. 2009; Civano et al. 2010; Robinson et al. 2010; Lusso et al. 2014). The key advantage of the method is its simplicity, as spectra that exhibit Doppler shift signatures are relatively straightforward to select from large archival data sets, such as the Sloan Digital Sky Survey (SDSS). Its main complication however is that the Doppler shift signature is not unique to these two physical scenarios and complementary observations are needed in order to determine the true nature of the observed candidates (Bogdanović et al. 2009b; Popović 2012).

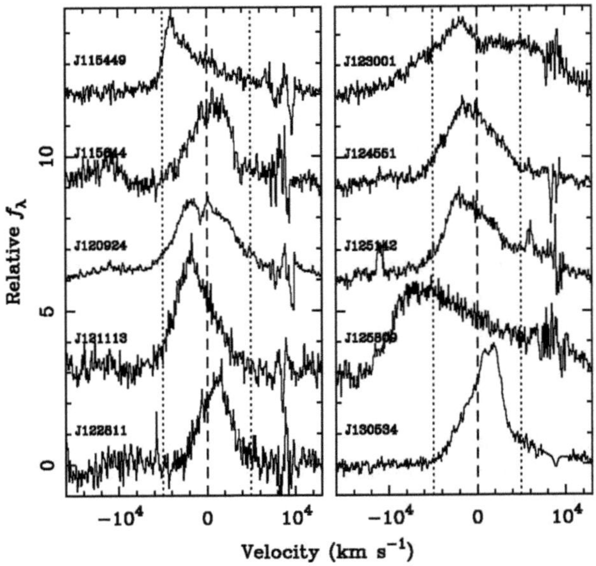

Fig. 3 Broad Hβ emission line profiles of selected SBHB candidates. Peakes of the asymmetric profiles are offset from the rest frame of the host galaxy (marked by the *dashed line*) by $\sim$ few $\times$ 10^3 km s^{-1}. In the context of the SBHB model the broad profile is attributed to the emission from the accretion disk of the smaller SBH, and the offset is attributed to its orbital motion. Figure adapted from Eracleous et al. (2012)

To address this ambiguity a new generation of spectroscopic searches has been designed to monitor the offset of the broad emission line profiles over multiple epochs and target sources in which modulations in the offset are consistent with the binary motion (Eracleous et al. 2012; Bon et al. 2012; Decarli et al. 2013; Liu et al. 2013; Shen et al. 2013; Ju et al. 2013). For example, Eracleous et al. (2012) searched for $z < 0.7$ SDSS quasars whose broad $H\beta$ lines are offset by $\gtrsim$1,000 km s^{-1}. Using this criterion they selected 88 quasars for observational followup from the initial catalog of $\sim$16,000 objects. After the second and third epoch of observations of this sample, statistically significant changes in the velocity offset were found in 14 (Eracleous et al. 2012) and 9 objects (Mathes et al. 2014), respectively, in broad agreement with theoretical predictions (Volonteri et al. 2009).

Figure 3 shows several representative broad $H\beta$ profiles selected in this search. The profiles are asymmetric and have peaks offset by $\sim$ few $\times$ 10^3 km s^{-1} from the rest frame of the galaxy, as inferred from from the narrow emission lines (the narrow emission line components were subsequently removed from these profiles for clarity). If velocity offset of 1,000 km s^{-1} is interpreted as a signature of the orbital motion of a binary with mass $M \sim 10^8 M_\odot$, it follows that the targeted population of SBHBs has the average orbital separation of $\sim$0.1 pc and orbital period of $\sim$300 years. Given a long average orbital period and observational campaigns which typically span a time line about 10 years, it follows that searches of this type are in principle capable of monitoring a SBHB during a fraction of its orbital cycle but are in general not expected to record multiple cycles.

A detection of an incomplete orbital cycle still leaves a possibility that other astrophysical processes may masquerade as binaries. For example, an orbiting hotspot produced by a local instability in the BLR of a galaxy, or outflows associated with the accretion disk may produce similar signatures and cannot be excluded. Consequently, the spectroscopic searches for SBHBs still require validation by another complementary observational technique. Thanks to their efficiency in selection of the SBHB candidate samples, spectroscopic searches can be combined with the direct imaging of the binary nuclei (discussed in Sect. 2.1) and imaging of the host galaxy, in order to search for any signs of interaction. This two-step approach can in principle also be used to distinguish SBHBs from spatially and kinematically offset recoiling SBHs.

3 Future Theoretical and Observational Prospects

While selection of a well defined sample of SBHBs remains a principal goal in this research field, an equally important consideration is *what can be learned once such sample is available*. For example, the ongoing searches are based on the monitoring of the broad optical emission lines, which are expected to encode some information about the kinematics of the binary BLRs. It is thus plausible, although it remains to be demonstrated, that by analysis of these line profiles one can learn about the structure and thermodynamics of the circumbinary accretion flow, the very questions that are hampering the progress of theoretical models. More generally, a comparison with the spectroscopic data can provide a test of the underlying SBHB-in-a-circumbinary disk model as well as a constraint on the time scale for evolution of the gravitationally bound SBHBs.

There are also other, fully developed observational techniques that can in principle be utilized to study SBHBs, once such a sample is defined. For example, *velocity-resolved reverberation mapping* is a method that has been successfully used to study the structure of the BLRs in several "conventional", single black hole AGNs (Bentz et al. 2008, 2009, 2010a, b; Pancoast et al. 2011; Barth et al. 2011; Dietrich et al. 2012; Pancoast et al. 2013). The method relies on the measurement of the propagation of a light echo from the central source of the continuum radiation across the BLR. This is achieved through monitoring of the response of the permitted broad optical emission lines to the variations in the continuum. The approach yields constrains on the structure of orbits occupied by the emitting gas (i.e., circular or eccentric, inflowing or outflowing), as well as the SBH mass and inclination of the BLR relative to the observer's line of sight. Because it is observationally intensive, this technique can initially be applied to a subset of selected SBHB candidates with the goal to study the structure of their BLR regions. A typical monitoring campaign over several months of time would allow to capture the kinematics of the gas and factor out the variability due to the orbital motion of the binary, which for the parsec scale binaries occurs on much longer time scales (see Sect. 2.3).

Over the past 2 years a detection of time- and energy-resolved reverberation lags also became possible in the X-ray band and specifically, in the FeKα emission line region (Zoghbi et al. 2012; Kara et al. 2013a, b, 2014; Cackett et al. 2014; Uttley et al. 2014). While key ideas are similar to the method used in the optical band, the

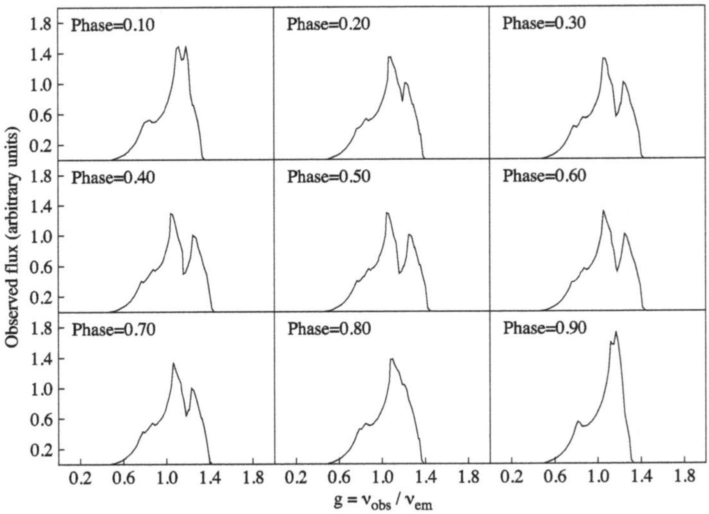

Fig. 4 Composite FeKα emission line profiles calculated for $q = 1$ eccentric binary with orbital separation of 0.01 pc and accretion disks of equal luminosity. Profiles are shown as a function of the binary orbital phase in arbitrary flux units against frequency normalized to the rest frame frequency. Figure adapted from Jovanović et al. (2013)

main difference is that the FeKα line emission originates from the BLR very close to the SBH, within $\sim 10\,r_g$[3] (Fabian et al. 1989). The relativistically broadened profiles probe the structure of the innermost accretion disk region as well as the the spin of the SBH. In the context of SBHB model, this technique can in principle be applied to very close binaries at orbital separations of $\sim 30 - 10^3\,r_g$, assuming that they can maintain the bright, X-ray emitting accretion disks within their orbit (Sesana et al. 2013).

In reality, the X-ray emission properties of such close SBHB systems are virtually unknown beyond several theoretical models that predict the shape of the composite FeKα emission lines from binary BLRs (McKernan et al. 2013; Jovanović et al. 2013). Figure 4 shows a sequence of the FeKα line profiles as a function of the binary orbital phase calculated for an equal-mass binary with two disks of equal luminosity. This and other models illustrate that if the FeKα emission from SBHB systems can be observed, that line profiles will be distinct from those observed in the single SBH systems. Their detection however may be challenging due to the fact that these SBHB are relatively short lived and are expected to merge due to the emission of GWs on time scales of 10^4–10^6 years (Sesana et al. 2013).

Another promising channel for detection of this population of close SBHBs is with the pulsar timing arrays (PTAs) (Sesana et al. 2008; Shannon et al. 2013; Sesana 2013; Burke-Spolaor 2013; Arzoumanian et al. 2014). This observational

[3] $r_g = GM/c^2$ is the gravitational radius.

technique is based on the monitoring of a network of stable millisecond pulsars with the goal to measure a characteristic departure in the time of arrival of their radio pulses caused by GWs in the frequency range ($10^{-9} - 10^{-7}$ Hz). While the PTA signal is expected to be dominated by the stochastic background of GWs emitted by a population of SBHBs with mass $> 10^8 M_\odot$ and $z < 1$, detection of individual, particularly bright GW sources cannot be excluded (Babak and Sesana 2012; Ravi et al. 2012; Tanaka and Haiman 2013; Petiteau et al. 2013). A detection of the PTA SBHBs can constrain the dominant evolutionary scenario and possibly the accretion history of the most massive black holes in the universe. It conjunction with observations of the FeKα line profiles, this approach could provide the only opportunity to study the GW emitting SBHBs before the GW space-based observatories come online.

Conclusions

Past 10 years have marked a period of very active research on SBH pairs and binaries. The interest in them has been driven by a realization that SBHs play an important role in evolution of their host galaxies and also, by intention to understand the parent population of merging binaries because these are the prime targets of the long anticipated space-based GW observatories. While the future space-based GW detectors will undoubtedly transform our understanding of SBHBs, the electromagnetic signatures remain the only way to learn about this population of objects in the next two decades.

Investigation of SBHBs has been spearheaded by theoretical studies which have investigated how black holes grow, form pairs and interact with their environment. They matured into a field that now faces questions rooted in thorough understanding of an extensive parameter space of SBHBs, their accretion flows and stellar environments. It became apparent that if further advances are to be made, observations of SBHBs must follow hand in hand, so to illuminate which portion of this parameter space is relevant for increasingly sophisticated theoretical models and simulations.

Observational searches for SBHB have however been challenging and while they unearthed a handful of binary candidates early on, their nature remained inconclusive in absence of the systematic multi-wavelength followup. They still provided valuable lessons as the present landscape of observational searches is represented by better designed, multi-wavelength and multi-year observational campaigns. The spectroscopic searches for SBHBs seem capable of delivering a statistically significant sample of binaries and their first results appear broadly consistent with theoretical predictions. In combination with other complementary observational techniques, they are well positioned to pave the way to a discovery of SBHBs.

Acknowledgements Tamara Bogdanović acknowledges the support from the Alfred P. Sloan Foundation under Grant No. BR2013-016. This research was supported in part by the National Science Foundation under Grant No. NSF AST-1211677 and NSF AST-1333360.

References

D. Alic, P. Moesta, L. Rezzolla, O. Zanotti, J.L. Jaramillo, ApJ **754**, 36 (2012). doi:10.1088/0004-637X/754/1/36

P. Amaro-Seoane, S. Aoudia, S. Babak, P. Binétruy, E. Berti, A. Bohé, C. Caprini, M. Colpi, N.J. Cornish, K. Danzmann, J.F. Dufaux, J. Gair, I. Hinder, O. Jennrich, P. Jetzer, A. Klein, R.N. Lang, A. Lobo, T. Littenberg, S.T. McWilliams, G. Nelemans, A. Petiteau, E.K. Porter, B.F. Schutz, A. Sesana, R. Stebbins, T. Sumner, M. Vallisneri, S. Vitale, M. Volonteri, H. Ward, B. Wardell, GW Notes **6**(4), 4–110 (2013)

P.J. Armitage, P. Natarajan, ApJ **634**, 921 (2005). doi:10.1086/497108

P. Artymowicz, S.H. Lubow, ApJ **467**, L77 (1996). doi:10.1086/310200

Z. Arzoumanian, A. Brazier, S. Burke-Spolaor, S.J. Chamberlin, S. Chatterjee, J.M. Cordes, P.B. Demorest, X. Deng, T. Dolch, J.A. Ellis, R.D. Ferdman, N. Garver-Daniels, F. Jenet, G. Jones, V.M. Kaspi, M. Koop, M. Lam, T.J.W. Lazio, A.N. Lommen, D.R. Lorimer, J. Luo, R.S. Lynch, D.R. Madison, M. McLaughlin, S.T. McWilliams, D.J. Nice, N. Palliyaguru, T.T. Pennucci, S.M. Ransom, A. Sesana, X. Siemens, I.H. Stairs, D.R. Stinebring, K. Stovall, J. Swiggum, M. Vallisneri, R. van Haasteren, Y. Wang, W.W. Zhu, ArXiv e-prints (2014)

S. Babak, A. Sesana, Phys. Rev. D **85**(4), 044034 (2012). doi:10.1103/PhysRevD.85.044034

J.E. Barnes, L. Hernquist, ARA&A **30**, 705 (1992). doi:10.1146/annurev.aa.30.090192.003421

J.E. Barnes, L. Hernquist, ApJ **471**, 115 (1996). doi:10.1086/177957

R.S. Barrows, C.H.S. Lacy, D. Kennefick, J. Kennefick, M.S. Seigar, New A **16**, 122 (2011). doi:10.1016/j.newast.2010.08.004

A.J. Barth, A. Pancoast, S.J. Thorman, V.N. Bennert, D.J. Sand, W. Li, G. Canalizo, A.V. Filippenko, E.L. Gates, J.E. Greene, M.A. Malkan, D. Stern, T. Treu, J.H. Woo, R.J. Assef, H.J. Bae, B.J. Brewer, T. Buehler, S.B. Cenko, K.I. Clubb, M.C. Cooper, A.M. Diamond-Stanic, K.D. Hiner, S.F. Hönig, M.D. Joner, M.T. Kandrashoff, C.D. Laney, M.S. Lazarova, A.M. Nierenberg, D. Park, J.M. Silverman, D. Son, A. Sonnenfeld, E.J. Tollerud, J.L. Walsh, R. Walters, R.L. da Silva, M. Fumagalli, M.D. Gregg, C.E. Harris, E.Y. Hsiao, J. Lee, L. Lopez, J. Rex, N. Suzuki, J.R. Trump, D. Tytler, G. Worseck, H.M. Yesuf, ApJ **743**, L4 (2011). doi:10.1088/2041-8205/743/1/L4

M.C. Begelman, R.D. Blandford, M.J. Rees, Nature **287**, 307 (1980). doi:10.1038/287307a0

M.C. Bentz, J.L. Walsh, A.J. Barth, N. Baliber, N. Bennert, G. Canalizo, A.V. Filippenko, M. Ganeshalingam, E.L. Gates, J.E. Greene, M.G. Hidas, K.D. Hiner, N. Lee, W. Li, M.A. Malkan, T. Minezaki, F.J.D. Serduke, J.H. Shiode, J.M. Silverman, T.N. Steele, D. Stern, R.A. Street, C.E. Thornton, T. Treu, X. Wang, J.H. Woo, Y. Yoshii, ApJ **689**, L21 (2008). doi:10.1086/595719

M.C. Bentz, J.L. Walsh, A.J. Barth, N. Baliber, V.N. Bennert, G. Canalizo, A.V. Filippenko, M. Ganeshalingam, E.L. Gates, J.E. Greene, M.G. Hidas, K.D. Hiner, N. Lee, W. Li, M.A. Malkan, T. Minezaki, Y. Sakata, F.J.D. Serduke, J.M. Silverman, T.N. Steele, D. Stern, R.A. Street, C.E. Thornton, T. Treu, X. Wang, J.H. Woo, Y. Yoshii, ApJ **705**, 199 (2009). doi:10.1088/0004-637X/705/1/199

M.C. Bentz, J.L. Walsh, A.J. Barth, Y. Yoshii, J.H. Woo, X. Wang, T. Treu, C.E. Thornton, R.A. Street, T.N. Steele, J.M. Silverman, F.J.D. Serduke, Y. Sakata, T. Minezaki, M.A. Malkan, W. Li, N. Lee, K.D. Hiner, M.G. Hidas, J.E. Greene, E.L. Gates, M. Ganeshalingam, A.V. Filippenko, G. Canalizo, V.N. Bennert, N. Baliber, ApJ **716**, 993 (2010a). doi:10.1088/0004-637X/716/2/993

M.C. Bentz, K. Horne, A.J. Barth, V.N. Bennert, G. Canalizo, A.V. Filippenko, E.L. Gates, M.A. Malkan, T. Minezaki, T. Treu, J.H. Woo, J.L. Walsh, ApJ **720**, L46 (2010b). doi:10.1088/2041-8205/720/1/L46
P. Berczik, D. Merritt, R. Spurzem, H.P. Bischof, ApJ **642**, L21 (2006). doi:10.1086/504426
E. Berti, M. Volonteri, ApJ **684**, 822 (2008). doi:10.1086/590379
L. Blecha, A. Loeb, MNRAS **390**, 1311 (2008). doi:10.1111/j.1365-2966.2008.13790.x
L. Blecha, T.J. Cox, A. Loeb, L. Hernquist, MNRAS **412**, 2154 (2011). doi:10.1111/j.1365-2966.2010.18042.x
T. Bode, R. Haas, T. Bogdanović, P. Laguna, D. Shoemaker, ApJ **715**, 1117 (2010). doi:10.1088/0004-637X/715/2/1117
T. Bode, T. Bogdanović, R. Haas, J. Healy, P. Laguna, D. Shoemaker, ApJ **744**, 45 (2012). doi:10.1088/0004-637X/744/1/45
T. Bogdanović, C.S. Reynolds, M.C. Miller, ApJ **661**, L147 (2007). doi:10.1086/518769
T. Bogdanović, B.D. Smith, S. Sigurdsson, M. Eracleous, ApJS **174**, 455 (2008). doi:10.1086/521828
T. Bogdanović, M. Eracleous, S. Sigurdsson, ApJ **697**, 288 (2009a). doi:10.1088/0004-637X/697/1/288
T. Bogdanović, M. Eracleous, S. Sigurdsson, New A Rev. **53**, 113 (2009b). doi:10.1016/j.newar.2009.09.005
E. Bon, P. Jovanović, P. Marziani, A.I. Shapovalova, N. Bon, V. Borka Jovanović, D. Borka, J. Sulentic, L.Č. Popović, ApJ **759**, 118 (2012). doi:10.1088/0004-637X/759/2/118
E.W. Bonning, G.A. Shields, S. Salviander, ApJ **666**, L13 (2007). doi:10.1086/521674
T.A. Boroson, T.R. Lauer, Nature **458**, 53 (2009). doi:10.1038/nature07779
S. Burke-Spolaor, MNRAS **410**, 2113 (2011). doi:10.1111/j.1365-2966.2010.17586.x
S. Burke-Spolaor, Class. Quantum Grav. **30**(22), 224013 (2013). doi:10.1088/0264-9381/30/22/224013
S. Burke-Spolaor, A. Brazier, S. Chatterjee, J. Comerford, J. Cordes, T.J.W. Lazio, X. Liu, Y. Shen, ArXiv e-prints (2014)
E.M. Cackett, A. Zoghbi, C. Reynolds, A.C. Fabian, E. Kara, P. Uttley, D.R. Wilkins, MNRAS **438**, 2980 (2014). doi:10.1093/mnras/stt2424
S. Callegari, L. Mayer, S. Kazantzidis, M. Colpi, F. Governato, T. Quinn, J. Wadsley, ApJ **696**, L89 (2009). doi:10.1088/0004-637X/696/1/L89
S. Callegari, S. Kazantzidis, L. Mayer, M. Colpi, J.M. Bellovary, T. Quinn, J. Wadsley, ApJ **729**, 85 (2011). doi:10.1088/0004-637X/729/2/85
M. Campanelli, C.O. Lousto, Y. Zlochower, D. Merritt, Physical Review Letters **98**(23), 231102 (2007). doi:10.1103/PhysRevLett.98.231102
S. Chandrasekhar, ApJ **97**, 255 (1943). doi:10.1086/144517
D. Chapon, L. Mayer, R. Teyssier, MNRAS **429**, 3114 (2013). doi:10.1093/mnras/sts568
F. Civano, M. Elvis, G. Lanzuisi, K. Jahnke, G. Zamorani, L. Blecha, A. Bongiorno, M. Brusa, A. Comastri, H. Hao, A. Leauthaud, A. Loeb, V. Mainieri, E. Piconcelli, M. Salvato, N. Scoville, J. Trump, C. Vignali, T. Aldcroft, M. Bolzonella, E. Bressert, A. Finoguenov, A. Fruscione, A.M. Koekemoer, N. Cappelluti, F. Fiore, S. Giodini, R. Gilli, C.D. Impey, S.J. Lilly, E. Lusso, S. Puccetti, J.D. Silverman, H. Aussel, P. Capak, D. Frayer, E. Le Floch, H.J. McCracken, D.B. Sanders, D. Schiminovich, Y. Taniguchi, ApJ **717**, 209 (2010). doi:10.1088/0004-637X/717/1/209
J. Condon, J. Darling, Y.Y. Kovalev, L. Petrov, ArXiv e-prints (2011)
L.R. Corrales, Z. Haiman, A. MacFadyen, MNRAS **404**, 947 (2010). doi:10.1111/j.1365-2966.2010.16324.x
J. Cuadra, P.J. Armitage, R.D. Alexander, M.C. Begelman, MNRAS **393**, 1423 (2009). doi:10.1111/j.1365-2966.2008.14147.x
R. Decarli, M. Dotti, C. Montuori, T. Liimets, A. Ederoclite, ApJ **720**, L93 (2010). doi:10.1088/2041-8205/720/1/L93
R. Decarli, M. Dotti, M. Fumagalli, P. Tsalmantza, C. Montuori, E. Lusso, D.W. Hogg, J.X. Prochaska, MNRAS **433**, 1492 (2013). doi:10.1093/mnras/stt831

F. De Paolis, G. Ingrosso, A.A. Nucita, A&A **388**, 470 (2002). doi:10.1051/0004-6361:20020519
B. Devecchi, E. Rasia, M. Dotti, M. Volonteri, M. Colpi, MNRAS **394**, 633 (2009). doi:10.1111/j.1365-2966.2008.14329.x
M. Dietrich, B.M. Peterson, C.J. Grier, M.C. Bentz, J. Eastman, S. Frank, R. Gonzalez, J.L. Marshall, D.L. DePoy, J.L. Prieto, ApJ **757**, 53 (2012). doi:10.1088/0004-637X/757/1/53
T. Di Matteo, J. Colberg, V. Springel, L. Hernquist, D. Sijacki, ApJ **676**, 33 (2008). doi:10.1086/524921
D.J. D'Orazio, Z. Haiman, A. MacFadyen, MNRAS **436**, 2997 (2013). doi:10.1093/mnras/stt1787
M. Dotti, M. Colpi, F. Haardt, MNRAS **367**, 103 (2006). doi:10.1111/j.1365-2966.2005.09956.x
M. Dotti, M. Colpi, F. Haardt, L. Mayer, MNRAS **379**, 956 (2007). doi:10.1111/j.1365-2966.2007.12010.x
M. Dotti, C. Montuori, R. Decarli, M. Volonteri, M. Colpi, F. Haardt, MNRAS **398**, L73 (2009a). doi:10.1111/j.1745-3933.2009.00714.x
M. Dotti, M. Ruszkowski, L. Paredi, M. Colpi, M. Volonteri, F. Haardt, MNRAS **396**, 1640 (2009b). doi:10.1111/j.1365-2966.2009.14840.x
M. Dotti, M. Volonteri, A. Perego, M. Colpi, M. Ruszkowski, F. Haardt, MNRAS **402**, 682 (2010). doi:10.1111/j.1365-2966.2009.15922.x
M. Dotti, A. Sesana, R. Decarli, Adv. Astron. **2012**, 940568 (2012). doi:10.1155/2012/940568
M. Dotti, M. Colpi, S. Pallini, A. Perego, M. Volonteri, ApJ **762**, 68 (2013). doi:10.1088/0004-637X/762/2/68
M. Eracleous, T.A. Boroson, J.P. Halpern, J. Liu, ApJS **201**, 23 (2012). doi:10.1088/0067-0049/201/2/23
A. Escala, R.B. Larson, P.S. Coppi, D. Mardones, ApJ **607**, 765 (2004). doi:10.1086/386278
A. Escala, R.B. Larson, P.S. Coppi, D. Mardones, ApJ **630**, 152 (2005). doi:10.1086/431747
A.C. Fabian, M.J. Rees, L. Stella, N.E. White, MNRAS **238**, 729 (1989)
J.H. Fan, G.Z. Xie, E. Pecontal, A. Pecontal, Y. Copin, ApJ **507**, 173 (1998). doi:10.1086/306301
B.D. Farris, Y.T. Liu, S.L. Shapiro, Phys. Rev. D **81**(8), 084008 (2010). doi:10.1103/PhysRevD.81.084008
B.D. Farris, Y.T. Liu, S.L. Shapiro, Phys. Rev. D **84**(2), 024024 (2011). doi:10.1103/PhysRevD.84.024024
B.D. Farris, P. Duffell, A.I. MacFadyen, Z. Haiman, ApJ **783**, 134 (2014). doi:10.1088/0004-637X/783/2/134
L. Ferrarese, D. Merritt, ApJ **539**, L9 (2000). doi:10.1086/312838
D. Fiacconi, L. Mayer, R. Roškar, M. Colpi, ApJ **777**, L14 (2013). doi:10.1088/2041-8205/777/1/L14
P.C. Fragile, P. Anninos, ApJ **623**, 347 (2005). doi:10.1086/428433
P.C. Fragile, O.M. Blaes, P. Anninos, J.D. Salmonson, ApJ **668**, 417 (2007). doi:10.1086/521092
K. Gebhardt, R. Bender, G. Bower, A. Dressler, S.M. Faber, A.V. Filippenko, R. Green, C. Grillmair, L.C. Ho, J. Kormendy, T.R. Lauer, J. Magorrian, J. Pinkney, D. Richstone, S. Tremaine, ApJ **539**, L13 (2000). doi:10.1086/312840
D. Gerosa, A. Sesana, ArXiv e-prints (2014)
B. Giacomazzo, J.G. Baker, M.C. Miller, C.S. Reynolds, J.R. van Meter, ApJ **752**, L15 (2012). doi:10.1088/2041-8205/752/1/L15
R. Gold, V. Paschalidis, Z.B. Etienne, S.L. Shapiro, H.P. Pfeiffer, Phys. Rev. D **89**(6), 064060 (2014). doi:10.1103/PhysRevD.89.064060
A. Gould, H.W. Rix, ApJ **532**, L29 (2000). doi:10.1086/312562
A. Gualandris, D. Merritt, ApJ **678**, 780 (2008). doi:10.1086/586877
J. Guedes, P. Madau, L. Mayer, S. Callegari, ApJ **729**, 125 (2011). doi:10.1088/0004-637X/729/2/125
K. Hayasaki, PASJ **61**, 65 (2009). doi:10.1093/pasj/61.1.65
K. Hayasaki, S. Mineshige, H. Sudou, PASJ **59**, 427 (2007). doi:10.1093/pasj/59.2.427
K. Holley-Bockelmann, K. Gültekin, D. Shoemaker, N. Yunes, ApJ **686**, 829 (2008). doi:10.1086/591218
P. Jovanović, V. Borka Jovanović, D. Borka, T. Bogdanović, ArXiv e-prints (2013)

W. Ju, J.E. Greene, R.R. Rafikov, S.J. Bickerton, C. Badenes, ApJ **777**, 44 (2013). doi:10.1088/0004-637X/777/1/44
E. Kara, A.C. Fabian, E.M. Cackett, J.F. Steiner, P. Uttley, D.R. Wilkins, A. Zoghbi, MNRAS **428**, 2795 (2013a). doi:10.1093/mnras/sts155
E. Kara, A.C. Fabian, E.M. Cackett, G. Miniutti, P. Uttley, MNRAS **430**, 1408 (2013b). doi:10.1093/mnras/stt024
E. Kara, E.M. Cackett, A.C. Fabian, C. Reynolds, P. Uttley, MNRAS **439**, L26 (2014). doi:10.1093/mnrasl/slt173
J.S. Key, N.J. Cornish, Phys. Rev. D **83**(8), 083001 (2011). doi:10.1103/PhysRevD.83.083001
F.M. Khan, A. Just, D. Merritt, ApJ **732**, 89 (2011). doi:10.1088/0004-637X/732/2/89
F.M. Khan, I. Berentzen, P. Berczik, A. Just, L. Mayer, K. Nitadori, S. Callegari, ApJ **756**, 30 (2012a). doi:10.1088/0004-637X/756/1/30
F.M. Khan, M. Preto, P. Berczik, I. Berentzen, A. Just, R. Spurzem, ApJ **749**, 147 (2012b). doi:10.1088/0004-637X/749/2/147
F.M. Khan, K. Holley-Bockelmann, P. Berczik, A. Just, ApJ **773**, 100 (2013). doi:10.1088/0004-637X/773/2/100
A. King, C. Nixon, Class. Quantum Grav. **30**(24), 244006 (2013). doi:10.1088/0264-9381/30/24/244006
B. Kocsis, Z. Haiman, A. Loeb, MNRAS **427**, 2660 (2012a). doi:10.1111/j.1365-2966.2012.22129.x
B. Kocsis, Z. Haiman, A. Loeb, MNRAS **427**, 2680 (2012b). doi:10.1111/j.1365-2966.2012.22118.x
S. Komossa, Adv. Astron. **2012**, 364973 (2012). doi:10.1155/2012/364973
S. Komossa, D. Merritt, ApJ **689**, L89 (2008). doi:10.1086/595883
S. Komossa, H. Zhou, H. Lu, ApJ **678**, L81 (2008). doi:10.1086/588656
D.A. Kornreich, R.V.E. Lovelace, ApJ **681**, 104 (2008). doi:10.1086/587511
D.N.C. Lin, J. Papaloizou, MNRAS **186**, 799 (1979)
X. Liu, Y. Shen, F. Bian, A. Loeb, S. Tremaine, ArXiv e-prints (2013)
F.K. Liu, S. Li, S. Komossa, ApJ **786**, 103 (2014). doi:10.1088/0004-637X/786/2/103
G. Lodato, D. Gerosa, MNRAS **429**, L30 (2013). doi:10.1093/mnrasl/sls018
A. Loeb, Physical Review Letters **99**(4), 041103 (2007). doi:10.1103/PhysRevLett.99.041103
C.O. Lousto, Y. Zlochower, Physical Review Letters **107**(23), 231102 (2011). doi:10.1103/PhysRevLett.107.231102
C.O. Lousto, Y. Zlochower, M. Dotti, M. Volonteri, Phys. Rev. D **85**(8), 084015 (2012). doi:10.1103/PhysRevD.85.084015
R.V.E. Lovelace, D.A. Kornreich, MNRAS **402**, 2753 (2010). doi:10.1111/j.1365-2966.2009.16095.x
E. Lusso, R. Decarli, M. Dotti, C. Montuori, D.W. Hogg, P. Tsalmantza, M. Fumagalli, J.X. Prochaska, MNRAS **441**, 316 (2014). doi:10.1093/mnras/stu572
A.I. MacFadyen, M. Milosavljević, ApJ **672**, 83 (2008). doi:10.1086/523869
H.L. Maness, G.B. Taylor, R.T. Zavala, A.B. Peck, L.K. Pollack, ApJ **602**, 123 (2004). doi:10.1086/380919
G. Mathes, M. Eracleous, S. Sigurdsson, J.C. Runnoe, T. Bogdanovic, in *American Astronomical Society Meeting Abstracts*, vol. 223 (2014), p. 250.15
L. Mayer, S. Kazantzidis, P. Madau, M. Colpi, T. Quinn, J. Wadsley, Science **316**, 1874 (2007). doi:10.1126/science.1141858
L. Mayer, Class. Quantum Grav. **30**(24), 244008 (2013). doi:10.1088/0264-9381/30/24/244008
B. McKernan, K.E.S. Ford, B. Kocsis, Z. Haiman, MNRAS **432**, 1468 (2013). doi:10.1093/mnras/stt567
J.C. McKinney, A. Tchekhovskoy, R.D. Blandford, Science **339**, 49 (2013). doi:10.1126/science.1230811
D. Merritt, J.D. Schnittman, S. Komossa, ApJ **699**, 1690 (2009). doi:10.1088/0004-637X/699/2/1690
M.C. Miller, J.H. Krolik, ApJ **774**, 43 (2013). doi:10.1088/0004-637X/774/1/43

M. Milosavljević, D. Merritt, ApJ **563**, 34 (2001). doi:10.1086/323830
P. Moesta, D. Alic, L. Rezzolla, O. Zanotti, C. Palenzuela, ApJ **749**, L32 (2012). doi:10.1088/2041-8205/749/2/L32
P. Mösta, C. Palenzuela, L. Rezzolla, L. Lehner, S. Yoshida, D. Pollney, Phys. Rev. D **81**(6), 064017 (2010). doi:10.1103/PhysRevD.81.064017
R. Morganti, B. Emonts, T. Oosterloo, A&A **496**, L9 (2009). doi:10.1051/0004-6361/200911705
C.J. Nixon, P.J. Cossins, A.R. King, J.E. Pringle, MNRAS **412**, 1591 (2011). doi:10.1111/j.1365-2966.2010.17952.x
C.J. Nixon, MNRAS **423**, 2597 (2012). doi:10.1111/j.1365-2966.2012.21072.x
S.C. Noble, B.C. Mundim, H. Nakano, J.H. Krolik, M. Campanelli, Y. Zlochower, N. Yunes, ApJ **755**, 51 (2012). doi:10.1088/0004-637X/755/1/51
S.M. O'Neill, M.C. Miller, T. Bogdanović, C.S. Reynolds, J.D. Schnittman, ApJ **700**, 859 (2009). doi:10.1088/0004-637X/700/1/859
E.C. Ostriker, ApJ **513**, 252 (1999). doi:10.1086/306858
C. Palenzuela, M. Anderson, L. Lehner, S.L. Liebling, D. Neilsen, Phys. Rev. Lett. **103**(8), 081101 (2009). doi:10.1103/PhysRevLett.103.081101
C. Palenzuela, L. Lehner, S. Yoshida, Phys. Rev. D **81**(8), 084007 (2010a). doi:10.1103/PhysRevD.81.084007
C. Palenzuela, L. Lehner, S.L. Liebling, Science **329**, 927 (2010b). doi:10.1126/science.1191766
A. Pancoast, B.J. Brewer, T. Treu, ApJ **730**, 139 (2011). doi:10.1088/0004-637X/730/2/139
A. Pancoast, B.J. Brewer, T. Treu, D. Park, A.J. Barth, M.C. Bentz, J.H. Woo, ArXiv e-prints (2013)
A. Petiteau, S. Babak, A. Sesana, M. de Araújo, Phys. Rev. D **87**(6), 064036 (2013). doi:10.1103/PhysRevD.87.064036
M. Ponce, J.A. Faber, J.C. Lombardi, ApJ **745**, 71 (2012). doi:10.1088/0004-637X/745/1/71
L.Č. Popović, New A Rev. **56**, 74 (2012). doi:10.1016/j.newar.2011.11.001
M. Preto, I. Berentzen, P. Berczik, R. Spurzem, ApJ **732**, L26 (2011). doi:10.1088/2041-8205/732/2/L26
V. Ravi, J.S.B. Wyithe, G. Hobbs, R.M. Shannon, R.N. Manchester, D.R.B. Yardley, M.J. Keith, ApJ **761**, 84 (2012). doi:10.1088/0004-637X/761/2/84
C.S. Reynolds, Class. Quantum Grav. **30**(24), 244004 (2013). doi:10.1088/0264-9381/30/24/244004
F.M. Rieger, K. Mannheim, A&A **359**, 948 (2000)
C. Rodriguez, G.B. Taylor, R.T. Zavala, A.B. Peck, L.K. Pollack, R.W. Romani, ApJ **646**, 49 (2006). doi:10.1086/504825
C. Rodriguez, G.B. Taylor, R.T. Zavala, Y.M. Pihlström, A.B. Peck, ApJ **697**, 37 (2009). doi:10.1088/0004-637X/697/1/37
A. Robinson, S. Young, D.J. Axon, P. Kharb, J.E. Smith, ApJ **717**, L122 (2010). doi:10.1088/2041-8205/717/2/L122
C. Roedig, M. Dotti, A. Sesana, J. Cuadra, M. Colpi, MNRAS **415**, 3033 (2011). doi:10.1111/j.1365-2966.2011.18927.x
C. Roedig, A. Sesana, M. Dotti, J. Cuadra, P. Amaro-Seoane, F. Haardt, A&A **545**, A127 (2012). doi:10.1051/0004-6361/201219986
J.D. Schnittman, ApJ **667**, L133 (2007). doi:10.1086/522203
J.D. Schnittman, Class. Quantum Grav. **30**(24), 244007 (2013). doi:10.1088/0264-9381/30/24/244007
J.D. Schnittman, J.H. Krolik, ApJ **684**, 835 (2008). doi:10.1086/590363
F. Schweizer, Science **231**, 227 (1986). doi:10.1126/science.231.4735.227
A. Sesana, MNRAS **382**, L6 (2007). doi:10.1111/j.1745-3933.2007.00375.x
A. Sesana, MNRAS **433**, L1 (2013). doi:10.1093/mnrasl/slt034
A. Sesana, A. Vecchio, C.N. Colacino, MNRAS **390**, 192 (2008). doi:10.1111/j.1365-2966.2008.13682.x
A. Sesana, A. Gualandris, M. Dotti, MNRAS **415**, L35 (2011). doi:10.1111/j.1745-3933.2011.01073.x

A. Sesana, C. Roedig, M.T. Reynolds, M. Dotti, MNRAS **420**, 860 (2012). doi:10.1111/j.1365-2966.2011.20097.x
R.M. Shannon, V. Ravi, W.A. Coles, G. Hobbs, M.J. Keith, R.N. Manchester, J.S.B. Wyithe, M. Bailes, N.D.R. Bhat, S. Burke-Spolaor, J. Khoo, Y. Levin, S. Oslowski, J.M. Sarkissian, W. van Straten, J.P.W. Verbiest, J.B. Want, Science **342**, 334 (2013)
Y. Shen, X. Liu, A. Loeb, S. Tremaine, ApJ **775**, 49 (2013). doi:10.1088/0004-637X/775/1/49
J.M. Shi, J.H. Krolik, S.H. Lubow, J.F. Hawley, ApJ **749**, 118 (2012). doi:10.1088/0004-637X/749/2/118
G.A. Shields, E.W. Bonning, ApJ **682**, 758 (2008). doi:10.1086/589427
G.A. Shields, E.W. Bonning, ApJ **772**, L5 (2013). doi:10.1088/2041-8205/772/1/L5
G.A. Shields, D.J. Rosario, K.L. Smith, E.W. Bonning, S. Salviander, J.S. Kalirai, R. Strickler, E. Ramirez-Ruiz, A.A. Dutton, T. Treu, P.J. Marshall, ApJ **707**, 936 (2009). doi:10.1088/0004-637X/707/2/936
K.A. Sorathia, J.H. Krolik, J.F. Hawley, ApJ **777**, 21 (2013). doi:10.1088/0004-637X/777/1/21
N. Stone, A. Loeb, MNRAS **422**, 1933 (2012). doi:10.1111/j.1365-2966.2012.20577.x
T.L. Tanaka, Z. Haiman, Class. Quantum Grav. **30**(22), 224012 (2013). doi:10.1088/0264-9381/30/22/224012
S. Tang, J. Grindlay, ApJ **704**, 1189 (2009). doi:10.1088/0004-637X/704/2/1189
S. Tremaine, K. Gebhardt, R. Bender, G. Bower, A. Dressler, S.M. Faber, A.V. Filippenko, R. Green, C. Grillmair, L.C. Ho, J. Kormendy, T.R. Lauer, J. Magorrian, J. Pinkney, D. Richstone, ApJ **574**, 740 (2002). doi:10.1086/341002
C.W. Tsai, T.H. Jarrett, D. Stern, B. Emonts, R.S. Barrows, R.J. Assef, R.P. Norris, P.R.M. Eisenhardt, C. Lonsdale, A.W. Blain, D.J. Benford, J. Wu, B. Stalder, C.W. Stubbs, F.W. High, K.L. Li, A.K.H. Kong, ApJ **779**, 41 (2013). doi:10.1088/0004-637X/779/1/41
P. Tsalmantza, R. Decarli, M. Dotti, D.W. Hogg, ApJ **738**, 20 (2011). doi:10.1088/0004-637X/738/1/20
P. Uttley, E.M. Cackett, A.C. Fabian, E. Kara, D.R. Wilkins, ArXiv e-prints (2014)
M.J. Valtonen, H.J. Lehto, K. Nilsson, J. Heidt, L.O. Takalo, A. Sillanpää, C. Villforth, M. Kidger, G. Poyner, T. Pursimo, S. Zola, J.H. Wu, X. Zhou, K. Sadakane, M. Drozdz, D. Koziel, D. Marchev, W. Ogloza, C. Porowski, M. Siwak, G. Stachowski, M. Winiarski, V.P. Hentunen, M. Nissinen, A. Liakos, S. Dogru, Nature **452**, 851 (2008). doi:10.1038/nature06896
J.R. van Meter, J.H. Wise, M.C. Miller, C.S. Reynolds, J. Centrella, J.G. Baker, W.D. Boggs, B.J. Kelly, S.T. McWilliams, ApJ **711**, L89 (2010). doi:10.1088/2041-8205/711/2/L89
E. Vasiliev, F. Antonini, D. Merritt, ApJ **785**, 163 (2014). doi:10.1088/0004-637X/785/2/163
M. Volonteri, ApJ **663**, L5 (2007). doi:10.1086/519525
M. Volonteri, P. Madau, ApJ **687**, L57 (2008). doi:10.1086/593353
M. Volonteri, F. Haardt, P. Madau, ApJ **582**, 559 (2003). doi:10.1086/344675
M. Volonteri, J.M. Miller, M. Dotti, ApJ **703**, L86 (2009). doi:10.1088/0004-637X/703/1/L86
M. Volonteri, K. Gültekin, M. Dotti, MNRAS **404**, 2143 (2010). doi:10.1111/j.1365-2966.2010.16431.x
S.D.M. White, MNRAS **184**, 185 (1978)
A. Zoghbi, A.C. Fabian, C.S. Reynolds, E.M. Cackett, MNRAS **422**, 129 (2012). doi:10.1111/j.1365-2966.2012.20587.x

General Relativistic Simulations of Accretion Disks Around Tilted Kerr Black Holes

Vassilios Mewes, Pedro J. Montero, Nikolaos Stergioulas, Filippo Galeazzi, and José A. Font

Abstract We simulate the dynamics of self-gravitating accretion disks around tilted Kerr black holes (BH) in full 3D general relativity. For this purpose we employ the EinsteinToolkit, using the thorn McLachlan for the evolution of the spacetime via the BSSN formalism of the Einstein equations and the thorn GRHydro for the evolution of the hydrodynamics, using a 3D Cartesian mesh with adaptive mesh refinement. We investigate the effects of the tilt angle between the disk angular momentum and BH spin vector on the dynamics of these systems as the disk evolves in the tilted spacetime. By evolving the spacetime and matter fields, we are able to observe how both BH and disk react and evolve in the tilted configuration. For the very light disk with a constant specific angular momentum profile that we study, we observe a significant change of the disk shape after a few orbits. This change is more pronounced the larger the initial inclination angle is.

1 Introduction

Stellar BH-torus systems are end states of binary neutron star (BNS) mergers and gravitational collapse of massive stars, and are linked to short- and long-duration gamma-ray bursts (GRBs). The 3D simulations of Rezzolla et al. (2010) (see also

V. Mewes (✉) • J.A. Font
Departamento de Astronomía y Astrofísica, Universitat de València, Dr. Moliner 50, 46100 Burjassot, València, Spain
e-mail: vassilios.mewes@uv.es

P.J. Montero
Max-Planck-Institut für Astrophysik, Karl-Schwarzschild-Strasse 1, 81748 Garching bei München, Germany

N. Stergioulas
Department of Physics, Aristotle University of Thessaloniki, Thessaloniki 54124, Greece

F. Galeazzi
Institut für Theoretische Physik, J.W. Goethe - Universität, Max-von-Laue-Str. 1, 60438 Frankfurt am Main, Germany

C.F. Sopuerta (ed.), *Gravitational Wave Astrophysics*, Astrophysics and Space Science Proceedings, DOI 10.1007/978-3-319-10488-1_10

references therein) show the self-consistent formation of massive accretion tori (or thick disks) around spinning black holes. Those systems do not manifest signs of any dynamical instability on short dynamical timescales but, however, on longer timescales, non-axisymmetric instabilities may set in Kiuchi et al. (2011), Korobkin et al. (2011). Moreover, for spinning BNS the angular momentum vector of the resulting accretion torus does not necessarily have to be aligned with that of the central BH. This kind of generic situation has been studied numerically in a series of seminal papers by Fragile and collaborators (see e.g. Fragile and Anninos 2005 and Fragile et al. 2007), who simulated the hydrodynamical evolution of thick accretion tori in the fixed background spacetime of a *tilted* Kerr BH. Building upon this work, we present in this paper preliminary results that incorporate the effects of the self-consistent evolution of the spacetime via the solution of the full 3+1 Einstein equations coupled to the hydrodynamics equations. A complete account of our study will be presented elsewhere (Mewes et al. xxxx).

We use the publicly available `Einstein Toolkit` (ET) (Löffler et al. 2012), to carry out our simulations. The ET is a toolkit for relativistic astrophysics simulations, which uses the `Cactus` framework (Goodale et al. 2003) and provides adaptive mesh refinement (AMR) via the `Carpet` driver (Schnetter et al. 2004). Integration in time is provided by the thorn `MoL` which implements the method of lines. The toolkit contains components to evolve the 3D Einstein equations,

$$G^{\mu\nu} = 8\,\pi\,T^{\mu\nu}, \tag{1}$$

coupling the evolution of the spacetime to the dynamic evolution of the matter fields through the stress-energy tensor $T_{\mu\nu}$. The methods to evolve the Einstein Equations are based on the 3+1 ADM (Arnowitt-Deser-Misner) (Arnowitt et al. 2008) splitting of spacetime, in which the metric becomes

$$ds^2 = \left(-\alpha^2 + \beta_i\beta^i\right)dt^2 + 2\beta_i\,dt\,dx^i + \gamma_{ij}\,dx^i\,dx^j\,, \tag{2}$$

where α is the lapse function, β^i is the shift vector and γ_{ij} is the spatial metric tensor.

The spacetime is evolved using the `MacLachlan` thorn (Brown et al. 2009; Reisswig et al. 2011), which solves the Einstein equations in the BSSN reformulation of the ADM formalism (Baumgarte and Shapiro 1998; Nakamura et al. 1987; Shibata and Nakamura 1995).

Correspondingly, the evolution of the general relativistic hydrodynamics equations is performed by the thorn `GRHydro` (Baiotti et al. 2005; Löffler et al. 2012; Mösta et al. 2014), which solves those equations in flux-conservative form in the so-called Valencia formulation (Banyuls et al. 1997; Font 2008), using high-resolution shock-capturing methods. In this formulation, the *conserved variables* are the generalised density, momenta and internal energy defined by:

$$D = \sqrt{\gamma}\,\rho W\,, \tag{3}$$

$$S_i = \rho h W^2 v_i\,, \tag{4}$$

$$\tau = \sqrt{\gamma}\,(\rho h W^2 - p) - D\,, \tag{5}$$

where γ is the determinant of the spatial metric and the *primitive variables* are the rest-mass density ρ, internal energy ϵ, three-velocity v^i, Lorentz factor W, pressure p and specific enthalpy h.

We measure the irreducible mass of the BH during the evolution using the `AHFinderDirect` thorn which implements the apparent horizon (AH) finder described in Thornburg (2003). The BH spin and Christodoulou mass are measured using the `QuasiLocalMeasures` thorn (Dreyer et al. 2003; Schnetter et al. 2006).

Our initial setup is a thick, self-gravitating axisymmetric accretion disk with a constant specific angular momentum profile around a Kerr BH. The main disk properties are given in the table below:

Model	ρ_{max} [$G=c=M_\odot=1$]	l	M_{torus}/M_{BH}	f_{orb} [Hz]
D2	1.05e-05	3.75, constant	4.4e-02	1,360

where ρ_{max} is the maximum rest mass density, l the specific angular momentum and f_{orb} the orbital frequency of the fluid at the location of ρ_{max}.

The initial data (ID) is generated with the `RNSID` code in spherical polar coordinates, and then interpolated onto the 3D Cartesian grid and evolved with the `Einstein Toolkit`. The ID describes a disk around a Schwarzschild BH, so for our tilted disk simulations we use the matter content of the ID with a Kerr spacetime in quasi-isotropic coordinates. Figure 1 shows contours of the intimal rest mass density profile in the disk, together with the initial grid structure.

The Kerr BH is set up to be initially tilted about the x-axis by an angle β_0, using the improved quasi-isotropic coordinates of Liu et al. (2009). In those coordinates, the radius of the AH of the BH does not shrink to 0 in the extreme Kerr limit, making them more suitable for evolutions of Kerr BHs with high initial spins. We use geometrised units ($G = c = M_\odot = 1$) throughout the text.

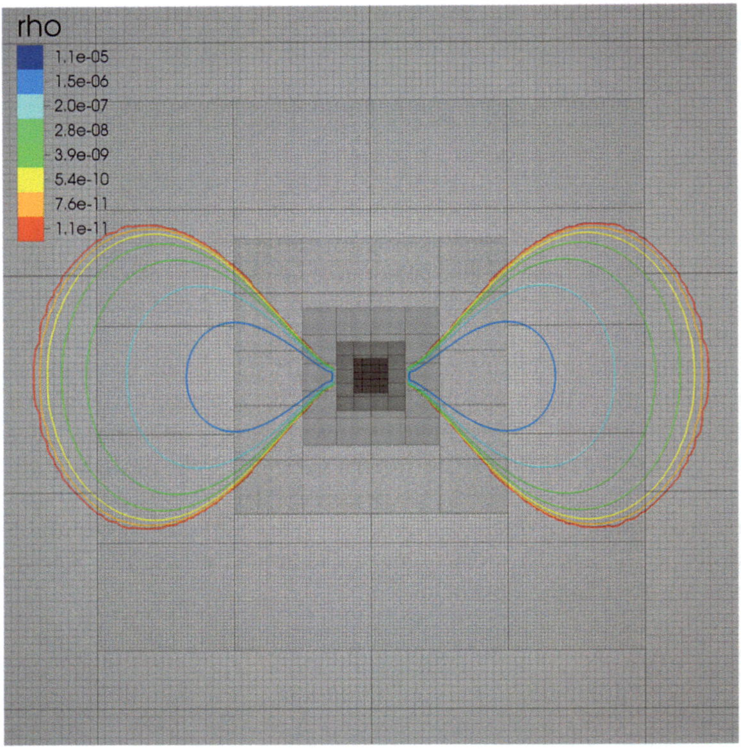

Fig. 1 Contour plot of the initial density profile in the disk. The mesh structure, showing the six innermost mesh refinement levels (out of 13), is shown as well

2 Results

We performed simulations of six different initial configurations: initial Kerr parameters of $a = 0.1$ and $a = 0.5$, combined with three different initial tilt angles β_0 : 5°, 15° and 30°. We plan to extend the study to higher initial BH spins in a future work as well (Mewes et al. xxxx). Simulations with higher spin are complicated by the fact that the conservation of spin decreases non-linearly with increasing spin (Marronetti et al. 2008). Figure 2 shows the fractional error in the Kerr parameter a of a tilted Kerr BH ($\beta_0 = 15°$) in vacuum for different initial spin magnitudes. The curves show satisfactory conservation of the initial spin through the evolution. The wavelike features in the curves are due to the interaction of the BH with spurious reflections of noise from the mesh refinement boundaries.

Figure 3 shows a surface plot of the rest mass density ρ for a simulation with initial Kerr parameter $a = 0.5$ and initial BH tilt angle $\beta_0 = 15°$ at $t = 0$ (left panel) and after $t = 6.5$ orbits. The BH spin axis has initially been tilted to the left about the x-axis. The disk has undergone a significant change in shape. In the central region around the BH, we can see two accretion streams reaching the BH.

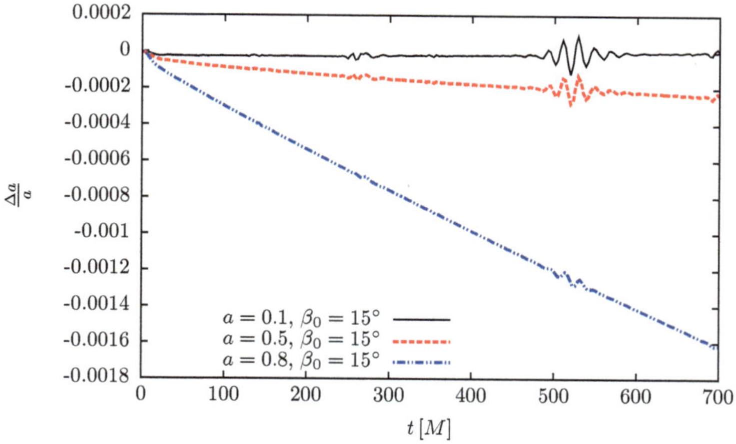

Fig. 2 Fractional error in Kerr spin parameter a for different initial spin magnitudes

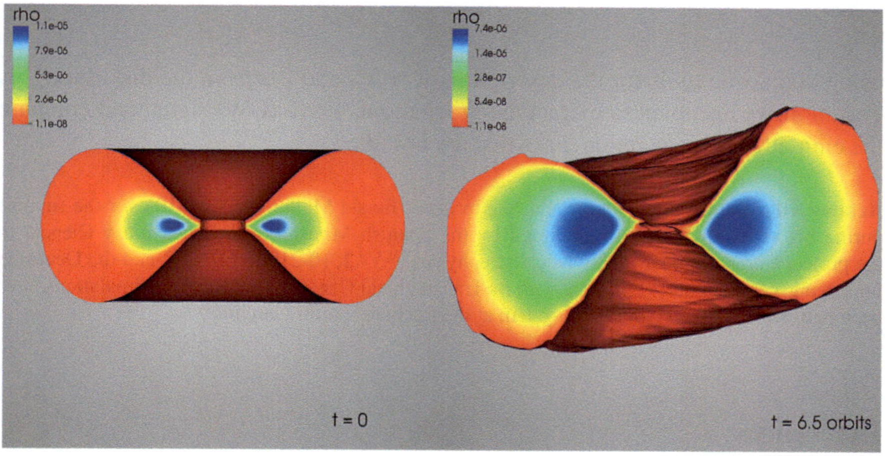

Fig. 3 Surface plot of ρ at $t = 0$ (*left*) and after 6.5 orbits for Kerr parameter $a = 0.5$ and initial BH tilt $\beta_0 = 15°$

We note that this has already been observed in simulations using the Cowling (fixed spacetime) approximation (Fragile et al. 2007).

Figure 4 shows a surface plot for a simulation with initial Kerr parameter $a = 0.5$ and initial BH tilt angle $\beta_0 = 30°$ at $t = 0$ (left panel) and after 6.5 orbits. The BH was initially tilted to the left about the x-axis as well. The change in shape is even more pronounced for a larger initial tilting angle. We note that in both simulations, the inner region of the disk does not align with the equatorial plane of the Kerr BH at the time shown. It therefore seems that the two simulations do not show evidence for the occurrence of the Bardeen-Petterson effect (Bardeen and Petterson 1975). This will be investigated more thoroughly elsewhere (Mewes et al. xxxx),

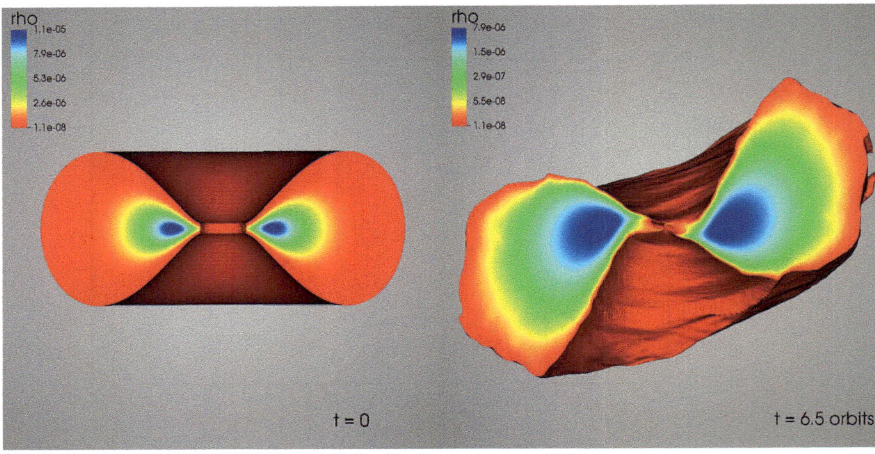

Fig. 4 Surface plot of ρ at $t = 0$ (*left*) and after 6.5 orbits for Kerr parameter $a = 0.5$ and initial BH tilt $\beta_0 = 30°$

using diagnostic tools for the description of *twists* and *warps* in the disk during its evolution.We also plan to extract the gravitational wave (GW) signal and compare the emitted GW to accretion disks that are not tilted.

Acknowledgements V. Mewes would like to thank Ewald Müller and the MPA for the support during his visit. This work is supported by the Spanish Ministerio de Educación y Ciencia in the project Computational Relativistic Astrophysics (AYA2010-21097-C03-01) and the Deutsche Forschungsgemeinschaft (DFG) through its Transregional Center SFB/TR7. The simulations have been performed on the Hydra HPC system at the Rechenzentrum Garching.

References

R. Arnowitt, S. Deser, C.W. Misner, Gen. Relativ. Gravit. **40**, 1997 (2008)
L. Baiotti, I. Hawke, P. Montero, F. Löffler, L. Rezzolla, N. Stergioulas, J. Font, E. Seidel, Phys. Rev. D **71**, 024035 (2005)
F. Banyuls, J.A. Font, J.M. Ibanez, J.M. Marti, J. A. Miralles, Astrophys. J. **476**, 221 (1997)
J.M. Bardeen, J.A. Petterson, Astrophys. J. **195**, L65 (1975)
T.W. Baumgarte S.L. Shapiro, Phys. Rev. D **59**, 024007 (1998)
D. Brown, P. Diener, O. Sarbach, E. Schnetter, M. Tiglio, Phys. Rev. D **79**, 044023 (2009)
O. Dreyer, B. Krishnan, D. Shoemaker, E. Schnetter, Phys. Rev. D **67**, 024018 (2003)
J.A. Font, Living Rev. Relativ. **11**, 7 (2008)
P.C. Fragile, P. Anninos, Astrophys. J. **623**, 347 (2005)
P.C. Fragile, O.M. Blaes, P. Anninos, J.D. Salmonson, Astrophys. J. **668**, 417 (2007)
T. Goodale, G. Allen, G. Lanfermann, J. Massó, T. Radke, E. Seidel, J. Shalf, VECPAR'2002, 5th International Conference, Springer (2003)
K. Kiuchi, M. Shibata, P. Montero, J. Font, Phys. Rev. Lett. **106**, 251102 (2011)
O. Korobkin, E. Abdikamalov, E. Schnetter, N. Stergioulas, B. Zink, Phys. Rev. D **83**, 043007 (2011)

Y.T. Liu, Z.B. Etienne, S.L. Shapiro, Phys. Rev. D **80**, 121503 (2009)
F. Löffler, J. Faber, E. Bentivegna, T. Bode, P. Diener, R. Haas, I. Hinder, B.C. Mundim, C.D. Ott, E. Schnetter, G. Allen, M. Campanelli, P. Laguna, Class. Quantum Grav. **29**, 115001 (2012)
P. Marronetti, W. Tichy, B. Bruegmann, J. Gonzalez, U. Sperhake, Phys. Rev. D **77**, 064010 (2008)
V. Mewes et al. (in preparation)
P. Mösta, B. C. Mundim, J. A. Faber, R. Haas, S. C. Noble, T. Bode, F. Löffler, C. D. Ott, C. Reisswig, E. Schnetter, Class. Quantum Grav. **31**, 015005 (2014)
T. Nakamura, K. Oohara, Y. Kojima, Prog. Theor. Phys. Suppl. **90**, 1 (1987)
C. Reisswig, C.D. Ott, U. Sperhake, E. Schnetter, Phys. Rev. D **83**, 064008 (2011)
L. Rezzolla, L. Baiotti, B. Giacomazzo, D. Link, J.A. Font, Class. Quantum Grav. **27**, 114105 (2010)
E. Schnetter, S.H. Hawley, I. Hawke, Class. Quantum Grav. **21**, 1465 (2004)
E. Schnetter, B. Krishnan, F. Beyer, Phys. Rev. D **74**, 024028 (2006)
M. Shibata, T. Nakamura, Phys. Rev. D **52**, 5428 (1995)
J. Thornburg, AIP Conf. Proc. **686**, 247–252 (2003)

Prompt Flare and Disk Formation in Tidal Disruptions by Massive Black Holes

Christopher Evans and Pablo Laguna

Abstract The ubiquitous observational signature in the disruption of a star by a massive black hole is a bright UV/X-ray flare with a light curve decaying at late times as $t^{-5/3}$. The flare is the product of the stellar debris falling back towards the black hole, self-intersecting, and eventually forming an accretion disk. The time elapsing between the disruption of the star as it passes periapsis, the onset of the subsequent flare from material falling back, and the ultimate formation of the accretion disk depends on the black hole mass, the characteristics of the star, and the strength of the encounter. We present preliminary results from a class of tidal disruption events that until now have not been considered in which the disruption-flare-disk sequence develops promptly. These tidal disruption events are from ultra-close encounters between main-sequence stars and intermediate mass black holes.

1 Introduction

Tidal stellar disruptions are fascinating cosmic events resulting from the encounter of a star with a massive black hole (BH). The transient flares produced by tidal disruption events (TDEs) make them a perfect target for time-domain astronomy, and thus for learning about the growth of massive BHs in non-active galaxies. Perhaps the cleanest TDE candidates so far are the transients J1644+57 (Levan et al. 2011) and J2058+0516 (Cenko et al. 2012) observed by *Swift*. The key TDE signature sought in the observational data is the onset of a flare and the subsequent decline in its luminosity, which theoretical models predict should drop as $t^{-5/3}$ (Phinney 1989; Evans and Kochanek 1989). The flare and decay in luminosity are consequences of the return of bound stellar debris to the neighborhood of the BH, which in time self-intersects and settles into an accretion disk.

C. Evans (✉) • P. Laguna
Center for Relativistic Astrophysics, School of Physics, Georgia Institute of Technology, Atlanta, GA 30332, USA
e-mail: cevans43@gatech.edu; plaguna@gatech.edu

C.F. Sopuerta (ed.), *Gravitational Wave Astrophysics*, Astrophysics and Space Science Proceedings, DOI 10.1007/978-3-319-10488-1_11

Interest in TDEs was sparked in the 1970s with the suggestion that active galactic nuclei (AGN) could contain massive BHs; of course, we know now that most galaxies host supermassive BHs. In 1975, Hills (1975) pointed out that TDEs could potentially provide not only the required matter to explain the growth of the massive BH, but they could also fuel the observed AGN luminosity. The observational challenge to demonstrate this lies in separating TDE signatures from the intrinsic variability observed in AGNs. This explains why the strongest observational evidence for TDEs has been in association with non-active galaxies.

In the 1980s, with the accumulation of evidence in support of massive BHs in galactic centers (Tonry 1984), the interest for TDE studies increased, with the tidal disruption paper by Rees (1988) considered the springboard for subsequent studies. Rees provided a magnificent description of the TDE astrophysical phenomena, estimating in particular the timescales for the bound debris to be swallowed by the hole and the decay rate $t^{-5/2}$ for the material to shower back into the BH. Notice that this is not the famed $t^{-5/3}$ decay rate. A discrepancy first corrected by Phinney (1989). Nonetheless, Rees elaborated that the accretion rate decay would be correlated with the luminosity and thus provide the smoking gun to identify TDEs.

Because of the complexities involved, most of the TDE studies have been computational (Evans and Kochanek 1989; Laguna et al. 1993; Haas et al. 2012; Lodato et al. 2009; Guillochon et al. 2009). The drawback to this approach is the limited dynamical range covered by the simulations. In particular, it has not yet been possible to simulate a TDE from the time of disruption to full accretion disk formation; only semi-analytic studies have been able to provide such insights (Shen and Matzner 2014). The first numerical study was carried out by Evans and Kochanek (1989) using smoothed particle hydrodynamics (SPH) (Hernquist and Katz 1989). Since in their encounters the star did not adventure areas of strong curvature, their results corroborated the back of the envelope estimates by Rees.

In his seminal paper (Rees 1988), Rees also noted that for ultra-close passages relativistic effects would be dynamically significant and potentially trigger interesting observational effects. An example is the multiple compressions a star would experience in ultra-close encounters (Luminet and Marck 1985). The same compression could, in some instances according to Carter and Luminet (1982), potentially ignite a thermonuclear detonation of the star. The first general relativistic study of TDEs was done by one of us (Laguna et al. 1993), also using SPH. These simulations demonstrated key relativistic effects as a result of orbital precession, such as the multiple tidal compressions. At the same time, the study found that the rise in the central temperature of the star was not sufficient for detonation. On the other hand, more recent studies of the disruption of white dwarfs by intermediate mass BHs found that detonation is indeed potentially feasible (Rosswog et al. 2008). The most recent TDE study on ultra-close encounters that require a general relativistic description of gravity was also in connection with white dwarfs and intermediate mass BHs (Haas et al. 2012). This study provided insights on the effects of the spin of the BH on the geometry of the debris.

In order to improve connections with observations and to predict other unique signatures that identify TDEs, additional theoretical modeling is needed not only in terms of the physics included in the model, but also on the dynamical rage covered by the simulations. The present study identifies a region of TDE parameter space where the dynamical range is such that the simulation is able to reach the stage of disk formation. These tidal disruption events are from ultra-close encounters between main-sequence stars and intermediate mass BHs.

2 Tidal Disruption Events in Brief

For a star of mass M_* and radius R_* to be disrupted by a BH of mass M_h, it would have to approach the hole within an approximate distance of

$$R_t \equiv R_* \left(\frac{M_h}{M_*}\right)^{1/3}, \tag{1}$$

called the *tidal radius*. If we denote by R_p the distance of closest approach to the BH, the strength of the encounter is characterized by the *penetration factor* β defined as

$$\beta \equiv \frac{R_t}{R_p} = \frac{R_*}{R_p}\left(\frac{M_h}{M_*}\right)^{1/3}. \tag{2}$$

In addition to R_*, R_t and R_p, the fourth length scale in the problem is the BH horizon radius R_h. For a non-spinning BH in Schwarzschild coordinates $R_h = 2\,G\,M_h/c^2$, but it can be as small as $R_h = G\,M_h/c^2$ for a maximally rotating (Kerr) BH in Boyer-Lindquist coordinates.

As first suggested by Luminet and Pichon (1989), one can use β to identify the domain of astrophysical relevance of TDEs as a function of the BH mass. By definition, independent of the mass of the BH, disruption events take place for $\beta \geq 1$. In addition, if $R_p \leq R_h$, one effectively has the situation in which the *star enters the BH*. From Eq. (2) it is clear that in this case

$$\beta \geq \beta_h \equiv \frac{R_*}{G\,M_*/c^2}\left(\frac{M_h}{M_*}\right)^{-2/3}. \tag{3}$$

On the other hand, if $R_p \leq R_*$, then the *BH penetrates the star*, and from Eq. (2)

$$\beta \geq \beta_* \equiv \left(\frac{M_h}{M_*}\right)^{1/3}. \tag{4}$$

For a main-sequence star with $M_* = M_\odot$ and $R_* = R_\odot$, in which $R_\odot/(G\,M_\odot/c^2) = 4.7 \times 10^5$, notice that the maximum penetration factor for which the star does not enter the hole or vice versa (i.e. $\beta_h = \beta_*$) is $\beta = 78$, and this involves a BH with mass $M_h = 4.7 \times 10^5\,M_\odot$.

Most TDE studies of main-sequence stars have considered BH masses $M_h \simeq 10^{6-7}\,M_\odot$ and penetration factors $\beta \sim 1$. The exceptions are the few studies with $\beta \simeq 5 - 10$ which required accounting for general relativistic effects (Laguna et al. 1993; Rosswog 2010). These TDE studies with large β and high BH masses are near the *star enters BH* boundary (i.e. $\beta \simeq \beta_h$). For the present study, we are still interested in large penetrations but now in encounters near the *BH enters the star* boundary (i.e. $\beta \simeq \beta_*$). As a consequence, these are TDE involving BH with masses $M_h \leq 4.7 \times 10^5\,M_\odot$.

3 Preliminary Results

We will present and discuss preliminary results from a general relativistic TDE simulation involving a main-sequence star with $M_* = M_\odot$ and $R_* = R_\odot$ and a non-spinning BH with mass $M_h = 10^5\,M_\odot$. The tidal radius, in units of M_h, for this situation is

$$\frac{R_t}{M_h} \simeq 218 \left(\frac{M_*}{M_\odot}\right)^{-1/3} \left(\frac{R_*}{R_\odot}\right) \left(\frac{M_h}{10^5\,M_\odot}\right)^{-2/3} . \tag{5}$$

We inject the star in a parabolic orbit with penetration factor $\beta = 10$, which translates into a periapsis distance, in solar radius units, of

$$\frac{R_p}{R_\odot} \simeq 4.64 \left(\frac{\beta}{10}\right)^{-1} \left(\frac{M_*}{M_\odot}\right)^{-1/3} \left(\frac{R_*}{R_\odot}\right) \left(\frac{M_h}{10^5\,M_\odot}\right)^{-1/3} . \tag{6}$$

Although the R_p above seems to suggest that the star could potentially swing by without the *BH entering the star*, this is not exactly the case. By the time the star approaches periapsis, it has been effectively disrupted by the BH. Accretion into the BH occurs almost instantaneously. In Fig. 1, we show the accretion rate through the horizon of the BH, with time $t = 0$ denoting the time at periapsis passage. Instantaneous accretion is evident from the flare occurring at $t \sim 25$ s. The flare is followed by the standard $t^{-5/3}$ decay, but it only lasts ~700 s. At $t \sim 400$ s after periapsis passage, the material has already begun to circularize and is in the process of forming a disk. Figure 2 shows a 3D snapshot at $t = 465$ s where disk formation is evident. At around $t \sim 800$ s, the accretion rate increases, reaching what seems to be a steady rate of $100\,M_\odot\,\mathrm{yr}^{-1}$. For reference, the Eddington accretion rate, assuming 10 % efficiency, is $\dot{M}_{\rm Edd} = 0.002\,(M_h/10^5\,M_\odot)\,M_\odot\,\mathrm{yr}^{-1}$.

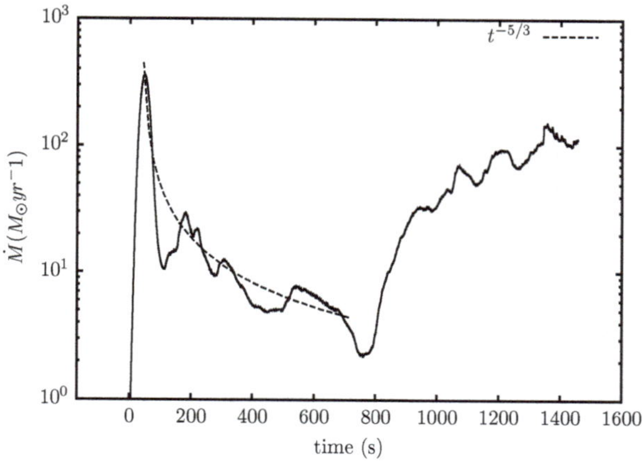

Fig. 1 Accretion rate of tidal debris through the BH horizon. Time has been chosen such that $t = 0$ s denotes periapsis passage. The *dashed line* is a reference to a $t^{-5/3}$ decay rate

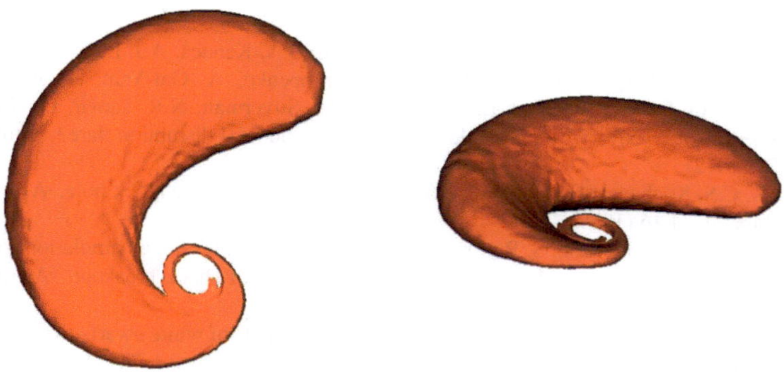

Fig. 2 3D snapshot of the debris at $t = 465$ s in Fig. 1. The *left panel* is face-on looking down into the orbital plane, and the *right panel* is at an inclination angle of 63°

Conclusions

We presented preliminary results from a class of TDEs that until now have not been considered. These are TDEs of main sequence stars disrupted by an intermediate mass BH with a mass $M_h = 10^5\,M_\odot$. The penetration factor is $\beta = 10$, putting the encounter near the *BH enters the star* limit. One of the main differences between the TDE encounter we have presented and the canonical encounters is that the time elapsed between disruption and the emergence of the flare is significantly reduced. That is, the disrupted material is accreted almost immediately after periapsis passage. Because general relativistic precession accelerates the circularization of the

(continued)

debris, the accretion disk forms few hundred seconds after disruption, cutting significantly the period of $t^{-5/3}$ flare-decay. The observed accretion rate through the BH horizon for these TDEs is also remarkably different from the canonical cases. If correlated with the luminosity, light curves from the TDE events we are considering provide clearly distinguishable, and thus unique, signatures for the presence of intermediate mass BHs.

Acknowledgements Work supported by NSF grants 1205864, 1212433, 1333360. Computations at XSEDE PHY120016 and the Cygnus cluster at Georgia Tech.

References

B. Carter, J.P. Luminet, Pancake detonation of stars by black holes in galactic nuclei. Nature **296**, 211–214 (1982)

S.B. Cenko, H.A. Krimm, A. Horesh, A. Rau, D.A. Frail, J.A. Kennea, A.J. Levan, S.T. Holland, N.R. Butler, R.M. Quimby, J.S. Bloom, A.V. Filippenko, A. Gal-Yam, J. Greiner, S.R. Kulkarni, E.O. Ofek, F. Olivares E., P. Schady, J.M. Silverman, N.R. Tanvir, D. Xu. Swift J2058.4+0516: discovery of a possible second relativistic tidal disruption flare? Astrophys. J. **753**, 77 (2012)

C.R. Evans, C.S. Kochanek, The tidal disruption of a star by a massive black hole. Astrophys. J. Lett. **346**, L13–L16 (1989)

J. Guillochon, E. Ramirez-Ruiz, S. Rosswog, D. Kasen, Three-dimensional simulations of tidally disrupted solar-type stars and the observational signatures of shock breakout. Astrophys. J. **705**, 844–853 (2009)

R. Haas, R.V. Shcherbakov, T. Bode, P. Laguna, Tidal disruptions of white dwarfs from ultra-close encounters with intermediate-mass spinning black holes. Astrophys. J. **749**, 117 (2012)

L. Hernquist, N. Katz, TREESPH - a unification of SPH with the hierarchical tree method. Astrophys. J. Suppl. **70**, 419–446 (1989)

J.G. Hills, Possible power source of Seyfert galaxies and QSOs. Nature **254**, 295–298 (1975)

P. Laguna, W.A. Miller, W.H. Zurek, M.B. Davies, Tidal disruptions by supermassive black holes - hydrodynamic evolution of stars on a Schwarzschild background. Astrophys. J. **410**, L83–L86 (1993)

A.J. Levan, N.R. Tanvir, S.B. Cenko, D.A. Perley, K. Wiersema, J.S. Bloom, A.S. Fruchter, A.D.U. Postigo, P.T. O'Brien, N. Butler, A.J. van der Horst, G. Leloudas, A.N. Morgan, K. Misra, G.C. Bower, J. Farihi, R.L. Tunnicliffe, M. Modjaz, J.M. Silverman, J. Hjorth, C. Thöne, A. Cucchiara, J.M.C. Cerón, A.J. Castro-Tirado, J.A. Arnold, M. Bremer, J.P. Brodie, T. Carroll, M.C. Cooper, P.A. Curran, R.M. Cutri, J. Ehle, D. Forbes, J. Fynbo, J. Gorosabel, J. Graham, D.I. Hoffman, S. Guziy, P. Jakobsson, A. Kamble, T. Kerr, M.M. Kasliwal, C. Kouveliotou, D. Kocevski, N.M. Law, P.E. Nugent, E.O. Ofek, D. Poznanski, R.M. Quimby, E. Rol, A.J. Romanowsky, R. Sánchez-Ramírez, S. Schulze, N. Singh, L. van Spaandonk, R.L.C. Starling, R.G. Strom, J.C. Tello, O. Vaduvescu, P.J. Wheatley, R.A.M.J. Wijers, J.M. Winters, D. Xu. An extremely luminous panchromatic outburst from the nucleus of a distant galaxy. Science **333**, 199 (2011)

G. Lodato, A.R. King, J.E. Pringle, Stellar disruption by a supermassive black hole: is the light curve really proportional to $t^{-5/3}$? MNRAS **392**, 332–340 (2009)

J.-P. Luminet, J.-A. Marck, Tidal squeezing of stars by Schwarzschild black holes. MNRAS **212**, 57–75 (1985)

J.-P. Luminet, B. Pichon, Tidal pinching of white dwarfs. Astron. Astrophys. **209**, 103–110 (1989)

E.S. Phinney, Manifestations of a massive black hole in the galactic center, in *The Center of the Galaxy*, ed. by M. Morris. IAU Symposium, vol. 136 (1989), p. 543

M.J. Rees, Tidal disruption of stars by black holes of 10 to the 6th-10 to the 8th solar masses in nearby galaxies. Nature **333**, 523–528 (1988)

S. Rosswog, Relativistic smooth particle hydrodynamics on a given background spacetime. Class. Quantum Grav. **27**(11), 114108 (2010)

S. Rosswog, E. Ramirez-Ruiz, W.R. Hix, Atypical thermonuclear supernovae from tidally crushed white dwarfs. Astrophys. J. **679**, 1385–1389 (2008)

R.-F. Shen, C.D. Matzner, Evolution of accretion disks in tidal disruption events. Astrophys. J. **784**, 87 (2014)

J.L. Tonry. Evidence for a central mass concentration in M32. Astrophys. J. Lett. **283**, L27 (1984)

Rival Families: Waveforms from Resonant Black-Hole Binaries as Probes of Their Astrophysical Formation History

Davide Gerosa

Abstract Astrophysical binary black holes formed following the successive core collapses of sufficiently massive binary stars are likely to be resonant. Post-Newtonian theory predicts the existence of two one-parameter families of equilibrium solutions ("resonances") in which the angular momentum and both spins share a common plane and precess at the same frequency. The two families are differentiated by either aligned or anti-aligned spin components in the orbital plane but both resonances have the capacity to attract generic non-resonant configurations. We develop astrophysical formation models showing that the fraction of binary black holes in each family is neatly related to the main stages of their formation history. Moreover, the gravitational-wave signals emitted from the two families are qualitatively different. Resonant binaries can be efficiently distinguished in events with signal-to-noise ratios ~ 10, typical of those expected for the first detections with Advanced LIGO/Virgo. Spin-orbit resonances thus consist in powerful, viable, probes of astrophysical processes in stellar-mass black-hole binary formation and evolution.

1 Stellar-Mass Black-Hole Binaries and the Key Role of Gravitational-Wave Astronomy

Stellar-mass black-hole (BH) binaries are the final endpoints of massive binary-star evolution. Such elusive objects have never been observed: stellar-mass BH binaries are electromagnetically invisible even in their merger phase. Unlike mergers involving neutron stars, where several solar masses of baryons are available to radiate after the collapse, accretion emission from BH binaries will be Eddington limited. The Eddington luminosity for several solar mass BHs ($\sim 10^{39}$ erg) is small enough to

D. Gerosa (✉)
Department of Applied Mathematics and Theoretical Physics, Centre for Mathematical Sciences, University of Cambridge, Wilberforce Road, Cambridge CB3 0WA, UK
e-mail: d.gerosa@damtp.cam.ac.uk

C.F. Sopuerta (ed.), *Gravitational Wave Astrophysics*, Astrophysics and Space Science Proceedings, DOI 10.1007/978-3-319-10488-1_12

make them very difficult to be observes at extragalactic distance, and the expected galactic merger rate is only $\sim$0.1 Myr^{-1} (Abadie et al. 2010). The first detection of a double stellar-mass BH will likely come from extragalactic gravitational-wave (GW) astronomy.

Gravitational radiation from stellar-mass BHs is mainly expected from coalescing binary systems, when one or both of the components is a BH. The larger mass of BH systems makes them visible in GWs from greater distances than neutron-star binaries: the horizon distance for the (Initial) Advanced LIGO/Virgo network is expected to be as large as ($\sim$160) $\sim$2,200 Mpc, twice the one expected for neutron-star binaries (Abadie et al. 2010). BHs require higher-mass progenitor stars and are therefore formed more rarely than neutron stars in Nature. However, the spatial abundance of binary systems with BHs is amplified relative to neutron-star binaries, because binary systems are much more easily disrupted by neutron star formation than by BH formation. When a neutron star forms, most of the progenitor-star mass ($6M_{\odot}$ or more) must be expelled from the system rapidly to produce a $\sim 1.5M_{\odot}$ object. Due to the stronger gravity of BHs, material is expected to fall back onto the core, which typically reduces the kick received by the newly formed compact object. Realistic expected rates for the Advanced LIGO/Virgo network for BH-BH binaries detections are as high as 20 year^{-1}, but even $\sim$1000 binaries per year for realistic astrophysical models (Abadie et al. 2010). With the upcoming advent of GW astronomy, statistical studies on binary BHs will be possible: information encoded in the dynamics of such binaries will be unveiled, thus opening a new window on the (largely unknown) physics of massive binary-star evolution.

Our contribution on the subject is mainly based on Gerosa et al. (2013, 2014). We tackle two questions:

- Do BH binaries retain memory of their formation history when they enter the GW sensitivity band of ground-based detectors? We develop an astrophysical model to describe the formation and the evolution of stellar-mass BH binaries. We find that precise astrophysical processes acting when the system is formed by massive stars are neatly related to the dynamics of the BH binaries as GW sources. We identify post-Newtonian spin-orbit resonances as the key dynamical features to encode information on the physical process that drive the binary evolution.
- Are such dynamical signatures observable with ground-based GW detectors? We analyze the GW signals emitted by BH binaries formed through different channels using the Advanced LIGO noise specifications. Waveforms from binaries in two different ("rival") resonant configurations are qualitatively different and can be distinguished in sources with sufficient signal-to-noise ratio (SNR).

2 Constraining Black-Hole Binary Formation

Astrophysical (hairless) BHs radiate away all their physical properties but their mass and their spins. This extreme simplicity implies that only those two quantities are available to store memory on how BHs formed. In order to finally become a compact binary, a binary star has to go through two supernova explosions. Mass and neutrino emission during supernova events is believed to be strongly asymmetric, thus imparting a natal kick on the newly formed compact object. Such kicks tilt the orbital plane: if rotating BHs are formed, the direction of their spins can be used to track the evolution of the binary orbital plane. This is the main idea behind the present contribution: spins (and spin directions in particular) encode valuable information on the evolutionary stage of compact binaries as GW sources.

The inspiral of compact binaries consists in an interplay between astrophysics and general relativity. GW emission causes a circular binary with a semimajor axis less than

$$a_{\rm GW} = 45\left(\frac{M}{10M_{\odot}}\right)^{3/4} \frac{q^{1/4}}{(1+q)^{1/2}}\, R_{\odot}\,, \tag{1}$$

to merge on a timescale less than the Hubble time $\sim 10\,\mathrm{Gyr}$ (Peters and Mathews 1963, where M is the binary total mass and $q \leq 1$ the mass ratio). If BH mergers occurs in Nature, astrophysical processes are required in order to shrink the binary down to separations smaller than $a_{\rm GW}$. From that point on, the evolution of the binary can be described using the post-Newtonian (PN) approximation. During the PN evolution, both the spins $\mathbf{S_i}$ ($i = 1, 2$) and the orbital angular momentum $\mathbf{L}$ of the binary precess about the total angular momentum $\mathbf{J}$. The orbital plane direction changes with time introducing modulations in the emitted GW signals. Such modulations are absent where non-spinning binaries are considered and consist in a precious richness to recover the binary parameters, including the spin orientations.

The manner in which spin precession alters the distribution of BH-binary spins can best be understood by appreciating the influence of PN spin-orbit resonances (Schnittman 2004). Those are special spin configurations in which the angular momentum and both spins share the same plane jointly precessing at the same frequency (hence the name "resonances"). Let us define θ_i to be the angle between $\mathbf{S}_i$ and $\mathbf{L}$; θ_{12} to be the angle between $\mathbf{S}_1$ and $\mathbf{S}_2$; and $\Delta\Phi$ to be the angle between the projections of the two spins onto the orbital plane. These spin-orbit resonances are divided between two rival families: resonances in which the spin components in the orbital plane are aligned ($\Delta\Phi = 0°$) and those in which these components are antialigned ($\Delta\Phi = \pm 180°$). Binaries near a resonant solution will be influenced by its presence, with $\Delta\Phi$ librating about either 0° or ±180° rather than evolving through the full range $[-180°, 180°]$. Resonant solutions act as attractors, and can profoundly modify the binary dynamics. PN spin-orbit couplings become important at separations (Kesden et al. 2010)

$$a_{\rm PNi} \sim 10^3 \frac{GM}{c^2} \simeq 10^{-2} \left(\frac{M}{10M_\odot}\right) R_\odot\,, \tag{2}$$

below which they can lock binaries into resonant configurations. The peak GW-frequency emitted from a compact binary is given by twice the orbital frequency $f_{\rm GW} = (GM/\pi^2 a^3)^{1/2}$. We expect resonant locking to be important at separations above

$$a_{\rm LIGO} \simeq 10^{-3} \left(\frac{M}{10M_\odot}\right)^{1/3} \left(\frac{f_{\rm GW}}{20\,{\rm Hz}}\right)^{-2/3} R_\odot\,, \tag{3}$$

at which the binary reaches the lower limit $f_{\rm GW} \simeq 10$–$20\,$Hz of the Advanced LIGO/Virgo sensitivity band. The final spin configuration at $a_{\rm LIGO}$ depends on the initial conditions with which binaries enter the PN regime at $\sim a_{\rm PNi}$. These are in turn related to the astrophysical processes that lead the binary to $a_{\rm GW}$. A unified treatment of the astrophysical initial conditions and the subsequent PN evolution is essential to determining which spin configurations are most relevant for GW detectors.

We develop a simple toy model to relate the spin angles at $a_{\rm PNi}$ with the main astrophysical processes that affect the spin directions at early stages; such angles are then used as initial data for numerical evolutions of the PN equation of motions to finally predict the spin configurations when binaries become detectable in GWs. Our spin evolutionary model is detailed in Gerosa et al. (2013). Three key processes leave neat imprints on the BH spin dynamics:

- When the system is formed by two massive stars, mass transfer events may change the binary mass ratio. The primary (more massive) star evolves more quickly; it can first fill its Roche lobe and transfer mass to the secondary. If a large amount of mass is transferred, the mass ratio of the binary may be reversed during the process: the initially less massive stars will end up forming the more massive BH and viceversa. On the other hand, moderate mass transfer preserves a "standard" mass ratio: the initially more massive star forms the more massive BH.
- Both SN explosions cause a tilt of the orbital plane. Angular-momentum conservation converts the asymmetry of the explosions into orbital plane tilts (Kalogera 2000). SN kicks can introduce an initial misalignment between the spins and the orbital angular momentum ($\theta_i \neq 0$), which will result in spin precession during the GW-driven inspiral phase.

- Between the two explosions, the system is naturally asymmetric: an already formed BH is orbiting together with a still-to-explode star. Such asymmetry is reflected by tidal interactions, which can only align the star's spin leaving the BH spin direction unaffected (Eggleton 2006). The system is likely to approach the second supernova, and therefore a second orbital-plane tilt, with only one spins misaligned with **L**. This can be either the primary or the secondary spin, depending on whether the mass ratio has previously been reversed. Binaries enter the PN evolution with different tilt angles ($\theta_1 \neq \theta_2$), which is exactly the condition to have strong resonant locking effects (Kesden et al. 2010). If tides are inefficient, binaries keep $\theta_1 \sim \theta_2$ at $a_{\rm PNi}$.

Our results are reported schematically in Fig. 1. The distribution of the azimuthal angle difference $\Delta\Phi$ is in a one-to-one correspondence with different BH formation channels. Whenever tides are efficient, the PN evolution attracts the spins and the orbital angular momentum in the same resonant plane with $\Delta\Phi \simeq 0°$ ($\Delta\Phi \simeq \pm 180°$) if the mass ratio has (has not) been reversed by mass transfers. When tidal interactions are inefficient, momenta precess freely with a moderate pile up at $\Delta\Phi \simeq \pm 90°$. Such findings are illustrated quantitatively in Fig. 2, where histograms of $\Delta\Phi$ are presented at different times during the PN evolutions. Distributions are flat in all scenarios at early times, where PN interactions are weak. As the evolution proceeds, *binary's memories grow stronger*: by the time binaries are detectable ($f_{\rm GW} = 20$ Hz for LIGO/Virgo, $f_{\rm GW} = 1$ Hz for the third-generation detectors), the evolved distribution of $\Delta\Phi$ becomes a neat indicator of the underlying formation scenarios.

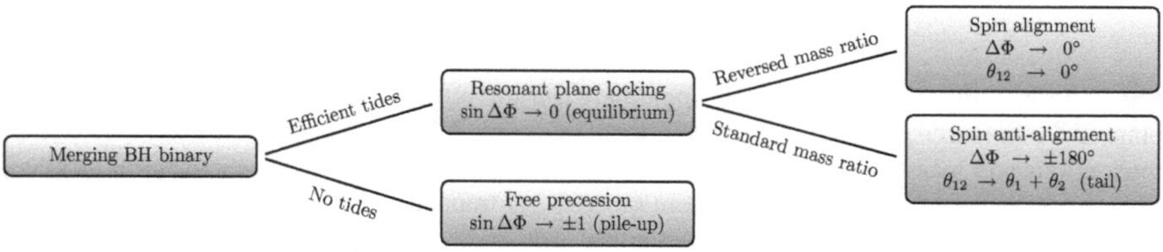

Fig. 1 Schematic summary of our predictions for the spin orientation of BH binaries as they become detectable in GWs. The efficiency of tidal interactions during the formation of the BH binaries cause an almost perfect resonant locking by the time binaries become detectable ($\sin\Delta\Phi = 0$, where $\Delta\Phi$ is the difference between the spin azimuthal angles). Only free precession is detected when tidal interactions are artificially turned off, with moderate pile-up around $\sin\Delta\Phi \to \pm 1$ at late times. The dynamics of resonant binaries is further constrained by mass transfer events which select one of the two resonant families ($\Delta\Phi = 0°$ or $\pm 180°$). A tail at large spin-spin angles θ_{12} is only present if the binary mass ratio has not been reversed during the binary evolution (Gerosa et al. 2013)

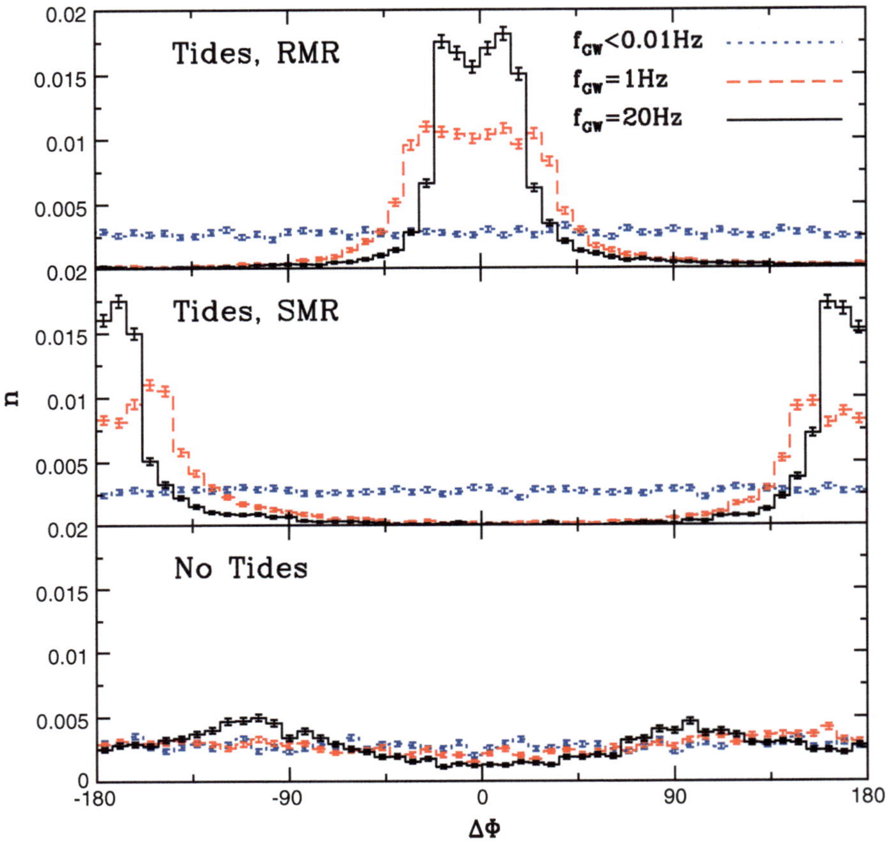

Fig. 2 Probability distribution of the angle between the projections of the spins onto the orbital plane $\Delta\Phi$. The three different scenarios outlined in our model (depending on the efficiency of tidal interactions and mass transfers), present orthogonal distributions in $\Delta\Phi$. As the inspiral proceeds to higher GW frequencies, more and more binaries get trapped into spin-orbit resonances in both "Tides" scenarios: binaries cluster around either $\Delta\Phi = 0°$ if their mass ratio has been reversed (RMR) or $\Delta\Phi = \pm 180°$ if it has been maintained standard (SMR). Such effect is completely absent in a formation scenario where tides are artificially removed: binaries freely precess spanning the whole range $\Delta\Phi$, with a moderate pile-up at $\Delta\Phi = \pm 90°$. This figure has been prepared considering maximally spinning BH binaries with $q = 0.8$ and $M = 13.5 M_{\odot}$ (Gerosa et al. 2013)

3 Telling the Two Resonant Families Apart

Information on the astrophysical formation mechanisms is encoded in the gravitational signals emitted by resonant binaries. As a second step of our study, we address the distinguishability of the two resonant families in Advanced LIGO-type detectors. If the two rival families are distinguishable, this would open a new window on probing astrophysical formation processes for stellar-mass BH binaries.

Figure 3 shows typical SNRs ρ per unit frequency as expected for detections by a single second-generation ground-based detector when source binaries are assumed to be in resonant configurations. Various panels show the expected SNR for the otherwise identical resonant binaries in each family (top/bottom row for $\Delta\Phi = 0°/180°$ respectively) when viewed at different inclination angles $\cos\iota = \hat{\mathbf{L}}\cdot\hat{\mathbf{n}}$ (where $\hat{\mathbf{n}}$ is the line of sight). Spinning-binary dynamics presents orbital plane precession, which is reflected in wide SNR modulations. However, these precessional oscillations are much more pronounced for binaries in the $\Delta\Phi = 0°$ family (where the orbital-plane components of the spins are aligned) than those in the $\Delta\Phi = 180°$ family. From this very simple kinematical property, we expect waveforms from the two rival families to be qualitatively different. As shown in the first column of Fig. 3, modulations are minimized if the line of sight $\hat{\mathbf{n}}$ is chosen to lie along the total angular momentum of the binary $\hat{\mathbf{J}}$ because its direction is almost conserved in the weak-field regime. We adopt this strongly conservative assumption where all geometrical effects due to the viewing directions cancel. Even in this unlucky case, we find that a sufficient percentage of binaries is distinguishable.

The key quantity needed to perform comparisons between a source waveform h_s and template h_t (which, in our case belong to either of the two different resonant families) is the *overlap*

$$\mathcal{O}(h_s, h_t) \equiv \max_{t_c,\phi_c} \frac{(h_t|h_s)}{\sqrt{(h_s|h_s)(h_t|h_t)}}, \quad (4)$$

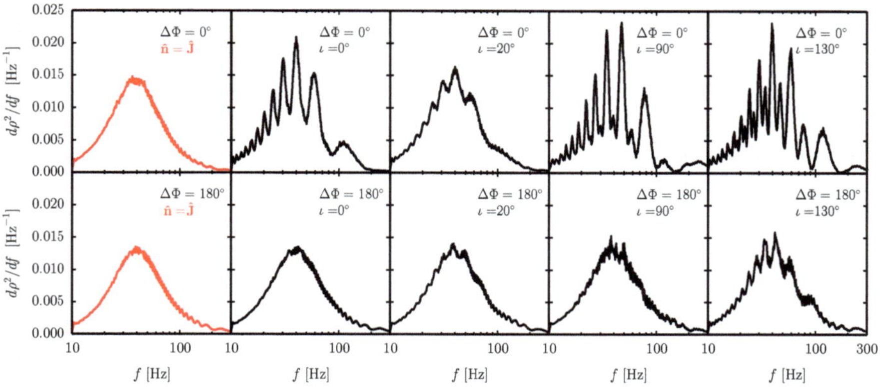

Fig. 3 Expected squared SNR per unit frequency for resonant binaries. The same resonant sources are viewed from different angles ι. Wider modulations are presents in the $\Delta\Phi = 0°$ resonance (*top row*) because of the greater precession of the orbital plane. The smaller orbital-plane component of the total spin $\mathbf{S_1} + \mathbf{S_2}$ suppress the SNR modulation in the $\Delta\Phi = 180°$ family (*bottom row*). This qualitative difference allow for the possibility to tell the two rivals apart. These (helpful) modulations are suppressed in the conservative assumption where the line of sight $\hat{\mathbf{n}}$ is taken to lie along the total angular momentum $\mathbf{J}$ (Gerosa et al. 2014)

the normalized inner product of the two responses maximized over the arrival time t_c and the phase at coalescence ϕ_c. Two waveforms can be distinguished when their overlap is sufficiently small $\mathcal{O} \lesssim 1-\rho^{-2}$ (Lindblom et al. 2008). Our final results are summarized in Fig. 4. For a given SNR ρ, we compute the number of binaries which present a maximum overlap with the other resonant family $\mathcal{O} < \mathcal{O}_{\max} = 1 - \rho^{-2}$. We use source distributions as predicted by our spin evolutionary models (here in Fig. 2). Figure 4 can therefore be interpreted as the fraction of binaries that will be distinguished with Advanced LIGO observation at SNR ρ. The three distributions are readily distinguishable by eye, although many sources have both spins too aligned with $\hat{\mathbf{L}}$ to be separated. When 100 sources will be detected at $\rho = 10$, about ~20 of them would be expected to be found in the $\Delta\Phi = 180°$ family for SMR-formed binaries; about ~15 would be found in the rival $\Delta\Phi = 180°$ family for the RMR scenario; but only a few would be found in *both* family in the "No Tides" case.

We are aware of the many limitations of this study, and much work still remains to be done: in particular, we are only using 1-dimensional template banks spanning over the resonant solutions, while all other intrinsic parameters are kept fixed. Nonetheless, our strongly conservative assumption $\hat{\mathbf{J}} = \hat{\mathbf{n}}$ eliminates all the geometrical effects like the precession of the orbital plane (Fig. 3), whose richness in the signal has already been claimed to be critical for GW physics.

To summarize, different astrophysical mechanisms in the progenitor systems (such as mass transfer, tidal interactions and supernova kicks) are neatly related to the spin dynamics of the remnant BH binaries. Such dynamics can be better

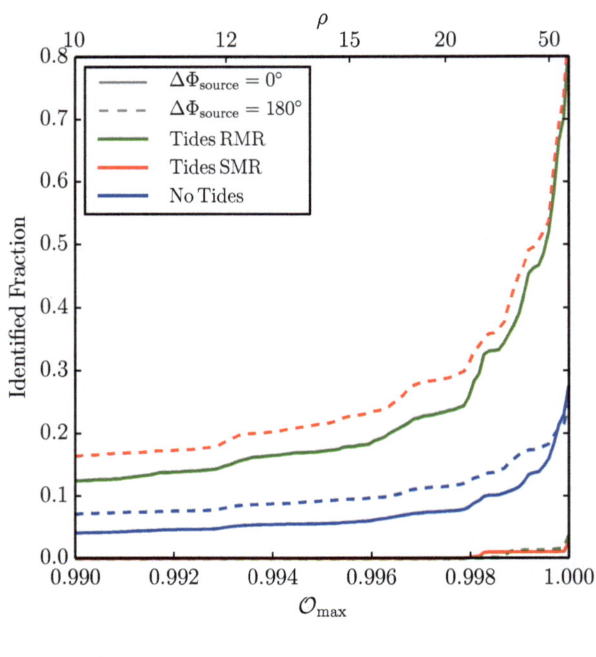

Fig. 4 Fraction of identified binaries from astrophysical distributions predicted by our evolutionary models (here in Fig. 2), as a function of their SNR ρ and maximum overlap $\mathcal{O}_{\max} = 1 - \rho^{-2}$. In both "Tides" cases, all the identified binaries belong to either the $\Delta\Phi = 0°$ family (RMR) or the $\Delta\Phi = 180°$ family (SMR) scenario. On the other hand, a comparable fraction of identified binaries belongs to both family in the "No Tides" scenario (Gerosa et al. 2014)

understood when Schnittman's (2004) resonant families are considered: two families are present, depending on the orientation of the orbital-plane components of the spins. Moreover, GW signals from the two rival resonant families are qualitatively different from each other because orbital plane precession is different. The Advanced LIGO/Virgo network is expected to detect ∼20 BH binaries per year (Abadie et al. 2010): our preliminary study shows that a few years of observations have the chance to revolutionize our understanding of astrophysical binary BH formation. With the upcoming advent of GW astronomy the electromagnetically invisible BH binaries will not only start *shining*, but the physics behind their formation will also be ready to be observed.

Acknowledgements The work contained in the present contribution has been done in collaboration with Emanuele Berti, Michael Kesden, Richard O'Shaughnessy and Ulrich Sperhake. DG is supported by the UK Science and Technology Facility Council and the Isaac Newton Studentship of the University of Cambridge; partial support to attend the Sant Cugat Forum on Astrophysics is also acknowledged from Darwin College, University of Cambridge.

References

J. Abadie, B.P. Abbott, R. Abbott, M. Abernathy, T. Accadia, F. Acernese, C. Adams, R. Adhikari, P. Ajith, B. Allen et al., TOPICAL REVIEW: predictions for the rates of compact binary coalescences observable by ground-based gravitational-wave detectors. Class. Quantum Grav. **27**(17), 173001 (2010)

P. Eggleton, *Evolutionary Processes in Binary and Multiple Stars* Cambridge University Press, ISBN 0521855578 (July 2006)

D. Gerosa, M. Kesden, E. Berti, R. O'Shaughnessy, U. Sperhake, Resonant-plane locking and spin alignment in stellar-mass black-hole binaries: a diagnostic of compact-binary formation. Phys. Rev. D **87**(10), 104028 (2013)

D. Gerosa, R. O'Shaughnessy, M. Kesden, E. Berti, U. Sperhake, Distinguishing black-hole spin-orbit resonances by their gravitational wave signature. **89**, 124025 (2014)

V. Kalogera, Spin-orbit misalignment in close binaries with two compact objects. ApJ **541**, 319–328 (2000)

M. Kesden, U. Sperhake, E. Berti, Final spins from the merger of precessing binary black holes. Phys. Rev. D **81**(8), 084054 (2010)

L. Lindblom, B.J. Owen, D.A. Brown, Model waveform accuracy standards for gravitational wave data analysis. Phys. Rev. D **78**(12), 124020 (2008)

P. C. Peters, J. Mathews, Gravitational radiation from point masses in a keplerian orbit. Phys. Rev. **131**, 435–440 (1963)

J.D. Schnittman, Spin-orbit resonance and the evolution of compact binary systems. Phys. Rev. D **70**(12), 124020 (2004)

Pulsar Timing Arrays and the Challenge of Massive Black Hole Binary Astrophysics

A. Sesana

Abstract Pulsar timing arrays (PTAs) are designed to detect gravitational waves (GWs) at nHz frequencies. The expected dominant signal is given by the superposition of all waves emitted by the cosmological population of supermassive black hole (SMBH) binaries. Such superposition creates an incoherent stochastic background, on top of which particularly bright or nearby sources might be individually resolved. In this contribution I describe the properties of the expected GW signal, highlighting its dependence on the overall binary population, the relation between SMBHs and their hosts, and their coupling with the stellar and gaseous environment. I describe the status of current PTA efforts, and prospect of future detection and SMBH binary astrophysics.

1 Introduction

Pulsar timing arrays (PTAs) provide a unique opportunity to obtain the very first low-frequency gravitational wave (GW) detection. The European Pulsar Timing Array (EPTA) (Kramer and Champion 2013), the North American Nanohertz Observatory for Gravitational Waves (NANOGrav) (McLaughlin 2013), and the Parkes Pulsar Timing Array (PPTA) (Hobbs 2013), joining together in the International Pulsar Timing Array (IPTA) (Hobbs et al. 2010; Manchester and IPTA 2013), are constantly improving their sensitivity in the frequency range of $\sim 10^{-9}$–10^{-6} Hz, where inspiralling supermassive black hole (SMBH) binaries populating merging galaxies throughout the Universe are expected to generate a strong signal (Rajagopal and Romani 1995; Jaffe and Backer 2003; Wyithe and Loeb 2003; Sesana et al. 2004).

Despite the fact that theoretical models of galaxy formation in the standard hierarchical framework predict a large population of SMBH binaries forming during galaxy mergers, to date there is only circumstantial observational evidence of their

A. Sesana (✉)
Max-Planck-Institut für Gravitationsphysik, Albert Einstein Institut, Am Mülenber 1, 14476 Golm, Germany
e-mail: alberto.sesana@aei.mpg.de

C.F. Sopuerta (ed.), *Gravitational Wave Astrophysics*, Astrophysics and Space Science Proceedings, DOI 10.1007/978-3-319-10488-1_13

existence. About 20 SMBH pairs with separations of $\sim$10 pc to $\sim$10 kpc are known as of today (see Dotti et al. 2012, for a comprehensive review), among which a new discovered SMBH triple (Deane et al. 2014). PTAs are sensitive to more compact systems, namely sub-parsec bound Keplerian SMBH binaries. Only few candidates of this class have been identified, based on peculiar broad emission line shifts (Tsalmantza et al. 2011; Eracleous et al. 2011); however, alternative explanations to the binary hypothesis exist (Dotti et al. 2012), and unquestionable observational evidence of their binary nature is still missing.

In this contribution I review the properties of the GW signals relevant to PTAs. I describe the status of current PTA efforts, future detection prospects and related astrophysical payouts. If as abundant as predicted, SMBH binaries are expected to form a low frequency background of GWs with a typical strain amplitude $A \sim 10^{-15}$ at a frequency $f = 1\,\text{year}^{-1}$ (Sesana et al. 2008; Ravi et al. 2012; McWilliams et al. 2014), with a considerable uncertainty of $\approx$ 0.5 dex.[1] The aforementioned studies indicate that the signal is expected to be dominated by a handful of sources, some of which might be individually resolvable; a situation similar to the foreground generated by WD-WD binaries in the mHz regime relevant to spaced based interferometers like eLISA (Consortium 2013). On the one hand, the unresolved background provides ways to probe the overall population of SMBH binaries in the low redshift universe ($z < 1$); on the other hand, electromagnetic counterparts to individually resolvable sources can be searched for with a number of facilities opening new avenues toward a multimessenger based understanding of these fascinating systems and their hosts.

The manuscript is organized as follows. In Sect. 2, I introduce the general response of a PTA to a passing GW, focusing in Sect. 3 on the overall stochastic background generated by a population of SMBH binaries. In Sect. 4 I describe the influence of SMBH astrophysics (dynamical coupling with the environment, eccentricity evolution) on the expected signal, and the related issues that a putative PTA detection might answer in the near future. Section 5 is devoted to individually resolved sources and multimessenger astronomy. A summary of the main results is given in the section "Conclusions". A concordance Λ–CDM universe with $\Omega_M = 0.27$, $\Omega_\lambda = 0.73$ and $h = 0.7$ is assumed. Unless otherwise specified, equations are expressed in geometric units where $G = c = 1$.

2 Pulsar Timing Array Response to Gravitational Waves

GWs affect the propagation of radio signals from the pulsar to the receiver on Earth (Sazhin 1978), leaving a characteristic fingerprint in the time of arrival of the radio pulses (Hellings and Downs 1983; Jenet et al. 2005). A binary system of masses

[1]In astronomy, the notation dex is commonly used for the $\log_{10}$ unit; therefore 0.5 dex$= 10^{0.5}$.

$M_1 > M_2$ in circular orbit at a Keplerian frequency $f_K = (1/2\pi)\sqrt{(M_1 + M_2)/a^3}$ (a is the semimajor axis of the binary) generates a time dependent GW at a frequency $f = 2f_K$. In the quadrupolar approximation (sufficient to our scopes) the wave can be described in terms of two independent polarization amplitudes in the form (see, e.g., Sesana and Vecchio 2010):

$$\begin{aligned} h_+(t) &= A_{gw}a(\iota)\cos\Phi(t)\,, \\ h_\times(t) &= A_{gw}b(\iota)\sin\Phi(t)\,, \end{aligned} \tag{1}$$

where

$$A_{gw}(f) = 2\frac{\mathcal{M}^{5/3}}{D}\,[\pi f(t)]^{2/3} \tag{2}$$

is the GW amplitude, $\mathcal{M} = (M_1M_2)^{3/5}/(M_1 + M_2)^{1/5}$ is the chirp mass of the system, D the luminosity distance to the GW source, $\Phi(t)$ is the GW phase $\Phi(t) = 2\pi\int^t f(t')dt'$, and $f(t)$ the instantaneous GW frequency.[2] The two functions $a(\iota) = 1 + \cos^2\iota$ and $b(\iota) = -2\cos\iota$ depend on the source inclination angle ι.

When crossing the Earth-pulsar line of sight, such GW induces an 'effective redshift' in the frequency of the received pulses, associated to the oscillatory change in the spacetime background metric, which is proportional to the amplitude h of the wave:

$$z(t,\Omega) \equiv \frac{\nu(t) - \nu_0}{\nu_0} = F_+(\Omega)\Delta h_+(t) + F_\times(\Omega)\Delta h_\times(t). \tag{3}$$

Here $F_+(\Omega)$ and $F_\times(\Omega)$ are response functions that depend on the angle defined by the relative pulsar-source orientations in the sky, and the quantities $\Delta h_{+,\times}(t) = h_{+,\times}(t_p) - h_{+,\times}(t_E)$ take into account of the 'double response' of the pulsar-Earth system to the passing GW: what matters is the difference between the metric perturbation at the pulsar and the metric perturbation on Earth. The integral over the observation time of Eq. (3) results in a dephasing of the received pulse: i.e. *the GW causes the photons to arrive a little earlier or later than expected, inducing an effective 'residual' in the pulse time of arrival:*

[2]The frequency evolution of nHz SMBH binaries over an observing time of ~10 year is negligible Sesana and Vecchio (2010), we therefore assume monochromatic non-evolving sources and drop the time dependence in the frequency.

$$r(t) = \int_0^t dt' z(t', \Omega) \,. \tag{4}$$

Being the integral of a sinusoidal wave of amplitude h at a frequency f, the residual has a typical magnitude $r \sim h/(2\pi f)$. This relation can be used to conveniently convert strain amplitudes h (which we use in this contribution) into residuals r. To get a sense of the magnitude of the effect, the residual can be normalized to fiducial SMBH binary values as

$$r \simeq 30 \left(\frac{\mathcal{M}}{10^9\,M_\odot}\right)^{5/3} \left(\frac{D}{100\,\mathrm{Mpc}}\right)^{-1} \times \left(\frac{f}{5\times10^{-8}\,\mathrm{Hz}}\right)^{-1/3} \mathrm{ns}, \tag{5}$$

i.e. 100 ns or better timing precision is needed in order to make a detection.

Millisecond pulsars are the most stable natural clocks in the Universe. Nevertheless, timing them to 100 ns precision is challenging and several sources of uncontrolled noise might affect the process (Verbiest et al. 2009; Shannon and Cordes 2010; Cordes 2013). In particular, red noise processes that cannot be perfectly modeled and subtracted can mimic a low frequency GW signal. It is therefore necessary to monitor an ensemble, or array, of pulsars—which is the very essence of the concept of pulsar timing array (Foster and Backer 1990). Only by observing a signal that is consistently cross-correlated among them (Hellings and Downs 1983) one can be sure of its GW nature as opposed to some intrinsic random noise process.

In observations with PTAs, radio-pulsars are monitored weekly for periods of years. The relevant frequency band is therefore between $1/T$—where T is the total observation time—and the Nyquist frequency $1/(2\Delta t)$—where Δt is the time between two adjacent observations—, corresponding to 3×10^{-9} Hz–few$\times10^{-7}$ Hz. The frequency resolution bin is $1/T$.

Given a cosmological population of SMBH binaries, the overall GW signal reaching the Earth is an incoherent superposition of monochromatic waves $\sum_i h_i$. Depending on the details of the population, such superposition can behave as an effective 'stochastic background', and can be dominated by few bright resolvable sources. This concept is represented in Fig. 1. For the selected realization of the SMBH population, the signal looks like a stochastic background at $f < 10^{-8}$ Hz; it is however dominated by bright resolvable sources at higher frequencies. We will first treat the overall signal as an effective background composed of unresolvable sources, we will then focus on the properties of individual sources.

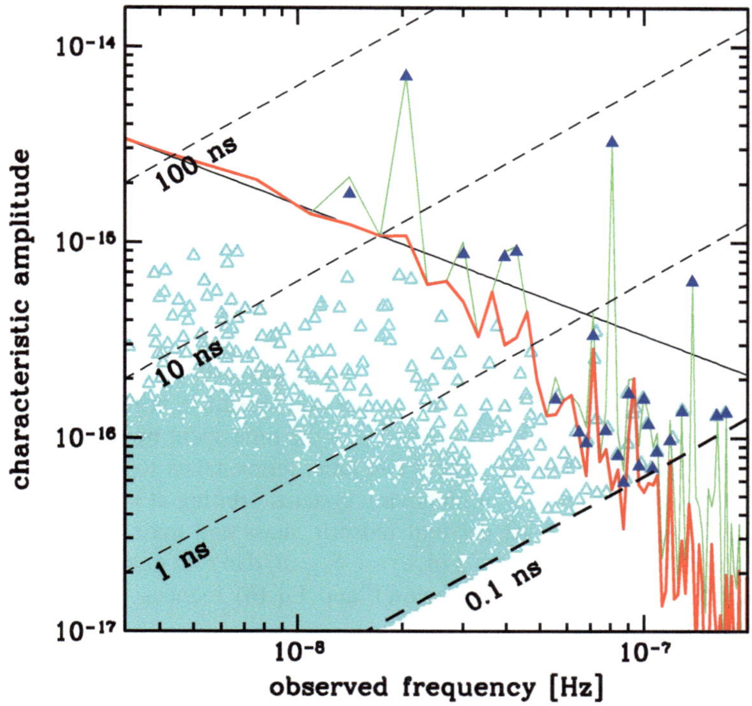

Fig. 1 Illustrative realization of the overall GW signal in the frequency domain; characteristic amplitude h_c vs frequency. Each *cyan triangle* corresponds to the contribution of an individual SMBH binary; among these, *blue triangles* identify bright, resolvable sources. The overall GW signal is given by the *jagged green line*, whereas the *red thick line* represents the unresolved signal after subtraction of the brightest sources. The *solid black line* represents the theoretical $h_c \propto f^{-2/3}$ behavior, and *dashed lines* mark characteristic residual levels according to the conversion $r = h/(2\pi f)$, as labeled in figure (Color figure online)

3 The Stochastic GW Background

We now consider a cosmological population of inspiralling SMBH binaries. Without making any restrictive assumption about the physical mechanism driving the binary semimajor axis—a—and eccentricity—e—evolution, we can write the characteristic amplitude h_c of the GW signal generated by the overall population as (Sesana 2013a):

$$h_c^2(f) = \int_0^\infty dz \int_0^\infty dM_1 \int_0^1 dq \frac{d^4N}{dz dM_1 dq dt_r} \frac{dt_r}{d\ln f_{K,r}} \times$$
$$h^2(f_{K,r}) \sum_{n=1}^\infty \frac{g[n, e(f_{K,r})]}{(n/2)^2} \delta\left[f - \frac{n f_{K,r}}{1+z}\right]. \quad (6)$$

Here $h(f_{K,r})$ is the strain emitted by a circular binary at a Keplerian rest frame frequency $f_{K,r}$,[3] averaged over source orientations

$$h(f_{K,r}) = \sqrt{\frac{32}{5}}\frac{\mathcal{M}^{5/3}}{D}(2\pi f_{K,r})^{2/3}. \tag{7}$$

The function $g(n,e)$ (Peters and Mathews 1963) accounts for the fact that the binary radiates GWs in the whole spectrum of harmonics $f_{r,n} = nf_{K,r}$ $(n = 1,2,\ldots)$, and is given by, e.g., Eqs. (5)–(7) in Amaro-Seoane et al. (2010). The δ function, ensures that each harmonic n contributes to the signal at an observed frequency $f = nf_{K,r}/(1+z)$, where the factor $1+z$ is given by the cosmological redshift. $d^4N/(dzdM_1dqdt_r)$ is the differential cosmological coalescence rate (number of coalescences per year) of SMBH binaries per unit redshift z, primary mass M_1, and mass ratio $q = M_2/M_1 < 1$. $dt_r/d\ln f_{K,r}$ is the time spent by the binary at each logarithmic frequency interval. These two latter terms, taken together, simply give the instantaneous population of comoving systems orbiting at a given logarithmic Keplerian frequency interval per unit redshift, mass and mass ratio. In the case of circular GW driven binaries, $g(n,e) = \delta_{n2}$, $dt/d\ln f$ is given by the standard quadrupole formula (see Eq. (9) below), and Eq. (6) reduces to the usual form (Phinney 2001)

$$h_c^2(f) = \frac{4f^{-4/3}}{3\pi^{1/3}}\int\int dzd\mathcal{M}\,\frac{d^2n}{dzd\mathcal{M}}\frac{1}{(1+z)^{1/3}}\mathcal{M}^{5/3}, \tag{8}$$

where we have introduced the differential merger remnant density (i.e. number of mergers remnants per co moving volume) $d^2n/(dzd\mathcal{M})$ (see Phinney 2001; Sesana 2013b for details). In this case, $h_c \propto f^{-2/3}$; it is therefore customary to write the characteristic amplitude in the form $h_c = A(f/\mathrm{year}^{-1})^{-2/3}$, where A is the amplitude of the signal at the reference frequency $f = 1\mathrm{year}^{-1}$. Observational limits on the GW background are usually given in terms of A.

4 The GW Background as a Tool for SMBH Binary Astrophysics

Equation (6), together with a prescription for the eccentricity distribution of the emitting SMBH binaries as a function of frequency, namely $e(M_1, q, f_{K,r})$, provides the most general description of the GW background generated by a population of SMBH binaries. The signal depends on three distinctive terms:

---

[3]For a source at redshift z, the general relation between rest frame and observed frequency is $f_r = f(1+z)$.

1. the cosmological SMBH binary coalescence rate, $d^4N/(dzdM_1dqdt_r)$;
2. the frequency evolution of each binary, $dt_r/d\ln f_{K,r}$;
3. the eccentricity evolution of the systems, which determines the emitted spectrum at any given binary Keplerian frequency.

In the following section, we will examine the impact of the items listed above on the GW signal, highlighting the enormous potential of PTA observations to enhance our understanding of the global population and the dynamical evolution of SMBH binaries in the Universe.

4.1 Spectral Amplitude: The Cosmic SMBH Binary Merger Rate

We first consider GW-driven, circular SMBH binaries only. It is clear by looking at Eq. (8) that the GW strain amplitude is proportional to the square root of the cosmic coalescence rate of SMBH binaries, and it is sensitive to their mass distributions. Therefore, the coalescence rate sets the *normalization* of the detectable signal. As discussed in Sesana (2013b) this, in practice, depends on four ingredients: (1) the galaxy merger rate; (2) the relation between SMBHs and their hosts, (3) the efficiency of SMBH coalescence following galaxy mergers and (4) when and how accretion is triggered during a merger event. Our limited knowledge of each of these ingredients is reflected in the uncertainty range of the predicted GW signal amplitude.

The expected level of the signal under the circular GW-driven approximation has been investigated by several authors (Sesana et al. 2008; Ravi et al. 2012; Sesana 2013b; Ravi et al. 2014; McWilliams et al. 2014), producing largely consistent results that cluster around the range $5 \times 10^{-16} < A < 2 \times 10^{-15}$. In particular Sesana and collaborators demonstrated that different methodologies that can be used in the computation of the signal yield consistent results (Sesana et al. 2008, 2009; Sesana 2013b). Ravi and collaborators corroborated those findings with independent investigations based on the Millennium Run (Springel et al. 2005), and by constructing an alternative observation-based model (Ravi et al. 2014). McWilliams et al. (2014) finds a somewhat larger signal, of the order $A \sim 4 \times 10^{15}$. This may be ascribed to a set of assumption, including a non evolving SMBH density with cosmic time and a merger driven only evolution of the SMBH mass function, that tend to favor more and more massive SMBH binary mergers.

Figure 2 compares the predicted GW amplitude to current observations. Here we plot 68, 95 and 99.7 % confidence level of the expected characteristic strain, extracted from a large compilation of models featuring different prescriptions for the SMBH binary population, as described in Sesana (2013b). In particular, we consider here the SMBH-bulge scaling relations derived in McConnell and Ma (2013), which

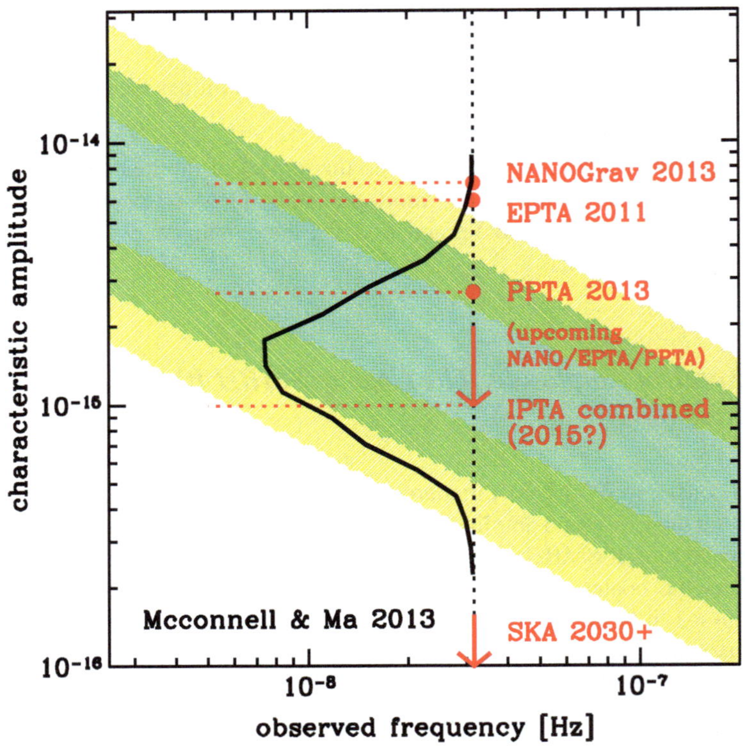

Fig. 2 Characteristic amplitude of the GW signal. *Shaded areas* represent the 68, 95 and 99.7 % confidence levels given by a representative set of observation-based models from Sesana (2013b), and the overlied *solid black line* traces the probability distribution function of A (i.e. h_c at $f = 1\text{year}^{-1}$). The *red dots* mark the progression of PTA limits from van Haasteren et al. (2011), Demorest et al. (2013), Shannon et al. (2013). The *two red arrows* anticipate the expected sensitivity of upcoming combined IPTA datasets, and of the full SKA (Dewdney et al. 2009) (Color figure online)

are representative of a recent upward revision of those relations (Kormendy and Ho 2013). It is worth mentioning how the most stringent upper limit published to date ($A > 2.4 \times 10^{15}$ at 95 % confidence, Shannon et al. 2013) is skimming the predicted 68 % confidence region. Upcoming limits from all the PTAs might soon push the numbers down by few dex to $A \approx 2 \times 10^{15}$, and the combined IPTA dataset might be sensitive to A as small as 10^{-15}, implying a good chance of detection in the next few years.

The first, obvious payout of a PTA detection is the *direct* confirmation of the existence of a vast population of sub-pc (to be precise, sub-0.01 pc) SMBH binaries. From this we learn that (1) binaries efficiently pair on pc scales following galaxy mergers and (2) stellar and/or gas dynamics is effective in removing energy and angular momentum from the binary, overcoming the 'last parsec problem'. It can still be the case that some level of stalling may occur, delaying the efficient energy loss due to GW emission. Nevertheless, a direct PTA detection will confirm that SMBH binaries eventually coalesce within an Hubble time. The exact level of the signal then depends on the combined ingredients outlined above, and those cannot be constrained all together by measuring one single number. It is even more dangerous to try to draw conclusions about SMBH coalescence rates based on PTA upper limits. In the following section we will discuss how environmental coupling can result in a loss of low frequency GW power; a non-detection at a given level might be either due to a particularly low coalescence rate, or to a particularly efficient coupling to the environment.

4.2 Spectral Shape: The SMBH Binary-Environment Coupling

SMBH binaries do not evolve in isolation, and although the circular GW-driven approximation has been widely employed to make predictions, GW emission alone cannot bring a SMBH binary to coalescence. This is clear by looking at the 'residence time' $dt_r/d\ln f_r$, i.e., the time a binary spend at each logarithmic frequency bin. In the circular GW-driven approximation, this is

$$\frac{dt_r}{d\ln f_r} = \frac{5}{64\pi^{8/3}}\mathcal{M}^{-5/3} f_r^{-8/3}. \tag{9}$$

—note that this, combined to Eq. (7), yields $h_c \propto M_1^{5/6} q^{1/2} f^{-2/3}$, i.e. the standard $f^{-2/3}$ power law given by Eq. (8). We note the steep frequency dependence of Eq. (9): as a matter of fact, typical binary residence times become longer than the Hubble time for $f < 10^{-10}$ Hz (roughly semimajor axis $a > 0.1$ pc for the $M \approx 10^9 M_\odot$ systems relevant here). It is therefore clear that some mechanism other then GW emission must bring the binary to subparsec separations.

Forming after galaxy mergers, SMBH binaries sit at the center of the stellar bulge of the remnant, and they are possibly surrounded by massive gas inflows triggered by dynamical instabilities related to the strong variations of the gravitational potential during the merger episode. Accordingly, two major routes for the SMBH binary dynamical evolution have been explored in the literature: (1) gas driven binaries, and (2) stellar driven binaries. A detailed description of both scenarios is beyond the scope of this contribution; here we consider simple evolutionary routes and assess their impact on the GW signal.

4.2.1 Change in the SMBH Binary Shrinking Rate

Let restrict ourselves to circular binaries first. A background of stars scattering off the binary drives its semimajor axis evolution according to the equation (Quinlan 1996)

$$\frac{da}{dt} = \frac{a^2 G\rho}{\sigma} H, \tag{10}$$

where ρ is the density of the background stars, σ is the stellar velocity dispersion and H is a numerical coefficient of order 15 (Quinlan 1996). Although Eq. (10) has been originally derived for static uniform backgrounds, recent numerical simulations (Khan et al. 2011; Preto et al. 2011) suggest that it can be applied to realistic systems providing that the density ρ is evaluated at the binary influence radius ($r_i \approx GM/\sigma^2$). If we consider, for simplicity, an isothermal sphere (i.e. $\rho \propto r^{-2}$), we substitute ρ_i in Eq. (10), and we assume $M_{BH} \propto \sigma^5$, we get $dt/d\ln f \propto f^{2/3} M_1^{2/3}$, which yields to a contribution of the single binary to the GW background of the form $h_c \propto M_1^2 q f$.

In the case of circumbinary disks, things are even more subtle, and the detailed evolution of the system depends on the complicated and uncertain dissipative physics of the disk itself. The simple case of a coplanar prograde disk, with a central cavity maintained by the torque exerted by the binary onto the disk admits a selfconsistent, non stationary solution that can be approximated as (Ivanov et al. 1999; Haiman et al. 2009)

$$\frac{da}{dt} = \frac{2\dot{M}}{\mu}(aa_0)^{1/2}. \tag{11}$$

Here, $\dot{M}$ is the mass accretion rate at the outer edge of the disk, a_0 is the semimajor axis at which the mass of the unperturbed disk equals the mass of the secondary black hole, and μ is the reduced mass of the binary. Considering a standard geometrically thin, optically thick disk model (Shakura and Sunyaev 1973), one finds $dt/d\ln f \propto f^{-1/3} M_1^{1/6}$, which yield to a contribution of the single binary to the GW background of the form $h_c \propto M_1^{7/4} q^{3/2} f^{1/2}$.

When compared to the GW driven case, $(da/dt)_{gw} \propto a^{-3}$, Eqs. (10) and (11) provide the transition frequency between the external environment driven and the GW driven regimes:

$$f_{\rm star/gw} \approx 5 \times 10^{-9} M_8^{-7/10} q^{-3/10}\,\mathrm{Hz}, \tag{12}$$

$$f_{\rm gas/gw} \approx 5 \times 10^{-9} M_8^{-37/49} q^{-69/98}\,\mathrm{Hz}, \tag{13}$$

where $M_8 = M/10^8 M_\odot$. We therefore see that if the signal is dominated by $10^9 M_\odot$ SMBH binaries, then the transition frequency is located around 10^{-9} Hz.

4.2.2 Eccentricity Evolution

It is well known that GW emission efficiently circularizes binaries, however things can be dramatically different in the star and gas dominated stages. If binaries get very eccentric in those phases, they can retain substantial eccentricity even during the GW dominated inspiral relevant to PTA observations, beyond the decoupling frequencies given by Eqs. (12) and (13). The eccentricity evolution in stellar environments has been tackled by several authors by means of full N-body simulations and semianalytic models. A consistent picture emerged according to which equal mass, circular binaries tend to stay circular or experience a mild eccentricity increase, while binaries that form already eccentric, or with $q \ll 1$ (regardless of their initial eccentricity) tend to grow more eccentric (Merritt et al. 2007; Matsubayashi et al. 2007; Preto et al. 2011; Quinlan 1996; Sesana et al. 2006; Sesana 2010). An important parameter here seems to be the eccentricity of the binary at the moment of formation e_0, which is often found to be larger than 0.6 in numerical studies (Preto et al. 2011). Large e_0 implies that systems emitting in the nHz regime can be highly eccentric, causing a significant suppression of the GW signal, as we will see in the next section. In the circumbinary disk scenario, excitation of eccentricity has been seen in several simulations (Armitage and Natarajan 2005; Cuadra et al. 2009). In particular, the existence of a limiting eccentricity $e_{\rm crit} \approx 0.6 - 0.8$ has been reported in Roedig et al. (2011) through a suite of high resolution smoothed particle hydrodynamics simulations, in the case of massive selfgravitating disks. Therefore, also in gaseous rich environments, eccentric binaries might be the norm—even though the extreme eccentricities ($e > 0.9$) that might be reached in the stellar driven case are unlikely.

4.2.3 Impact on the GW Signal

The main effect of the binary-environment coupling is to *suppress* the low frequency signal (Sesana et al. 2004; Enoki and Nagashima 2007; Kocsis and Sesana 2011; Ravi et al. 2014). On the one hand, the energy of the SMBH binary is transferred to the environment instead of going into GWs, the binary evolution is faster, and consequently there are less systems emitting at each frequency. On the other hand, if eccentricity grows significantly, GW power is moved from the second to higher harmonics, through the function $g(n, e)$ (Peters and Mathews 1963). However, since the energy carried by the wave is proportional to f^2h^2, shifting the emission to higher harmonics effectively *removes* power at low frequencies, without a significant enhancement (just marginal) of h at higher frequencies (Enoki and Nagashima 2007). Therefore, generally speaking, highly eccentric binaries might pose a threat to PTA GW detection (Sesana 2013b; Ravi et al. 2014).

This is clearly shown in Fig. 3, for selected systems evolving in stellar (left panel) and gaseous (right panel) environments. In both cases, the transition frequency is around 10^{-9} Hz (corresponding to 30 year timescale), and has only a marginalinfluence on detectability (assuming 10 year of observation). However, if the PTA baseline is further extended in time, departures from the standard $f^{-2/3}$ power law become significant. The situation may get problematic when binaries grow very eccentric. This is particularly true in the stellar driven case, where, if binaries form already quite eccentric (say $e_0 > 0.7$), eccentricity can easily grow to $e > 0.95$. Consequently, a large amount of power at low frequency is removed, as shown by the $e_0 = 0.9$ line in the left panel of Fig. 3. Note that in this case, the suppression of the signal at low frequency is so severe that even 10 year observations with a nominal array sensitivity of 100 ns would not be sufficient for a confident detection. This problem is generally less severe in gaseous driven systems, where the existence of a limiting eccentricity (at least in the prograde case) prevents significant loss of power at low frequencies (as shown by the blue curves in the right panel of Fig. 3).

We must caution, however, that these results are based on toy models that probably tend to overestimate the impact of the binary coupling to the environment. The isothermal sphere implies central stellar densities higher than those typically seen in large ellipticals (see, e.g., Terzić and Graham 2005), which accelerates

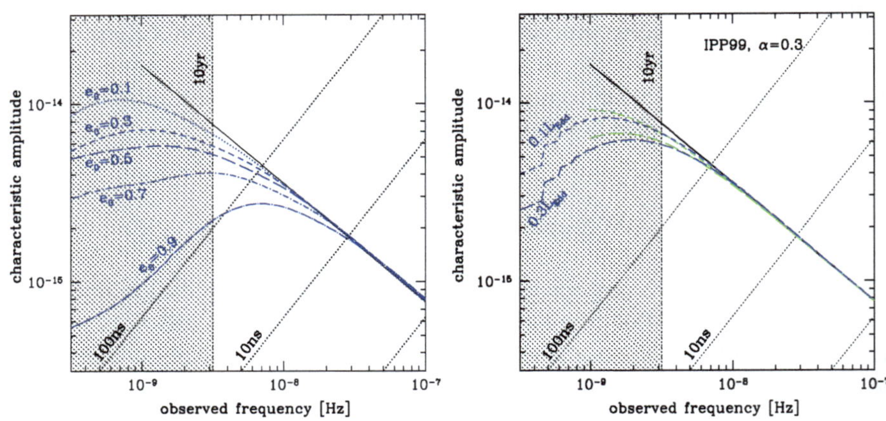

Fig. 3 Effect of the environmental coupling on the overall GW spectrum. *Left panel*: SMBH binary evolution in a stellar background. *Blue lines* represent the signal generated by a population of binaries evolving in a dense stellar background, according to the model presented in Sesana (2010); each line is for a specific fiducial eccentricity at binary formation—e_0, as labeled in figure. *Left panel*: SMBH binary evolution in circumbinary gaseous disks. Here the *two green curves* assume circular binaries, whereas the *blue curves* allow self-consistent eccentricity evolution (resulting in a slightly lower signal). The upper and lower pair of curves are for two different Eddington ratios (see text), as labeled in figure. In both panels, the *shaded area* cuts off the region at $f < 3\times10^{-9}$, to highlight the modified signal in a frequency range that is relevant to a ~10 year observation time. As in Fig. 1, the *solid black lines* represent the theoretical $h_c \propto f^{-2/3}$ behavior, and *dashed lines* mark residual levels according to the conversion $r = h/(2\pi f)$, as labeled in figure (Color figure online)

the pace at which the binary shrinks according to Eq. (10). Also in the gas driven case, the evolution described by Eq. (11) is function of the accretion rate. Here we considered binaries accreting at 0.1–0.3 of the Eddington rate.[4] However, observations of active SMBHs at low redshift indicate that the most massive systems (most relevant to PTA) are generally severely sub-Eddington (Kauffmann and Heckman 2009), which implies a weaker binary-environment coupling. It is, however, clear that the determination of the GW background spectral slope carries a lot of information about the dynamics of SMBH binaries. A well defined turnover frequency around 10^{-9} Hz will be the distinctive signature that strong coupling with the environment is the norm, whereas a plateau might be indicative of a population of highly eccentric systems. PTA detection will therefore provide important information about the dynamics of individual SMBH binaries, not only about the statistics of their collective population.

5 Resolvable Sources

As shown in Fig. 1, a handful of sources might be bright enough to be individually resolved, i.e., they stand out of the total signal level produced by the combination of all other systems. This can be seen by computing the signal-to-noise ratio (SNR) of a bright source with respect to the instrumental noise plus the unresolved GW background. Assuming a monochromatic source, the SNR in a single pulsar α can be approximated as (Sesana and Vecchio 2010)

$$\rho_\alpha^2 = (r_\alpha | r_\alpha) \approx \frac{2}{S_\alpha} \int_0^T r_\alpha^2(t)dt, \tag{14}$$

where the inner product $(*|*)$ is defined by the last approximate equality and S_0 in the noise spectral density evaluated at the frequency of the source, including both the pulsar noise and the unresolved GW background contribution. For an array of M pulsars, the single contribution of each pulsar can be coherently added in quadrature to get $\rho^2 = \sum_{\alpha=1}^{M} \rho_\alpha^2$.

[4]We recall that the Eddington luminosity is the maximum admitted luminosity for which the radiation pressure exerted by the photons emitted in the accretion process is smaller than the gravitational binding energy of the accreting material. If the contrary is true, radiation pressure blows away the reservoir of gas, suppressing the accretion process. For standard radiatively efficient accretion flows (Shakura and Sunyaev 1973), the Eddington luminosity corresponds to an Eddington accretion rate of $\dot{M}_{\rm Edd} \approx 2.5 M_8 M_\odot\,\mathrm{year}^{-1}$, where M_8 is the MBH mass normalized to $10^8 M_\odot$. Accretion rates are customarily given as fractions of the Eddington rate.

The statistics of resolvable sources has only been investigated by Sesana et al. (2009). They found that an handful of sources might be resolvable on top of the stochastic background, which seems to be corroborated by recent work from Ravi and collaborators (Ravi et al. 2014). This is an active area of research at the moment; several single source data analysis pipelines are under construction, and a better understanding of the typical properties of resolvable SMBH binaries will help setting the requirements for detection algorithm development.

5.1 Science with Resolvable PTA Sources

A pioneering investigation of the detectability of resolvable PTA sources was performed by Sesana and Vecchio (2010) assuming circular, non spinning monochromatic systems, and considering only the coherent 'Earth term' in the analysis. In this case the waveform is function of seven parameters only: the amplitude R,[5] sky location θ, ϕ, polarization ψ, inclination ι, frequency f and phase Φ_0, defining the parameter vector $\boldsymbol{\lambda} = \{R, \theta, \phi, \psi, \iota, f, \Phi_0\}$. Sesana and Vecchio (2010) finds that for SNR$=$ 10, the source amplitude is determined to a 20 % accuracy, whereas ϕ, ψ, ι are only determined within a fraction of a radian and sky location within few tens to few deg^2 (depending on the number of pulsars in the array). These results were confirmed and extended by Babak and Sesana (2012) and Petiteau et al. (2013). In particular (Petiteau et al. 2013) demonstrated that, providing there are enough pulsars in the array, multiple individual sources can be separated. This is shown in Fig. 4, which is taken from the original paper. Using a multi-search genetic algorithm, the authors succeeded to correctly identify all the eight sources blindly injected in the data, and to infer that there were no other sources with SNR$>$ 5. Experiments like this were conducted on simulated data including Gaussian noise only, and in a relatively high SNR regime. The effective capabilities of PTAs to identify and separate individual sources in more realistic situations is subject of active work in the field (Ellis et al. 2012; Taylor et al. 2014; Arzoumanian et al. 2014).

For bright enough sources (SNR$\approx$ 10) sky location within few tens to few deg^2 is possible (see also Ellis et al. 2012), and even sub deg^2 determination, under some specific conditions (Lee et al. 2011). Even though this is a large chunk of the sky, resolvable systems are extremely massive and at relatively low redshift ($z < 0.5$), making them suitable targets for electromagnetic follow-up. The possibility of performing multimessenger astronomy with PTA sources has been firstly investigated in Sesana et al. (2012) and Tanaka et al. (2012) (see Burke-Spolaor 2013; Tanaka and Haiman 2013 for a comprehensive review). They

[5]For monochromatic signals the two masses and the luminosity distance degenerate into a single amplitude parameter.

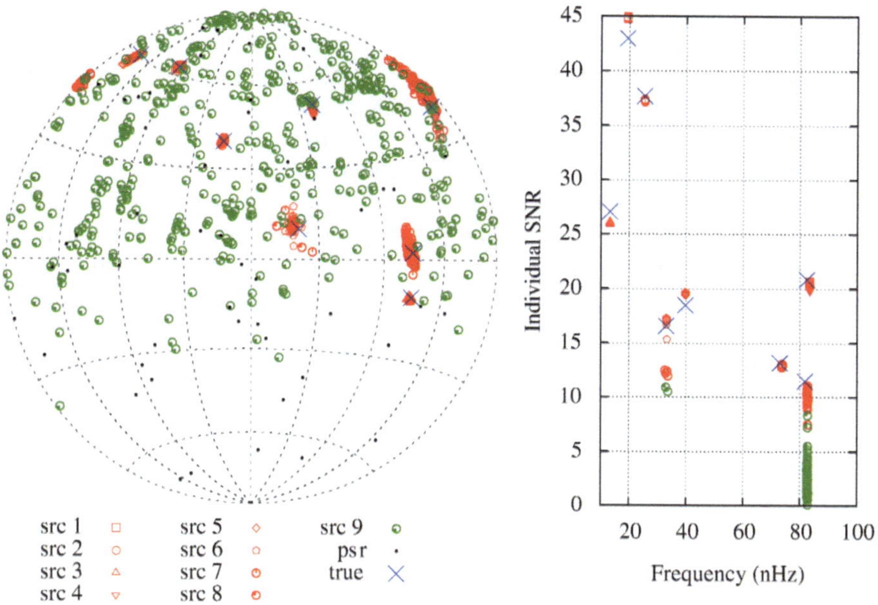

Fig. 4 Performance of the multi-search genetic algorithm in finding resolvable sources. The figure shows the source localization in the sky (sky map, *left panel*) and in the frequency-SNR space (*right panel*). *Blue crosses* (×) correspond to injected signals and *black dots* to the position of the millisecond pulsars forming the array. *Red marks* represent sources 1-to-8 found by all the organisms with $\mathrm{SNR}^2_{\mathrm{tot}} > 99\,\%\ \mathrm{SNR}^2_{\mathrm{best}}$, while *green circles* represent a putative 9th source. Note that this latter one does not have a defined position, and typically has SNR< 5 (from Petiteau et al. 2013) (Color figure online)

identified a number of possible signatures, including: periodicity related to the binary orbital period; peculiar emission spectra due to the presence of a central cavity; peculiar Kα line profiles; recurring flaring activity (see also Tanaka 2013). Given the relatively poor sky localization, suitable hosts must be selected among galaxies falling in a large error box. In general, bright ellipticals in galaxy cluster centers experience a large merger activity at low redshift, and might be the best candidates for periodic monitoring and dedicated pointings (Rosado and Sesana 2014; Simon et al. 2014).

Conclusions

PTA increasing sensitivities promise the detection of the predicted GW signal produced by a cosmological population of SMBH binaries in the near future. Beyond the obvious excitement of a direct GW observation, detection will carry an enormous wealth of information about these fascinating astrophysical

(continued)

systems, providing a *direct unquestionable* evidence of their existence and an empirical prove that the 'final parsec problem' is solved by nature. The determination of the overall signal amplitude and spectral slope will provide important information about the global properties of the SMBH binary population, giving, for example, insights about the relation between SMBH binaries and their hosts and will inform us about the dynamics of SMBH binaries and their stellar and/or gaseous environment, possibly constraining the efficiency of their mutual interaction. Identification and sky localization of individual sources, will add further excitement to the picture, making multimessenger studies of SMBH binaries and their hosts possible. Although the current focus is inevitably on the first detection, we should never forget that pulsar timing arrays will also be groundbreaking astrophysical probes.

Acknowledgements A.S. acknowledges the DFG grant SFB/TR 7 Gravitational Wave Astronomy and by DLR (Deutsches Zentrum fur Luft- und Raumfahrt), and wishes to thank Carlos F. Sopuerta and all the organizers of the 2014 Sant Cugat Forum on Astrophysics (Gravitational Wave Astrophysics).

References

P. Amaro-Seoane, A. Sesana, L. Hoffman, M. Benacquista, C. Eichhorn, J. Makino, R. Spurzem, Triplets of supermassive black holes: astrophysics, gravitational waves and detection. Mon. Not. R. Astron. Soc. **402**, 2308–2320 (2010)

P.J. Armitage, P. Natarajan, Eccentricity of supermassive black hole binaries coalescing from gas-rich mergers. Astrophys. J. **634**, 921–927 (2005)

Z. Arzoumanian et al., NANOGrav limits on gravitational waves from individual supermassive black hole binaries in circular orbits. ArXiv e-prints (April 2014)

S. Babak, A. Sesana, Resolving multiple supermassive black hole binaries with pulsar timing arrays. Phys. Rev. D **85**(4), 044034 (2012)

S. Burke-Spolaor, Multi-messenger approaches to binary supermassive black holes in the continuous-wave regime. Class. Quantum Grav. **30**(22), 224013 (2013)

T.E. Consortium, The gravitational universe. ArXiv e-prints (May 2013)

J.M. Cordes, Limits to PTA sensitivity: spin stability and arrival time precision of millisecond pulsars. Class. Quantum Grav. **30**(22), 224002 (2013)

J. Cuadra, P.J. Armitage, R.D. Alexander, M.C. Begelman, Massive black hole binary mergers within subparsec scale gas discs. Mon. Not. R. Astron. Soc. **393**, 1423–1432 (2009)

R.P. Deane, Z. Paragi, M.J. Jarvis, M. Coriat, G. Bernardi, R.P. Fender, S. Frey, I. Heywood, H.-R. Klöckner, K. Grainge, C. Rumsey, A close-pair binary in a distant triple supermassive black-hole system. ArXiv e-prints (June 2014)

P.B. Demorest, R.D. Ferdman, M.E. Gonzalez, D. Nice, S. Ransom, I.H. Stairs, Z. Arzoumanian, A. Brazier, S. Burke-Spolaor, S.J. Chamberlin, J.M. Cordes, J. Ellis, L.S. Finn, P. Freire, S. Giampanis, F. Jenet, V.M. Kaspi, J. Lazio, A.N. Lommen, M. McLaughlin, N. Palliyaguru, D. Perrodin, R.M. Shannon, X. Siemens, D. Stinebring, J. Swiggum, W.W. Zhu, Limits on the stochastic gravitational wave background from the North American nanohertz observatory for gravitational waves. Astrophys. J. **762**, 94 (2013)

P.E. Dewdney, P.J. Hall, R.T. Schilizzi, T.J.L.W. Lazio, The square kilometre array. IEEE Proc. **97**, 1482–1496 (2009)

M. Dotti, A. Sesana, R. Decarli, Massive black hole binaries: dynamical evolution and observational signatures. Adv. Astron. **2012** (2012)

J.A. Ellis, X. Siemens, J.D.E. Creighton, Optimal strategies for continuous gravitational wave detection in pulsar timing arrays. Astrophys. J. **756**, 175 (2012)

M. Enoki, M. Nagashima, The effect of orbital eccentricity on gravitational wave background radiation from supermassive black hole binaries. Prog. Theor. Phys. **117**, 241–256 (2007)

M. Eracleous, T.A. Boroson, J.P. Halpern, J. Liu, A large systematic search for recoiling and close supermassive binary black holes. ArXiv e-prints (June 2011)

R.S. Foster, D.C. Backer, Constructing a pulsar timing array. Astrophys. J. **361**, 300–308 (1990)

Z. Haiman, B. Kocsis, K. Menou, The population of viscosity- and gravitational wave-driven supermassive black hole binaries among luminous active galactic nuclei. Astrophys. J. **700**, 1952–1969 (2009)

R.W. Hellings, G.S. Downs, Upper limits on the isotropic gravitational radiation background from pulsar timing analysis. Astrophys. J. Lett. **265**, L39–L42 (1983)

G. Hobbs, The Parkes pulsar timing array. Class. Quantum Grav. **30**(22), 224007 (2013)

G. Hobbs et al., The International pulsar timing array project: using pulsars as a gravitational wave detector. Class. Quantum Grav. **27**(8), 084013 (2010)

P.B. Ivanov, J.C.B. Papaloizou, A.G. Polnarev, The evolution of a supermassive binary caused by an accretion disc. Mon. Not. R. Astron. Soc. **307**, 79–90 (1999)

A.H. Jaffe, D.C. Backer, Gravitational waves probe the coalescence rate of massive black hole binaries. Astrophys. J. **583**, 616–631 (2003)

F.A. Jenet, G.B. Hobbs, K.J. Lee, R.N. Manchester, Detecting the stochastic gravitational wave background using pulsar timing. Astrophys. J. Lett. **625**, L123–L126 (2005)

G. Kauffmann, T.M. Heckman, Feast and famine: regulation of black hole growth in low-redshift galaxies. Mon. Not. R. Astron. Soc. **397**, 135–147 (2009)

F.M. Khan, A. Just, D. Merritt, Efficient merger of binary supermassive black holes in merging galaxies. Astrophys. J. **732**, 89 (2011)

B. Kocsis, A. Sesana, Gas-driven massive black hole binaries: signatures in the nHz gravitational wave background. Mon. Not. R. Astron. Soc. **411**, 1467–1479 (2011)

J. Kormendy, L.C. Ho, Coevolution (or not) of supermassive black holes and host galaxies. Ann. Rev. Astron. Astrophys. **51**, 511–653 (2013)

M. Kramer, D.J. Champion, The European pulsar timing array and the large European array for pulsars. Class. Quantum Grav. **30**(22), 224009 (2013)

K.J. Lee, N. Wex, M. Kramer, B.W. Stappers, C.G. Bassa, G.H. Janssen, R. Karuppusamy, R. Smits, Gravitational wave astronomy of single sources with a pulsar timing array. Mon. Not. R. Astron. Soc. **414**, 3251–3264 (2011)

R.N. Manchester, IPTA, The International pulsar timing array. Class. Quantum Grav. **30**(22), 224010 (2013)

T. Matsubayashi, J. Makino, T. Ebisuzaki, Orbital evolution of an IMBH in the galactic nucleus with a massive central black hole. Astrophys. J. **656**, 879–896 (2007)

N.J. McConnell, C.-P. Ma, Revisiting the scaling relations of black hole masses and host galaxy properties. Astrophys. J. **764**, 184 (2013)

M.A. McLaughlin, The North American nanohertz observatory for gravitational waves. Class. Quantum Grav. **30**(22), 224008 (2013)

S.T. McWilliams, J.P. Ostriker, F. Pretorius, Gravitational waves and stalled satellites from massive galaxy mergers at z ¡= 1. Astrophys. J. **789**, 156 (2014)

D. Merritt, S. Mikkola, A. Szell, Long-term evolution of massive black hole binaries. III. Binary evolution in collisional nuclei. Astrophys. J. **671**, 53–72 (2007)

P.C. Peters, J. Mathews, Gravitational radiation from point masses in a Keplerian orbit. Phys. Rev. **131**, 435–440 (1963)

A. Petiteau, S. Babak, A. Sesana, M. de Araújo, Resolving multiple supermassive black hole binaries with pulsar timing arrays. II. Genetic algorithm implementation. Phys. Rev. D **87**(6), 064036 (2013)

E.S. Phinney, A practical theorem on gravitational wave backgrounds. ArXiv Astrophysics e-prints 0108028 (August 2001)

M. Preto, I. Berentzen, P. Berczik, R. Spurzem, Fast coalescence of massive black hole binaries from mergers of galactic nuclei: implications for low-frequency gravitational-wave astrophysics. Astrophys. J. Lett. **732**, L26 (2011)

G.D. Quinlan, The dynamical evolution of massive black hole binaries I. Hardening in a fixed stellar background. New Astron. **1**, 35–56 (1996)

M. Rajagopal, R.W. Romani, Ultra–low-frequency gravitational radiation from massive black hole binaries. Astrophys. J. **446**, 543 (1995)

V. Ravi, J.S.B. Wyithe, G. Hobbs, R.M. Shannon, R.N. Manchester, D.R.B. Yardley, M.J. Keith, Does a "stochastic" background of gravitational waves exist in the pulsar timing band? Astrophys. J. **761**, 84 (2012)

V. Ravi, J.S.B. Wyithe, R.M. Shannon, G. Hobbs, R.N. Manchester, Binary supermassive black hole environments diminish the gravitational wave signal in the pulsar timing band. Mon. Not. R. Astron. Soc. **442**, 56–68 (2014)

V. Ravi, J.S.B. Wyithe, R.M. Shannon, G. Hobbs, Prospects for gravitational-wave detection and supermassive black hole astrophysics with pulsar timing arrays. ArXiv e-prints (June 2014)

C. Roedig, M. Dotti, A. Sesana, J. Cuadra, M. Colpi, Limiting eccentricity of subparsec massive black hole binaries surrounded by self-gravitating gas discs. Mon. Not. R. Astron. Soc. **415**, 3033–3041 (2011)

P.A. Rosado, A. Sesana, Targeting supermassive black hole binaries and gravitational wave sources for the pulsar timing array. Mon. Not. R. Astron. Soc. **439**, 3986–4010 (2014)

M.V. Sazhin, Opportunities for detecting ultralong gravitational waves. Sov. Astron. **22**, 36–38 (1978)

A. Sesana, Self consistent model for the evolution of eccentric massive black hole binaries in Stellar environments: implications for gravitational wave observations. Astrophys. J. **719**, 851–864 (2010)

A. Sesana, Insights into the astrophysics of supermassive black hole binaries from pulsar timing observations. Class. Quantum Grav. **30**(22), 224014 (2013a)

A. Sesana, Systematic investigation of the expected gravitational wave signal from supermassive black hole binaries in the pulsar timing band. Mon. Not. R. Astron. Soc. **433**, 1 (2013b)

A. Sesana, A. Vecchio, Measuring the parameters of massive black hole binary systems with pulsar timing array observations of gravitational waves. Phys. Rev. D **81**(10), 104008 (2010)

A. Sesana, F. Haardt, P. Madau, M. Volonteri, Low-frequency gravitational radiation from coalescing massive black hole binaries in hierarchical cosmologies. The Astrophys. J. **611**, 623–632 (2004)

A. Sesana, F. Haardt, P. Madau, Interaction of massive black hole binaries with their Stellar environment. I. Ejection of hypervelocity stars. Astrophys. J. **651**, 392–400 (2006)

A. Sesana, A. Vecchio, C.N. Colacino, The stochastic gravitational-wave background from massive black hole binary systems: implications for observations with Pulsar Timing Arrays. Mon. Not. R. Astron. Soc. **390**, 192–209 (2008)

A. Sesana, A. Vecchio, M. Volonteri, Gravitational waves from resolvable massive black hole binary systems and observations with Pulsar Timing Arrays. Mon. Not. R. Astron. Soc. **394**, 2255–2265 (2009)

A. Sesana, C. Roedig, M.T. Reynolds, M. Dotti, Multimessenger astronomy with pulsar timing and X-ray observations of massive black hole binaries. Mon. Not. R. Astron. Soc. **420**, 860–877 (2012)

N.I. Shakura, R.A. Sunyaev, Black holes in binary systems. Observational appearance. Astron. Astrophys. **24**, 337–355 (1973)

R.M. Shannon, J.M. Cordes, Assessing the role of spin noise in the precision timing of millisecond pulsars. Astrophys. J. **725**, 1607–1619 (2010)

R.M. Shannon, V. Ravi, W.A. Coles, G. Hobbs, M.J. Keith, R.N. Manchester, J.S.B. Wyithe, M. Bailes, N.D.R. Bhat, S. Burke-Spolaor, J. Khoo, Y. Levin, S. Oslowski, J.M. Sarkissian, W. van Straten, J.P.W. Verbiest, J.-B. Want, Gravitational-wave limits from pulsar timing constrain supermassive black hole evolution. Science **342**, 334–337 (2013)

J. Simon, A. Polin, A. Lommen, B. Stappers, L.S. Finn, F.A. Jenet, B. Christy, Gravitational wave hotspots: ranking potential locations of single-source gravitational wave emission. Astrophys. J. **784**, 60 (2014)

V. Springel et al., Simulations of the formation, evolution and clustering of galaxies and quasars. Nature **435**, 629–636 (2005)

T.L. Tanaka, Recurring flares from supermassive black hole binaries: implications for tidal disruption candidates and OJ 287. Mon. Not. R. Astron. Soc. **434**, 2275–2288 (2013)

T.L. Tanaka, Z. Haiman, Electromagnetic signatures of supermassive black hole binaries resolved by PTAs. Class. Quantum Grav. **30**(22), 224012 (2013)

T. Tanaka, K. Menou, Z. Haiman, Electromagnetic counterparts of supermassive black hole binaries resolved by pulsar timing arrays. Mon. Not. R. Astron. Soc. **420**, 705–719 (2012)

S. Taylor, J. Ellis, J. Gair, Including the pulsar-term in continuous gravitational-wave searches using pulsar timing arrays: a blessing and a curse. ArXiv e-prints (June 2014)

B. Terzić, A.W. Graham, Density-potential pairs for spherical stellar systems with Sérsic light profiles and (optional) power-law cores. Mon. Not. R. Astron. Soc. **362**, 197–212 (2005)

P. Tsalmantza, R. Decarli, M. Dotti, D.W. Hogg, A systematic search for massive black hole binaries in the Sloan Digital Sky survey spectroscopic sample. Astrophys. J. **738**, 20 (2011)

R. van Haasteren et al., Placing limits on the stochastic gravitational-wave background using European Pulsar Timing Array data. Mon. Not. R. Astron. Soc. **414**, 3117–3128 (2011)

J.P.W. Verbiest, M. Bailes, W.A. Coles, G.B. Hobbs, W. van Straten, D.J. Champion, F.A. Jenet, R.N. Manchester, N.D.R. Bhat, J.M. Sarkissian, D. Yardley, S. Burke-Spolaor, A.W. Hotan, X.P. You, Timing stability of millisecond pulsars and prospects for gravitational-wave detection. Mon. Not. R. Astron. Soc. **400**, 951–968 (2009)

J.S.B. Wyithe, A. Loeb, Low-frequency gravitational waves from massive black hole binaries: predictions for LISA and pulsar timing arrays. Astrophys. J. **590**, 691–706 (2003)

Gravitational Wave Verification Sources

Mukremin Kilic, Warren R. Brown, J.J. Hermes, and A. Gianninas

Abstract Ultra-compact binary systems containing white dwarfs dominate the gravitational wave foreground in the mHz frequency range. Many of these binaries are weak gravitational wave sources. However, there are nine systems where the gravitational wave strain is strong enough to make them verification sources for gravitational wave missions, e.g. the evolved Laser Interferometer Space Antenna (*eLISA*). Here we present a summary of the physical characteristics of these nine systems, including the observed rate of orbital decay. In addition, we present the recent discoveries from the ELM Survey, which identified two verification sources in the last 3 years. We also discuss the future prospects for increasing the sample of verification binaries.

1 Ultra-Compact Binary Systems

The evolution of ultra-compact binaries ($P_{\rm orb} < 1$ h) containing white dwarfs (WDs), neutron stars, and stellar-mass black holes is dominated by gravitational waves. These systems lose angular momentum through gravitational wave radiation and evolve to shorter periods until they start mass transfer. The orbital decay of several ultra-compact binary systems (including double pulsars, AM Canum Venaticorum, and double WD systems) has already been detected, indicating that these systems must emit gravitational waves. Perhaps the best-known example is the Hulse Taylor pulsar where the orbital period is shrinking by -2.42×10^{-12} s s^{-1} (Weisberg et al. 2010), matching the predictions from general relativity.

M. Kilic (✉) • A. Gianninas
Department of Physics and Astronomy, University of Oklahoma, 440 W. Brooks St., Norman, OK 73019, USA
e-mail: kilic@ou.edu; alexg@nhn.ou.edu

W.R. Brown
Smithsonian Astrophysical Observatory, 60 Garden St, Cambridge, MA 02138, USA
e-mail: wbrown@cfa.harvard.edu

J.J. Hermes
Department of Physics, University of Warwick, Coventry CV4 7AL, UK
e-mail: j.j.hermes@warwick.ac.uk

C.F. Sopuerta (ed.), *Gravitational Wave Astrophysics*, Astrophysics and Space Science Proceedings, DOI 10.1007/978-3-319-10488-1_14

Nelemans (2013) predicts $\sim 10^8$ double WDs in the Galaxy. Given the large number of such sources, a fraction of them will be found at very short ($\sim$ min) orbital periods, providing verification binaries for gravitational wave missions in the mHz frequency range. In fact, Nelemans (2013) predicts several thousand sources that should be individually detected by *eLISA*. However, there are only nine *eLISA* verification sources currently known.

Here we discuss the current sample of gravitational wave verification binaries, including the latest discoveries in the ELM Survey (Brown et al. 2011; Kilic et al. 2014) in Sect. 2. The latest results on the rate of orbital period decay in these systems are also included in this section. We discuss the future prospects for increasing their numbers in Sect. 3.

2 Verification Sources

2.1 AM CVn Stars

AM CVn are interacting binary systems consisting of a WD accreting helium-rich material from a companion star. There are currently $\approx$40 AM CVn systems known with orbital periods ranging from 5 to 65 min. The shortest period systems, HM Cnc and V407 Vul, have orbital periods of 5.4 and 9.5 min, respectively (Roelofs et al. 2010). Seven of these systems, namely HM Cnc, V407 Vul, ES Cet, AM CVn, HP Lib, Cr Boo, and V803 Cen, are *eLISA* verification sources (Amaro-Seoane et al. 2012).

Strohmayer (2005) measure $\dot{P} = -3.75 \times 10^{-11}\,\mathrm{s\,s^{-1}}$ for HM Cnc. Similarly, Ramsay et al. (2005) find a spin-up rate of $\dot{P} = -3.17 \times 10^{-12}\,\mathrm{s\,s^{-1}}$ for V407 Vul. The orbital periods of these two shortest period AM CVn systems known are decreasing at a rate consistent with orbital angular momentum loss through gravitational wave radiation. However, the next shortest period AM CVn system, ES Cet, does not show any evidence of a change in orbital period so far (Copperwheat et al. 2011). Even though this may sound surprising, stellar evolution theory predicts that AM CVn should have stable mass transfer and evolve to longer periods. Hence, the inconsistency between the observed $\dot{P}$ and the expectations demonstrate the complicated orbital angular momentum evolution in short-period interacting systems. Regardless of these issues, the seven AM CVn systems mentioned above are excellent gravitational wave sources.

2.2 Double WDs

Double WDs outnumber all other known types of gravitational wave sources in the Galaxy and they form a gravitational wave foreground. Until recently, WD 0957-666 was the shortest period detached double WD system known with an orbital period of 1.46 h (Moran et al. 1997). Unfortunately, the gravitational wave strain from this system is below the *eLISA* sensitivity limit.

WD 0957-666 contains two low-mass WDs with masses 0.32–0.37 $M_{\odot}$. Marsh et al. (1995) demonstrated that low-mass He-core WDs with $M < 0.45 M_{\odot}$ come in pairs. The Galaxy is not old enough to produce He-core WDs from single stars. Hence, they must form when a companion strips the outer envelope from a post main-sequence star before the star reaches the tip of the red giant branch and ignites helium. Low mass WDs are, therefore, signposts of short period binary systems. The ELM Survey (Brown et al. 2013) takes advantage of this fact to identify binary systems containing Extremely-Low-Mass ($M \leq 0.3 M_{\odot}$) WDs. The ELM Survey has so far discovered 56 binaries, all with $P \leq 1$ d, including 33 that will merge within a Hubble time. Four of these have orbital periods less than an hour; J0651+2844, WD 0931+444, J0106−1000, and J1630+4233.

J0651 is the best system found so far. It is the shortest period detached WD binary yet discovered. Figure 1 shows the radial velocity observations of J0651 (Brown

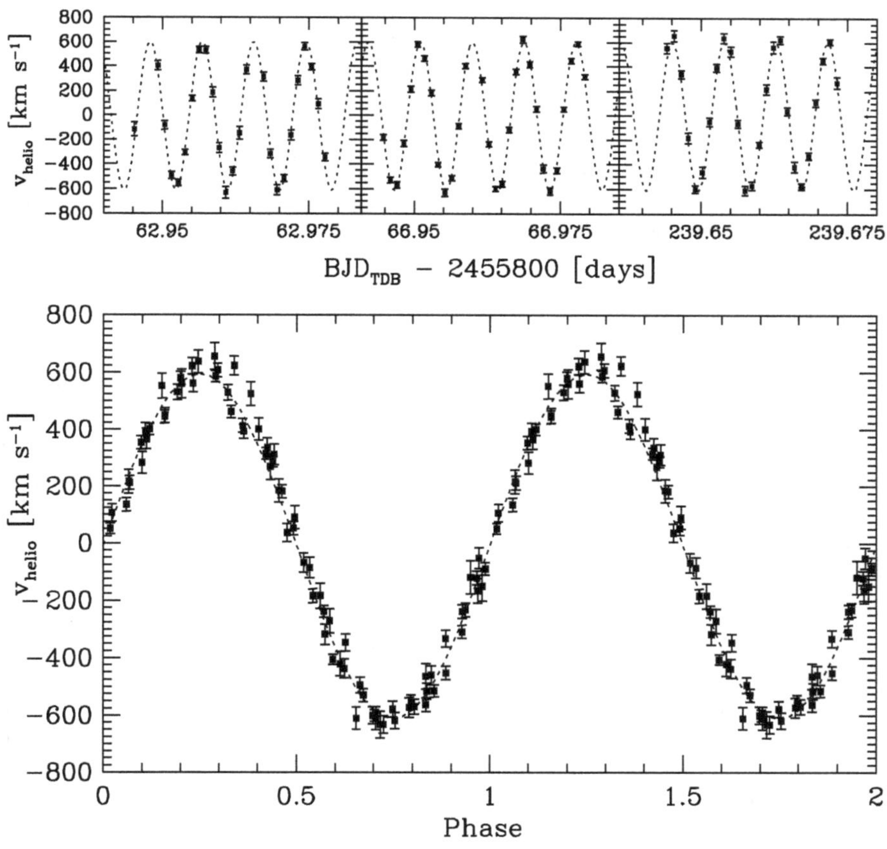

Fig. 1 Radial velocity observations of J0651 from three epochs (*top panel*), and those data phased to the orbital period (*bottom panel*, Hermes et al. 2012)

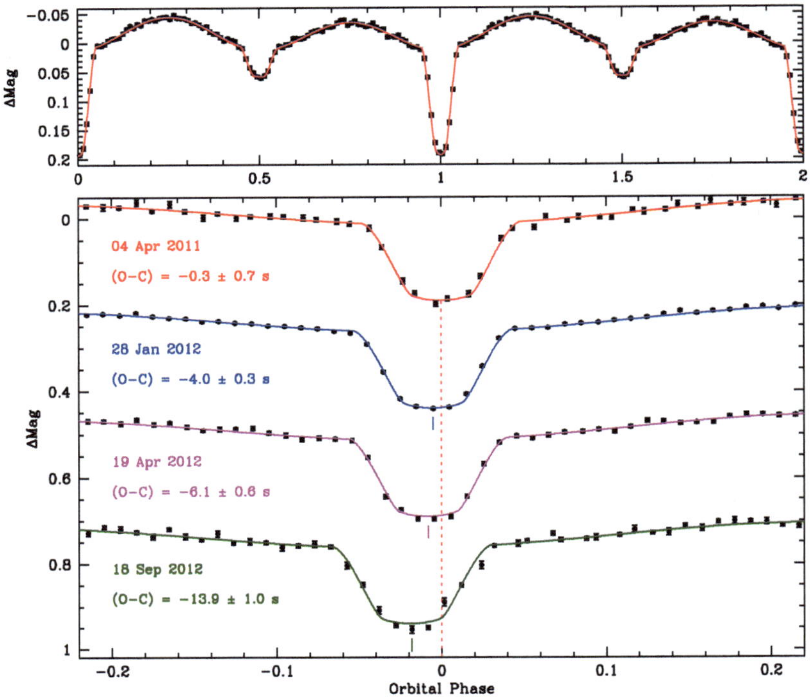

Fig. 2 High-speed photometry of J0651 folded at the orbital period (*top panel*). The *red solid line* shows the best-fit model. The *bottom panel* shows data from four different subsets. The decreasing orbital period is evident as the primary eclipses shift sooner (Color figure online)

et al. 2011; Hermes et al. 2012). J0651 has a best-fit orbital period of 765 s and a velocity semi-amplitude of 616.9 km s^{-1}. Luckily, J0651 is also an eclipsing system. Figure 2 shows its optical light curve, which reveals primary and secondary eclipses, ellipsoidal variations from the tidally distorted ELM WD, and doppler boosting. J0651 contains a 0.25 $M_{\odot}$ WD with a 0.49 $M_{\odot}$ secondary WD at $i = 84.4°$. There is no evidence for mass transfer, thus J0651 is one of the cleanest known strong gravitational wave sources.

We expect J0651 to merge in less than 1 Myr due to gravitational radiation. Hermes et al. (2012) measured the rate of orbital decay in this system using eclipse timing measurements over a year and found $\dot{P} = -9.8 \pm 2.8 \times 10^{-12}$ s s^{-1} (see Fig. 2). By now, the mid-eclipse times have actually shifted by almost 50 s (over the course of 3 years; Hermes et al. in prep). This should be compared to the $\approx$ 45 s cumulative shift of periastron time over three decades in the Hulse Taylor pulsar (Weisberg et al. 2010). The observed rate of orbital period change in J0651 is consistent with the predictions from general relativity within the errors. The short period of J0651 makes it one of the loudest known sources of gravitational wave radiation.

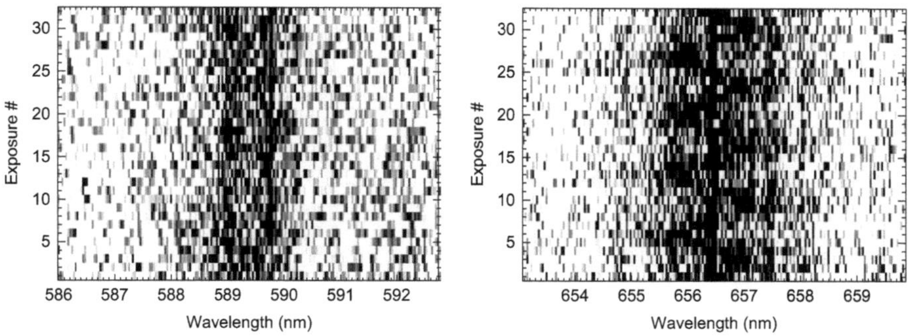

Fig. 3 Time-resolved spectroscopy of the Na I doublet (*left panel*) and the Hα line (*right panel*) in WD 0931+444 (J0935) over 90 min. The Na I lines and the Hα line from the M dwarf are stationary, whereas the Hα line from the WD clearly shows a 20 min periodicity (Kilic et al. 2014)

Not as spectacular as J0651, but still interesting, is the recent discovery of a 20 min detached double WD system in the ELM Survey. WD 0931+444 (SDSS J093506.93+441106.9) was originally identified as a UV excess object (Usher et al. 1982). The SDSS spectroscopy of WD 0931+444 clearly shows a composite spectrum of a DA WD plus an M dwarf. We identified WD 0931+444 as an interesting ELM WD target based on the model fits to the Balmer line profiles. Our follow-up time-resolved spectroscopy observations at the MMT and Gemini show that the low-mass WD in this system orbits an unseen companion every 20 min. Figure 3 shows these spectra, which reveal two components for the Hα line. The first one, from the M dwarf, is stationary. While the second one, from the WD, shows significant velocity variations. The Hα line and the other Balmer lines (not shown in this figure) from the WD clearly show a 20 min periodicity, whereas the Hα and Na I lines from the M dwarf are stationary. Hence, the M dwarf is not a member of the 20 min binary. In fact, the orbital period of this system is significantly shorter than the period minimum for cataclysmic variables (Hellier 2001). Hence, no main-sequence star can fit into this orbit.

There is no evidence of any significant photometric variability in WD 0931+444 above 0.4 % in a blue broad-bandpass filter. Assuming that the secondary star has a radius comparable to the ELM WD, the lack of eclipses constrain the inclination angle to $i \leq 70°$, which implies a $M \geq 0.14 M_{\odot}$ WD companion. The two WDs in this system will merge in $\leq$9 Myr. Hence, after J0651, WD 0931+444 becomes the second quickest WD merger system currently known. The expected gravitational wave strain at Earth $\log h \geq -22.17$ at $\log \nu$ (Hz) = -2.77 from WD 0931+444 makes it an *eLISA* verification source, only the ninth such system known. Given the unknown inclination angle and the non-detection of the WD companion in the spectroscopy and the photometry data, the gravitational wave strain from WD 0931+444 may be even higher than this estimate, potentially making it one of the best verification sources known.

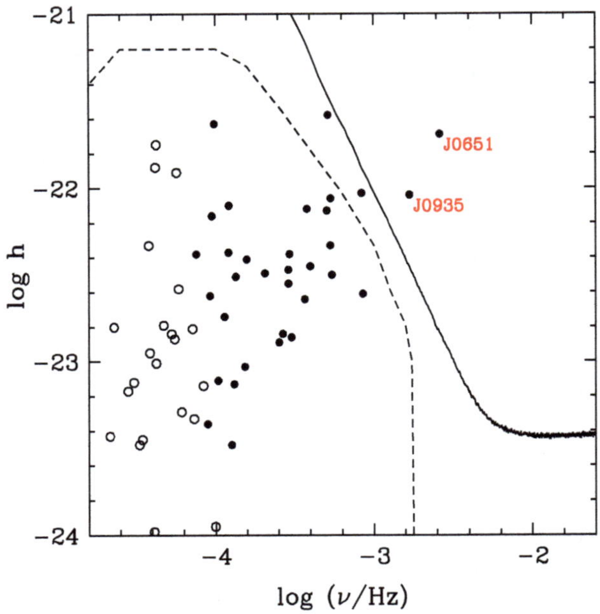

Fig. 4 Gravitational wave strain versus frequency for the binary systems found in the ELM survey. *Solid dots* are those systems that will merge within a Hubble time. We assume an average inclination angle of $i = 60°$, except where the inclination is known from eclipses or ellipsoidal variations. The *solid line* shows the design sensitivity of eLISA after 1 year. The *dashed line* shows the predicted Galactic foreground from Nelemans et al. (2001)

Figure 4 shows the predicted gravitational wave strain amplitudes of the binary systems found in the ELM Survey so far. There are two binaries, J0651 and J0935 (WD 0931+444) that should clearly be detected by an *eLISA*-like mission within the first year of operation. It may be possible to detect several other ELM WD binaries because we know their coordinates and physical parameters from the optical observations. The remaining several dozen sources form part of the Galactic foreground for gravitational wave detectors.

3 Future Prospects

Currently, there are nine *eLISA* verification sources known; seven AM CVn systems with orbital periods ranging from 5 to 27 min, and two detached double WDs with orbital periods of 12–20 min. The best verification binary is the 5.4 min period HM Cnc. The second best system is the detached 12 min binary J0651. The observed rapid orbital period decay in HM Cnc, J0651, and V407 Vul demonstrate that they are strong gravitational wave sources that should be detected by *eLISA* within the first few weeks of operation.

V803 Cen is the longest period verification binary currently known with an orbital period of 27 min. Identification of similar (or better) verification sources requires wide-field synoptic surveys. There are several groups searching for AM CVn in the SDSS and other ground-based surveys. Our group is continuing the ELM Survey to identify many more short period binary WDs. On the other hand, GAIA and the Large Synoptic Survey Telescope (LSST) perhaps have the best chance of discovering a large number of ultra-compact binary systems. GAIA will identify binary WDs down to a limiting magnitude of 20, whereas each LSST visit will have a limiting magnitude of >24. These should lead to at least an order of magnitude increase in the number of verification binaries over the next decade.

Acknowledgements We gratefully acknowledge the support of the NSF under grant AST-1312678. JJH acknowledges funding from the European Research Council under the European Union's Seventh Framework Programme (FP/2007–2013)/ERC Grant Agreement no. 320964 (WDTracer).

References

P. Amaro-Seoane, S. Aoudia, S. Babak et al., Class. Quantum Grav. **29**, 124016 (2012)
W.R. Brown, M. Kilic, J.J. Hermes et al., Astrophys. J. **737**, L23 (2011)
W.R. Brown, M. Kilic, C. Allende Prieto, A. Gianninas, S.J. Kenyon, Astrophys. J. **769**, 66 (2013)
C.M. Copperwheat, T.R. Marsh, V.S. Dhillon et al., Mon. Not. R. Astron. Soc. **413**, 3068 (2011)
C. Hellier, *Cataclysmic Variable Stars* (Springer, Berlin, 2001)
J.J. Hermes, M. Kilic, W.R. Brown et al., Astrophys. J. **757**, L21 (2012)
M. Kilic, W.R. Brown, A. Gianninas, J.J. Hermes, C. Allende Prieto, S.J. Kenyon, Mon. Not. R. Astron. Soc. MNRAS **444**, L1 (2014)
T.R. Marsh, V.S. Dhillon, S.R. Duck, Mon. Not. R. Astron. Soc. **275**, 828 (1995)
C. Moran, T.R. Marsh, A. Bragaglia, Mon. Not. R. Astron. Soc. **288**, 538 (1997)
G. Nelemans, 9th LISA Symposium, vol. 467 (2013), p. 27
G. Nelemans, L.R. Yungelson, S.F. Portegies Zwart, Astron. Astrophys. **375**, 890 (2001)
G. Ramsay, P. Hakala, K. Wu et al., Mon. Not. R. Astron. Soc. **357**, 49 (2005)
G.H.A. Roelofs, A. Rau, T.R. Marsh et al., Astrophys. J. **711**, L138 (2010)
T.E. Strohmayer, Astrophys. J. **627**, 920 (2005)
P.D. Usher, D. Mattson, A. Warnock III, Astrophys. J. Suppl. **48**, 51 (1982)
J.M. Weisberg, D.J. Nice, J.H. Taylor, Astrophys. J. **722**, 1030 (2010)

The Gravitational Wave Emission of White Dwarf Dynamical Interactions

Gabriela Aznar-Siguán, Enrique García-Berro, and Pablo Lorén-Aguilar

Abstract We compute the emission of gravitational waves of white dwarf dynamical interactions and close encounters in dense stellar environments and we compare it with the sensitivity curves of planned space-borne gravitational wave detectors, like eLISA and ALIA. We find that for the three possible outcomes of these interactions—which are the formation of an eccentric binary system, a lateral collision in which several mass transfer episodes occur, and a direct one in which just a single mass transfer episode takes place—only those in which an eccentric binary are formed are likely to be detected by the planned gravitational wave mission eLISA, while ALIA would be able to detect the gravitational wave signal emitted in lateral collisions.

1 Introduction

Close encounters of two white dwarfs in dense stellar environments, as the central regions of galaxies or the cores of globular clusters, is an interesting phenomenon that has several potential applications, and deserves close scrutiny. Among these applications we mention that it has been shown (Rosswog et al. 2009; Raskin et al. 2009) that head-on collisions of two white dwarfs is a viable mechanism to produce Type Ia Supernovae (SNIa), one of the most energetics events in the Universe. This, in turn, has obvious implications in Cosmology, as SNIa are one of the best

G. Aznar-Siguán (✉)
Departament de Física Aplicada, Universitat Politècnica de Catalunya, c/Esteve Terrades 5, 08860 Castelldefels, Spain
e-mail: gabriela@fa.upc.edu

E. García-Berro
Departament de Física Aplicada, Universitat Politècnica de Catalunya, c/Esteve Terrades 5, 08860 Castelldefels, Spain

Institute for Space Studies of Catalonia, c/Gran Capità 2–4, Edif. Nexus 104, 08034 Barcelona, Spain
e-mail: enrique.garcia-berro@upc.edu

P. Lorén-Aguilar
School of Physics, University of Exeter, Stocker Road, Exeter EX4 4QL, UK

C.F. Sopuerta (ed.), *Gravitational Wave Astrophysics*, Astrophysics and Space Science Proceedings, DOI 10.1007/978-3-319-10488-1_15

distance indicators. However, there are situations in which the two white dwarfs interact but the interaction does not result in the complete disruption of the two stars and a powerful thermonuclear outburst. Specifically, it has been recently shown (Lorén-Aguilar et al. 2010) that white dwarf close encounters have three different outcomes, depending on the initial conditions of the interaction—namely, the initial energy and angular momentum of the of the white dwarf pair. In particular, these interactions can lead to the formation of an eccentric binary system, to a lateral collision in which several mass transfer episodes between the disrupted less massive star and the more massive one occur, and finally to a direct collision, in which only one catastrophic mass transfer event occurs. According to these considerations, during the last few years this research topic has been the subject of renewed interest, and the dynamical behavior of these interactions has been simulated using in most cases Smoothed Particle Hydrodynamics techniques (Rosswog et al. 2009; Raskin et al. 2009; Lorén-Aguilar et al. 2010; Raskin et al. 2010; Hawley et al. 2012; Kushnir et al. 2013; Aznar-Siguán et al. 2013). It is important to mention that in Aznar-Siguán et al. (2013) a significantly broad range of initial conditions (initial velocities and separations) for a larger set of white dwarf masses and core chemical compositions, w as computed. In this work we adopt this set of simulations, and we calculate the gravitational waveforms produced in these interactions.

2 Numerical Setup

We compute the gravitational wave emission in the slow-motion, weak-field quadrupole approximation (Misner et al. 1973). Specifically, we follow closely the procedure outlined in Lorén-Aguilar et al. (2005). To assess the detectability of the gravitational waveforms we proceed as follows. For the well defined eccentric orbits, we first accumulate the power of the signal during 1 year. We then compute the characteristic frequencies and amplitudes, and the signal-to-noise-ratios (SNR) according to Zanotti et al. (2003). For the short-lived signals obtained in lateral collisions, we compute the SNR as in Giacomazzo et al. (2011). Finally, because in direct collisions only one powerful burst of gravitational waves is radiated it is quite unlikely that these interactions could be eventually detected.

3 Gravitational Wave Radiation

Some of the close encounters computed in Aznar-Siguán et al. (2013) result in the formation of a binary system that survives the interaction and no mass transfer between the two stars occurs. In these cases the eccentricity of the orbit is usually relatively large. Specifically, $e > 0.5$ in all cases. The top panels of Fig. 1 show the resulting gravitational dimensionless strains of a typical case, in units of 10^{-22}, for an edge-on orbit. The two polarizations of the gravitational radiation, h_+ and $h_\times$, are shown for a full period of the orbit. The corresponding dimensionless

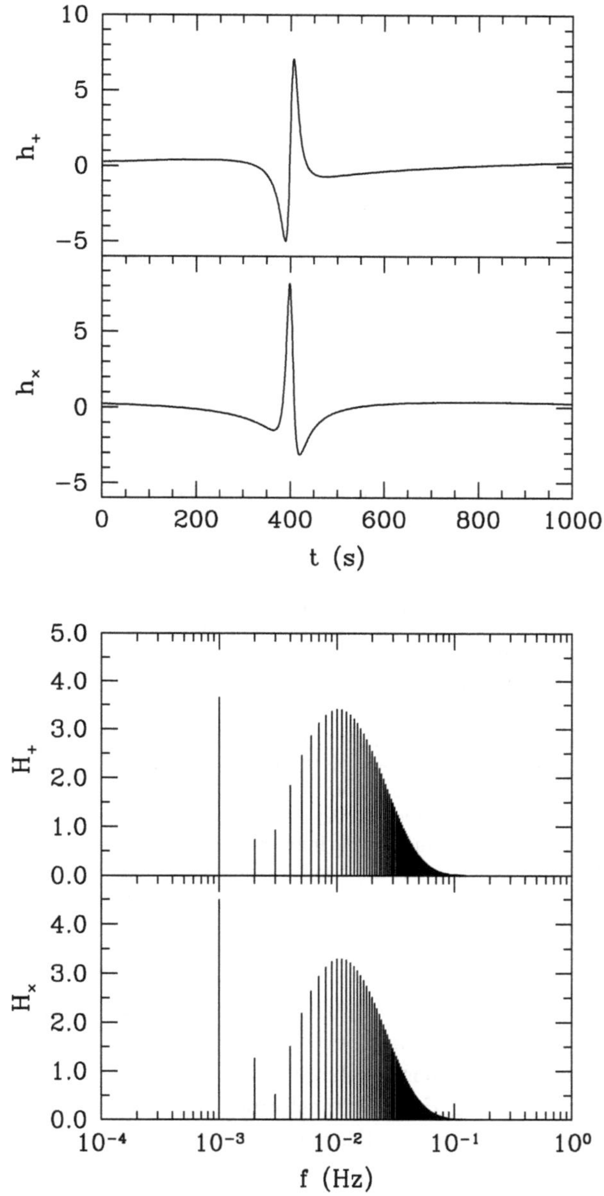

Fig. 1 Gravitational waveforms for a close encounter in which an eccentric binary is formed. For this particular case both white dwarfs have equal masses, $M = 0.8\,M_{\odot}$, whilst the adopted initial conditions lead to the formation of a binary system with eccentricity $e = 0.820$ and frequency $f = 0.001$ Hz. In the *top panels* we show the h_{+} and $h_{\times}$ waveforms for an inclination $i = 0°$ and a distance of 10 kpc, in units of 10^{-22}. The *bottom panels* display the Fourier transforms of the signals, in units of 10^{-23}

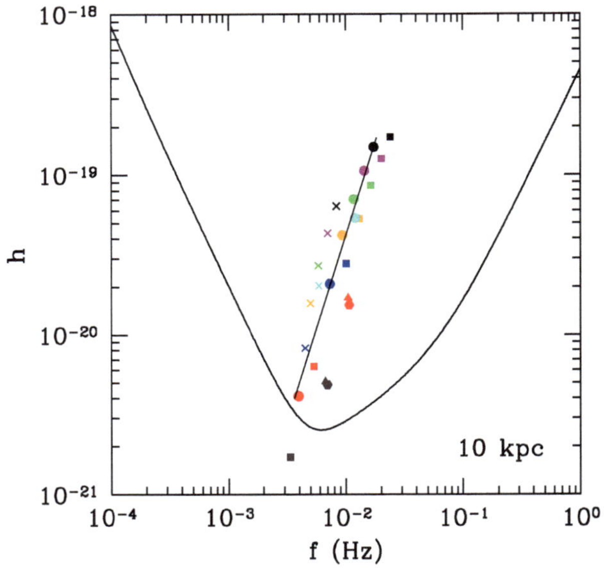

Fig. 2 A comparison of the signal produced by the close white dwarf binary systems studied here, when a distance of 10 kpc is adopted, with the spectral distribution of noise of eLISA for a 1 year integration period, for a null inclination. We assess the detectability of the h_+ dimensionless strain. The *different colors* denote different masses of the interacting white dwarfs, whereas the *different symbols* are used to distinguish between different initial conditions. *Black symbols* correspond to a $1.2\,M_\odot + 0.8\,M_\odot$ binary system, *magenta* ones to a $1.0\,M_\odot + 0.8\,M_\odot$, *green symbols* to a $0.8\,M_\odot + 0.8\,M_\odot$, *cyan* ones to a $1.2\,M_\odot + 0.4\,M_\odot$, *orange symbols* to a $0.8\,M_\odot + 0.6\,M_\odot$, *blue* ones to a $0.8\,M_\odot + 0.4\,M_\odot$, *red symbols* to a $0.4\,M_\odot + 0.4\,M_\odot$, and *grey* ones are used for a $0.4\,M_\odot + 0.2\,M_\odot$ binary system. On the other hand, *hexagons* are used for the case in which the initial relative velocity $v_{\rm ini} = 150$ km/s and the initial distance perpendicular to the relative velocity $\Delta y = 0.4\,R_\odot$, *triangles* for $v_{\rm ini} = 200$ km/s and $\Delta y = 0.3\,R_\odot$, *squares* for $v_{\rm ini} = 200$ km/s and $\Delta y = 0.4\,R_\odot$, *circles* for $v_{\rm ini} = 300$ km/s and $\Delta y = 0.3\,R_\odot$, *crosses* for $v_{\rm ini} = 300$ km/s and $\Delta y = 0.4\,R_\odot$ (Color figure online)

Fourier transforms, $H(f) = \tilde{h}/T$ where $T = 1$ year, of these gravitational wave patterns are displayed in the bottom panels of Fig. 1. The gravitational wave emission presents a single, narrow, large pulse, which occurs at the periastron, whereas the corresponding Fourier transforms, H_+ and $H_\times$, present a peak at the orbital frequency, and several Fourier components at larger frequencies. This was expected, as a binary system in a circular orbit produces a monochromatic gravitational waveform at twice the orbital frequency, whereas for eccentric orbits it is found—see, for instance, Wahlquist et al. (1987)—that the binary system radiates power at the fundamental mode and at several harmonics of the orbital frequency. The reason for this is clear, since it is at closest approach when the accelerations are larger, thus resulting in an enhanced emission of gravitational radiation. Moreover, it turns out that for a fixed energy, the larger the eccentricity, the smaller the distance at closest approach, and therefore the larger the radiated power. As can be seen in Fig. 2 all these events but one would be most likely detected by eLISA.

In lateral collisions the less massive white dwarf is tidally deformed by the more massive star at closest approach to such an extent that in the end some of its material is accreted by the massive companion. This occurs in several mass transfer episodes, leading to a final merger. A typical example of the gravitational wave pattern resulting in these cases is shown in Fig. 3. As can be seen, the time evolution of the dimensionless strains presents a series of peaks, each one corresponding to a passage through the periastron. It is to be noted that, depending on the masses of the stars and on the initial conditions of the close encounter, the eccentricity of the orbit and the distance at closest approach may be quite different for the several cases studied here. Hence, the number of periastron passages shows a wide range of variation. However, a general feature in all cases is that the emission of gravitational waves is largest for the first passages through the periastron and the amplitude of the narrow peaks of gravitational wave radiation decreases in subsequent passages. Note as well that the time difference between successive peaks of the gravitational wave pattern also decreases as time passes by. All this is a consequence of the fact that after every passage through the periastron the orbit is slightly modified, either because the less massive white dwarf is substantially deformed by tidal forces in those cases in which during the first passages through the periastron there is no mass transfer from the less massive white dwarf to the more massive one, or because mass transfer has happened, and the orbit becomes less eccentric. Finally, after several mass transfers the less massive star is totally destroyed, leading to a merger. This causes the sudden decrease of the gravitational wave emission, although some oscillations of the remnant still radiate gravitational waves—in a way very much similar to that of the ring-down phase found in mergers of two compact objects—but with a significantly smaller amplitude.

The dimensionless Fourier transforms of the gravitational wave patterns shown in the top panels of Fig. 3 are displayed in the bottom ones. In contrast with the discrete Fourier transform of the gravitational wave emission presented in Fig. 1, for this case the emission of gravitational presents a continuous spectrum, which has two peaks. The first one is a broad, smooth peak at around $f = 3\,\text{mHz}$, while the second one, which is of larger amplitude but more noisy, occurs at larger frequencies, $\sim$20 mHz. Since in our numerical simulations all lateral collisions have orbits which can be well approximated until near the first periastron passage by elliptical ones, the first of these peaks corresponds to the fundamental mode of the eccentric orbit, while the second one is due to the presence of higher order harmonics, as it occurs for the pure elliptical orbits previously analyzed, being the most significant difference the continuous shift in frequency due to orbit circularization, which noticeably broadens the fundamental mode. In Table 1 we list the SNRs of the simulated waveforms for both for eLISA and ALIA, as well as the number of periastron passages (N). As can be seen, eLISA will not be able to detect these dynamical interactions at a distance of 10 kpc. However, ALIA Crowder et al. (2005) would eventually be able to detect most of these gravitational signals.

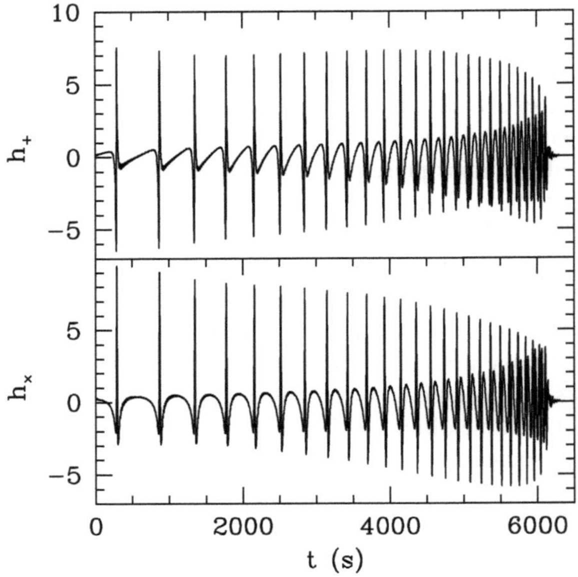

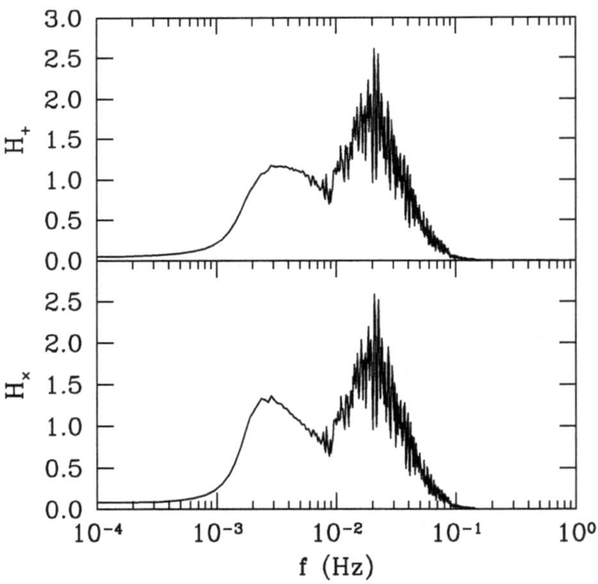

Fig. 3 Gravitational waveforms for a close encounter in which the outcome of the interaction is a lateral collision. For this particular case two white dwarfs of masses $0.8\,M_\odot$ and $0.6\,M_\odot$, respectively, collide. The adopted initial conditions of this interaction are $v_{\rm ini} = 200\,{\rm km\,s^{-1}}$ and $\Delta y = 0.3\,R_\odot$. *Left*: the waveforms h_+ (*top panel*) and $h_\times$ (*bottom panel*) for an inclination $i = 0°$ and a distance of 10 kpc are shown. The units of of the dimensionless stresses are 10^{-22}. *Right*: Fourier transforms of the signals in units of 10^{-23}

Table 1 Detectability of lateral collisions

$M_1 + M_2$	v_{ini}	Δy	N	SNR		$M_1 + M_2$	v_{ini}	Δy	N	SNR	
$(M_\odot)$	km/s	$(R_\odot)$		eLISA	ALIA	$(M_\odot)$	km/s	$(R_\odot)$		eLISA	ALIA
0.4+0.2	75	0.4	2	0.02	1.60	0.6+0.8	150	0.4	6	0.28	24.25
0.4+0.2	100	0.3	2	0.02	1.61	0.8+0.8	150	0.4	4	0.29	31.33
0.6+0.8	100	0.4	2	0.11	12.24	1.0+0.8	150	0.4	5	0.36	36.83
0.8+0.8	100	0.4	2	0.15	18.03	1.2+0.8	150	0.4	3	0.31	32.34
1.0+0.8	100	0.4	2	0.16	18.74	0.8+0.4	150	0.4	3	0.12	8.56
1.2+0.8	100	0.4	2	0.16	18.73	1.2+0.4	150	0.4	2	0.11	9.70
0.4+0.2	100	0.4	2	0.02	1.58	0.6+0.8	200	0.3	27	0.57	45.83
0.4+0.4	100	0.4	2	0.07	5.66	0.8+0.8	200	0.3	20	0.64	63.28
0.6+0.8	150	0.3	2	0.13	13.47	1.0+0.8	200	0.3	5	0.37	37.69
0.8+0.8	150	0.3	2	0.17	19.38	1.2+0.8	200	0.3	3	0.31	32.10
1.0+0.8	150	0.3	2	0.17	20.16	0.8+0.4	200	0.3	3	0.12	8.70
1.2+0.8	150	0.3	2	0.14	15.75	1.2+0.4	200	0.3	2	0.11	9.67
0.4+0.2	150	0.3	2	0.03	1.38	1.2+0.4	200	0.4	8	0.25	15.50
0.4+0.4	150	0.3	2	0.07	6.33						

The last outcome of the dynamical interactions studied here consists in a direct collision, in which a single mass transfer episode, of very short duration occurs. The physical conditions achieved in all these interactions are such that the densities and temperatures necessary to produce a powerful detonation are met, leading in nearly all the cases to the disruption of either the lightest white dwarf or of both stars and their material is ejected as a consequence of the interaction. Although it will not possible to detect the gravitational signal produced in these interactions, owing to their extremely short duration, for the sake of completeness in Fig. 4 a typical example of the gravitational wave pattern is displayed. As can be seen the signal is characterized by a single, very narrow pulse.

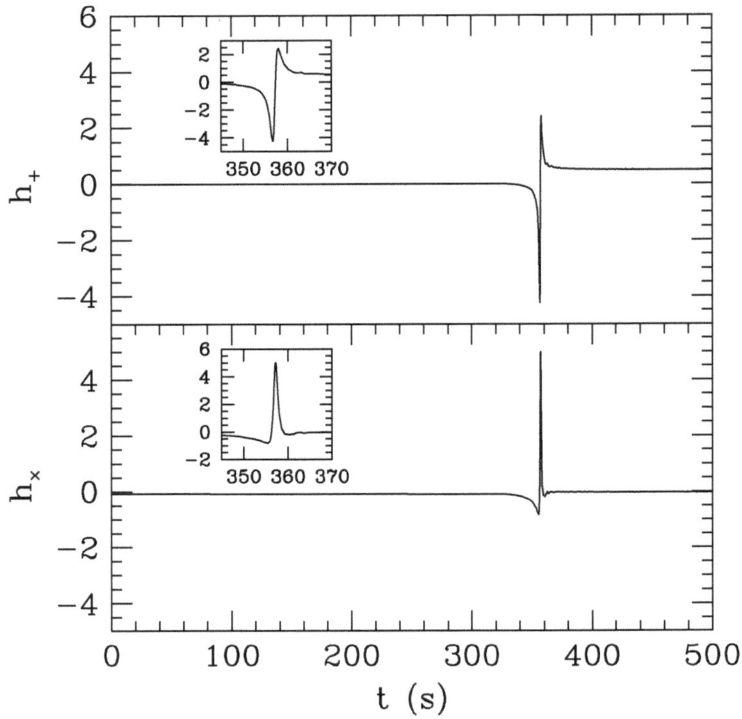

Fig. 4 Gravitational waveform for a typical direct collision, in units of 10^{-21}. The *inset* shows in greater detail the region of the burst of gravitational waves

Conclusions

We have computed the emission of gravitational waves in white dwarf dynamical interactions. This is of interest mainly for those systems which experience a lateral collision, in which several mass transfer episodes occur, or for those pairs of white dwarfs which as a result of the interaction end up forming an eccentric binary system, although we also computed the emission of gravitational waves for those events in which a direct collision, in which only one violent mass transfer event occurs. We have shown that these signals are likely to be detected by future spaceborne detectors like eLISA up to relatively long distances (larger than 10 kpc). This, however, is not the case of the gravitational wave radiation resulting from lateral collisions, since eLISA will not be able to detect them, although more advanced experiments, like ALIA, would be able to detect them. Finally, since for the case of direct collisions the emitted signal consists of a single, well-defined peak, of very short duration, there are no hopes to detect them.

Acknowledgements This work was partially supported by MCINN grant AYA2011–23102, by the AGAUR and by the European Union FEDER funds.

References

G. Aznar-Siguán, E. García-Berro, P. Lorén-Aguilar, J. José, J. Isern, MNRAS **434**, 2539 (2013)
J. Crowder, N.J. Cornish, Phys. Rev. D **72**(8), 083005 (2005)
B. Giacomazzo, L. Rezzolla, L. Baiotti, Phys. Rev. D **83**(4), 044014 (2011)
W.P. Hawley, T. Athanassiadou, F.X. Timmes, ApJ **759**, 39 (2012)
D. Kushnir, B. Katz, S. Dong, E. Livne, R. Fernández, ApJ **778**, L37 (2013)
P. Lorén-Aguilar, J. Guerrero, J. Isern, J.A. Lobo, E. García-Berro, MNRAS **356**, 627 (2005)
P. Lorén-Aguilar, J. Isern, E. García-Berro, MNRAS **406**, 2749 (2010)
C.W. Misner, K.S. Thorne, J.A. Wheeler, *Gravitation* (W.H. Freeman and Co., 1973)
C. Raskin, F.X. Timmes, E. Scannapieco, S. Diehl, C. Fryer, MNRAS **399**, L156 (2009)
C. Raskin, E. Scannapieco, G. Rockefeller, C. Fryer, S. Diehl, F.X. Timmes, ApJ **724**, 111 (2010)
S. Rosswog, D. Kasen, J. Guillochon, E. Ramirez-Ruiz, ApJ **705**, L128 (2009)
H. Wahlquist, Gen. Rel. Grav. **19**, 1101 (1987)
O. Zanotti, L. Rezzolla, J.A. Font, MNRAS **341**, 832 (2003)

Gravitational Recoil and Astrophysical Impact

Ulrich Sperhake

Abstract Asymmetric emission of gravitational waves from astrophysical sources leads to a net flux of linear momentum from the source and, by momentum conservation, imparts a gravitational recoil on the emitting source. Numerical relativity simulations have revealed that this effect can lead to astonishingly large kick velocities, so-called *superkicks*, of several thousand km/s in the inspiral and merger of black-hole binaries. We here discuss the calculation of the recoil in black-hole spacetimes and the astrophysical repercussions of such large kicks, in particular related to the possible displacement or ejection of supermassive black holes from their host galaxies. We also discuss possible mechanisms that would make superkicks less likely to occur in astrophysical binaries and thus explain why most, if not all, galaxies observed in this context appear to harbor a black hole at their center.

1 Introduction and Motivation

A dictionary definition of the term *recoil* is given as the act of "moving abruptly backward as a reaction on firing a bullet, shell, or other missile"; see e.g. Google Dictionary (2011). It may appear surprising at first glace that this effect plays an important role for astrophysical black holes (BHs). After all, BHs are (at classical level) black objects and not supposed to emit anything, not even light. And yet, the gravitational recoil or "kick" of BHs is generally regarded an important dynamical feature in a variety of astrophysical scenarios involving, in particular, supermassive BHs (SMBHs). In this case, the role of the "missile" mentioned in the above definition is played by the gravitational waves (GWs) generated in the inspiral

U. Sperhake (✉)
Department of Applied Mathematics and Theoretical Physics, University of Cambridge, Wilberforce Road, Cambridge CB3 0WA, UK

California Institute of Technology, 1200 E California Boulevard, Pasadena, CA 91125, USA

Department of Physics and Astronomy, The University of Mississippi, University, MS 38677, USA
e-mail: U.Sperhake@damtp.cam.ac.uk

C.F. Sopuerta (ed.), *Gravitational Wave Astrophysics*, Astrophysics and Space Science Proceedings, DOI 10.1007/978-3-319-10488-1_16

and merger of BH binaries. In addition to energy, GWs also carry away linear momentum from their source and if the emission of linear momentum is anisotropic, the net loss of momentum is compensated for by a recoil of the emitting source. We note in this context that kicks also arise in supernova core collapse, mostly through anisotropic emission of neutrinos; see e.g. Wongwathanarat et al. (2013). While these are also of high relevance in astrophysics, our focus here will be on gravitational recoil of BHs generated through GW emission.

The fact that anisotropic gravitational radiation will impart a kick on the emitting source was already realized in the early 1960s (Bonnor and Rotenberg 1961; Peres 1962); see also Bekenstein (1973). The determination of accurate values for the magnitude of the resulting kick velocities, however, represents a major challenge given the high complexity of the dynamical processes responsible for the GW generation. It is only through the use of complex numerical codes, that a quantitative exploration of BH kicks has become possible in the last ~10 years. As we shall discuss in more detail below in Sect. 2, kick velocities range over several orders of magnitude depending on the binary constituents' parameters. For interpreting these numbers, we need some reference of astrophysically relevant velocities. Such reference numbers are given by the escape velocities from the astrophysical objects hosting the BHs. These escape velocities vary with the mass of the host and typical values are $v_{\rm esc} \approx 30$ km/s for globular clusters, 20–100 km/s for dwarf galaxies and ~1,000 km/s for giant elliptic galaxies; cf. Fig. 2 in Merritt et al. (2004). Recoils in the range of these numbers would then be able to displace or even eject the merger remnant BH from the centre of its host with wide ranging astrophysical implications.

Observations of galaxies provide strong evidence that most, if not all, galaxies harbor massive dark objects which are most plausibly interpreted as BHs (Kormendy and Richstone 1995; Magorrian et al. 1998). Observations furthermore demonstrate that the properties of the BHs are correlated with properties of the host galaxies such as the luminosity, mass and velocity dispersion (Kormendy and Richstone 1995; Magorrian et al. 1998; Ferrares and Merritt 2000; Gebhardt et al. 2000). The formation of SMBHs by redshift $z \approx 6$, as suggested by the presence of quasars at such rather early times in the universe, is often described in terms of hierarchical or "bottom-up" growth through accretion and BH mergers; see for example Volonteri et al. (2003). At high redshift $z \geq 10$, the dark matter halos hosting the BHs have escape velocities of less than about 150 km/s, so that even moderate kicks would be sufficient to eject BHs (Micic et al. 2006). Efficient ejection of BHs does not only have consequences for the merger rate of BHs (Holley-Bockelmann et al. 2008; Miller and Lauburg 2009) but also leads to BH depleted globular clusters (Mandel et al. 2008) and may necessitate accretion above the Eddington limit to ensure a sufficiently rapid assembly of SMBHs (Haiman 2004).

BHs ejected or displaced from their hosts' centers may also be directly observable in electromagnetic radiation. Several candidates have indeed been identified, although the interpretation of the data is not unambiguous in these cases. Komossa et al. (2008), for example, have observed a blue shift corresponding to 2,650 km/s of the broad-line region relative to the narrow-line region in the quasar J0927+2943. As we shall see in Sect. 2 below, this magnitude is about half of the maximum recoil velocities generated in BH inspiral and merger and, thus, may represent a BH kicked out of its host. Alternatively, however, the observed features could be explained in terms of a superposition of two active galactic nuclei (AGN) or a binary BH system (Bogdanović et al. 2009; Shields et al. 2009). Similarly, the galaxy CID-42 shows features compatible with an inspiraling AGN pair or one AGN with kick velocity $\sim$1,300 km/s. Using hydrodynamic galaxy merger simulations coupled to radiative transfer, Blecha et al. (2012) find both a double-AGN and a recoil model compatible with the observations and a clear identification of the system is not possible with the present data. A similar example is given by the galaxy merger remnant COSMOS J100043+020637 which contains two optical nuclei about 2 kpc apart (Wrobel et al. 2014). These can be interpreted as a BH ejected or displaced from the galactic centre. Further electromagnetic signatures of BH kicks include tidal disruptions of stars by the moving BH and resulting X ray flares (Komossa and Bade 1999; Komossa and Merritt 2008a; Bloom et al. 2011) and repeated flares caused by a displaced BH oscillating on a scale comparable to the accretion torus (Komossa and Merritt 2008b). Also, the relative motion between a recoiling BH and the accreting material would introduce vibrations in the density of the shock cone with frequencies similar to those observed in quasi-periodic oscillations of X ray sources (Lora-Clavijo and Guzmán 2013). For a more comprehensive discussion of the electromagnetic signatures resulting from recoiling BHs, the reader is referred to Komossa's review (Komossa 2012).

A clear understanding of the kick magnitude and its dependency on the BH binary parameters is clearly necessary for the interpretation of the observations as well as theoretical modeling of galaxies, structure formation in the universe and the assembly of SMBHs. In the next section we discuss the progress made in this direction focusing on numerical relativity applications.

2 Calculation of the Recoil

By simple symmetry considerations the inspiral and merger of a binary system composed of identical compact objects cannot generate a net recoil. Consider for example two non-spinning BHs of equal mass and let us denote the kick velocity resulting from their coalescence by $\mathbf{v}_{\text{kick}}$. Because the two BHs are identical, the configuration obtained by reflection of the system across the center of mass, will be identical to the original one but invert the recoil velocity so that $\mathbf{v}_{\text{kick}} = -\mathbf{v}_{\text{kick}}$ which is only satisfied by $v_{\text{kick}} = 0$. In order to obtain a non-vanishing recoil, we therefore need to break the symmetry of the system under consideration. In the case of BH binaries in vacuum, this can be achieved either by considering BHs of unequal masses or rotating BHs with different spins.

The recoil of unequal-mass but non-spinning BH binary systems was first studied systematically by Fitchett (1983) who used a quasi-Newtonian approach modeling the binary as two point masses in Keplerian orbit. For binaries with zero eccentricity, Fitchett's calculation suggests a functional dependency $v_{\text{kick}} \sim \eta^4 \sqrt{1-4\eta}$ of the kick velocity on the mass ratio which we here define in terms of the individual BH masses m_1 and m_2 through

$$q \equiv \frac{m_2}{m_1} \leq 1\,, \qquad \eta \equiv \frac{m_1 m_2}{(m_1 + m_2)^2} = \frac{q}{(1+q)^2}\,. \tag{1}$$

Note that there is no general convention in the literature but here we label the BHs such that $m_2 \leq m_1$. η is often referred to as the "dimensionless mass-ratio parameter". Fitchett's calculation also includes a generalization to eccentric orbits, but due to the circularizing effect of GW emission (Peters 1964) the eccentricity of BH binaries is typically expected to vanish with high precision in the late inspiral stage which dominates the net emission of linear momentum.

The main question left open by these calculation is the actual magnitude of the kick or, in other words, the proportionality constant in the functional relation between the kick and mass ratio. The determination of these numbers only became possible with the breakthroughs of numerical relativity (Pretorius 2005; Baker et al. 2006a; Campanelli et al. 2006a) which opened up the regime of dynamic, strong-field interaction of BH binaries for precision studies. The recoil generated in the merger of non-spinning, unequal-mass binaries was indeed one of the first applications of numerical relativity (Baker et al. 2006b; González et al. 2007a; Herrmann et al. 2007a) after the breakthroughs. The most extended study of the η parameter space was performed by González et al. (2007a) who simulated a large number of binaries with mass ratios ranging from $q = 1$ to $q = 1/4$. Based on Fitchett's calculations, they fit their data using the relation

$$v_{\text{kick}} = A\eta^2\sqrt{1-4\eta}\,(1 + B\eta)\,, \tag{2}$$

and determine the parameters $A = 1.2 \times 10^4$ km/s, $B = -0.93$. The numerical results of their study, augmented by data for $q = 1/10$ (Gonzalez et al. 2009), are displayed in Fig. 1 together with the fit (2) as well as an alternative fit from Baker et al. (2006b) and an effective-one-body prediction from Schnittman and Buonanno (2007). The maximum kick of non-spinning BH binaries determined from these simulations is 175.2 ± 11 km/s realized for $\eta = 0.195 \pm 0.005$. This value is large enough to eject BHs from globular clusters or dwarf galaxies but well below the escape velocity from giant galaxies.

The inclusion of BH spins significantly complicates the calculation of the recoil. In addition to the mass ratio, there are now six spin parameters, three for each BH. A first exploration of the spins' impact on the emission of linear momentum through GWs was made by Kidder (1995) using post-Newtonian (PN) calculations. The resulting linear momentum is composed of a contribution due to the unequal masses, the Fitchett contribution, and a term due to the spin-orbit coupling. Again, the determination of precise numbers became possible through numerical relativity calculations. These first considered spins parallel to the orbital angular momentum $\boldsymbol{L}$, i.e. spins aligned or anti-aligned with $\boldsymbol{L}$. As discussed above, a non-zero kick is only realized in these cases when the BHs have different masses or different

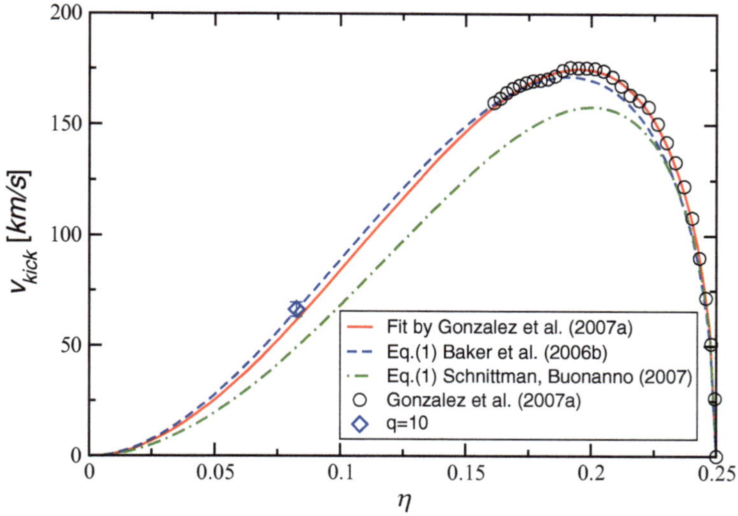

Fig. 1 Recoil velocity resulting from the merger of non-spinning BH binaries with dimensionless mass-ratio parameter η. The *circle* and *diamond symbols* represent numerical relativity results and the *curves* various fits or predictions. Figure taken from Gonzalez et al. (2009)

spins. A variety of configurations were analysed in Herrmann et al. (2007b), Koppitz et al. (2007), and Pollney et al. (2007) and predict kicks of up to 500 km/s. This maximum is obtained by extrapolating from the simulations to the maximal spin magnitude $\chi \equiv a/m = 1$, where a is the Kerr spin parameter and m the BH mass. Kidder's work, however, suggests a separate contribution to the total kick pointing out of the orbital plane. Based on this motivation, González et al. (2007b) and Campanelli et al. (2007a, b) simulated equal-mass BH binaries with spins oriented perpendicular to $\boldsymbol{L}$, i.e. in the orbital plane, but opposite to each other. The astonishing result of these simulations were kick velocities of a few thousand km/s with an extrapolated (to $\chi = 1$) value of 4,000 km/s, often referred to as *superkick*. Figure 2 illustrates the BH trajectories in an inspiral lasting ~2 orbits; during inspiral, the orbital plane moves up and down along the direction of the orbital angular momentum (pointing upwards in the figure) and, after merger, the single BH moves in that direction with v_{kick}. Closer investigation of these superkicks revealed that v_{kick} has a sinusoidal dependence on the angle between the spin of either hole and the line-of-sight between the holes at the start of the simulation which corresponds to a dependency on the orbital phase; cf. Fig. 4 in Brügmann et al. (2008). An intuitive interpretation of the superkick phenomenon is provided in Fig. 5 of Pretorius (2009). The motion of the orbital plane up and down along the direction of $\boldsymbol{L}$ is a consequence of the frame dragging exerted on each binary member by the other hole. This motion introduces a Doppler shift to the emitted gravitational radiation and, thus, an asymmetry in the amount of GWs emitted in the corresponding directions. The net effect is an asymmetry in the $l = 2, m = \pm 2$ GW

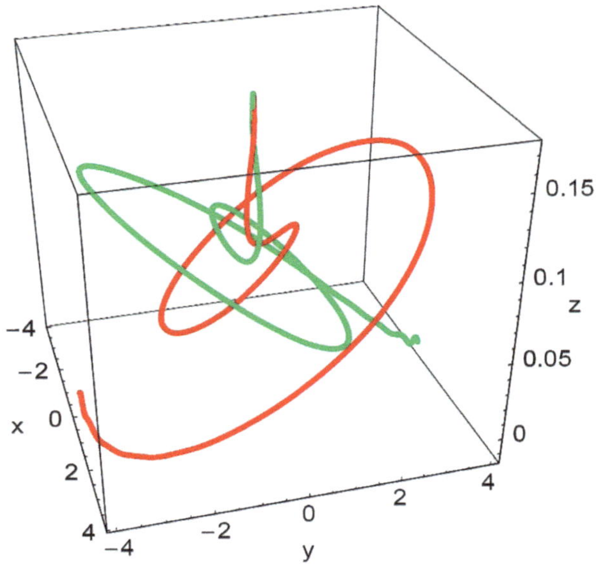

Fig. 2 Trajectories of the BHs in a superkick configuration with individual spins oriented in the orbital plane but opposite to one another

multipoles (Brügmann et al. 2008) and a net amount of linear momentum emitted in GWs whose magnitude depends on the precise phase in this cyclic process when the merger puts an end to it.

A further increase in the maximum kick in BH binary inspiral has been realized by Lousto and Zlochower (2011) and Lousto et al. (2012) by combining the superkick with the "orbital hang up" effect. The orbital hang up arises in the inspiral of BHs with spins aligned with the orbital angular momentum and results in a larger number of orbits at small separation shortly before merger and, consequently, increased emission of GW energy by about a factor of two compared with the non-spinning counterpart (Campanelli et al. 2006b; Lovelace et al. 2012). For equal masses and spins, these configurations do not result in a net linear momentum due to the symmetry argument mentioned above. For BH spins *partially* aligned with $\boldsymbol{L}$ such that the spin projections onto the orbital plane are equal in magnitude and opposite in direction, however, Lousto and Zlochower were able to maximize the increased GW emission of the hang-up effect with the net-emission of linear momentum and observed even larger kicks of up to 5,000 km/s realized for inclination angles of about 45° and extrapolated to maximal spins $\chi = 1$.

A main goal for the modeling of gravitational recoil in BH binaries is the generation of a "black box" or fitting formula that takes as input the parameters of the binary and returns the kick velocity as output. We have already seen such a formula for the simple case of non-spinning binaries in Eq. (2). Various suggestions of varying degrees of complexity have been made to generalize such a formula for spinning binaries. Campanelli et al. (2007a) have proposed the following expression for modeling superkicks, but not including the hang-up kicks. Here we again denote the dimensionless spin parameters by χ with index 1 or 2 for either hole. Boldface characters represent vectorial quantities and we use sub or super scripts $\perp$ and $||$ to denote vector components perpendicular and parallel to the orbital angular momentum $\boldsymbol{L}$, respectively. With this notation we have[1]

$$\begin{aligned}
\boldsymbol{v}_{\text{kick}}(q, \boldsymbol{\chi}_i) &= v_m \boldsymbol{e}_1 + v_\perp (\cos\xi\, \boldsymbol{e}_1 + \sin\xi\, \boldsymbol{e}_2) + v_{||} \boldsymbol{e}_{||}\,, \\
v_m &= A\eta^2 \frac{1-q}{1+q} (1 + B\eta)\,, \\
v_\perp &= H \frac{\eta^2}{1+q} \left(q \left|\chi_2^{||}\right| - \left|\chi_1^{||}\right| \right), \\
v_{||} &= K \frac{\eta^2}{1+q} \cos(\Theta - \Theta_0) \left| q\boldsymbol{\chi}_2^\perp - \boldsymbol{\chi}_1^\perp \right|. \qquad (3)
\end{aligned}$$

Here $(\boldsymbol{e}_1, \boldsymbol{e}_2, \boldsymbol{e}_{||})$ is an orthonormal basis with $\boldsymbol{e}_{||}$ pointing along $\boldsymbol{L}$, ξ is an angle found to be $\sim 145°$ for a wide variety of quasi-circular configurations (Lousto and Zlochower 2009) and Θ is the angle between $q\boldsymbol{\chi}_2^\perp - \boldsymbol{\chi}_1^\perp$ and the line of sight between the BHs at the "time of merger". This angle is determined by the phase angle mentioned above and enters the superkick in the sinusoidal form discussed in this context. For most astrophysical applications, this angle is only known in the form of a statistical distribution commonly chosen to be flat in $[0, 180°]$. The kick magnitude is then determined by the fitting coefficients which are $A = 1.2 \times 10^4$ km/s, $B = -0.93$ (González et al. 2007a), $H = (6.9 \pm 0.5) \times 10^3$ km/s (Lousto and Zlochower 2008), $K = (6.0 \pm 0.1) \times 10^4$ km/s (Campanelli et al. 2007b).

Various attempts have been made to extend Eq. (3). A systematic spin expansion of the recoil has been explored by Boyle et al. (2008) and Boyle and Kesden (2008) exploiting all possible symmetry conditions. The resulting expansion is yet to be calibrated by numerical relativity simulations. An extension of Eq. (3) that includes the hang-up kick is given in Lousto et al. (2012) and further elaboration through inclusion of various higher-order terms has resulted in a further $\sim 10\,\%$ increase in the maximum kick, an effect dubbed *cross kick* (Lousto and Zlochower 2013). A detailed discussion of these formulae is beyond the scope of this work and we refer the interested reader to the articles for further information.

---

[1] We use here the notation of Kesden et al. (2010b) which differs from that of Campanelli et al. (2007a) by the convention that $q \equiv m_2/m_1 \leq 1$.

3 Open Questions

The accurate determination of kick velocities generated through GW emission in the merger of BH binaries has made tremendous progress over the last decade and the discovery of the superkicks remains one of the most astonishing results obtained with numerical relativity simulations to date. Still, a good deal of work needs to be done before the kick phenomenon can be regarded as comprehensively understood. In this section we will list our pick of the most important questions that need to be resolved in future work.

*(1) **A complete kick formula:*** We have seen that asymmetries in the BH spins introduce significantly larger recoil velocities than unequal mass parameters do for non-spinning BHs. The inclusion of spins and the associated six parameters is therefore absolutely necessary for obtaining astrophysically helpful kick formulas and, as we have seen, plenty of work has already been invested in this direction. Nevertheless, a full exploration of the parameter space will require more numerical simulations. The majority of simulations used to calibrate existing formulae for spinning binaries have been performed for equal-mass systems. While there have been explorations of spinning-unequal mass BHs (Baker et al. 2008a, b; Campanelli et al. 2007a; Lousto and Zlochower 2009), the dependence of the superkick and hang-up kicks, in particular, is not yet known with sufficient precision. The good news is that this question can be addressed by simply performing more numerical simulations. Even though this requires considerable resources given the high dimensionality of the parameter space, it appears straightforward to fill in the missing gaps in the kick formulae.

*(2) **Parameter evolution during the inspiral:*** A conceptually more delicate problem arises from the BH parameters that are actually used as input for existing (or future) kick formulae. Numerical simulations typically start a few orbits prior to merger of the binary. For a given system this provides accurate estimates for the recoil magnitude because the emission of linear momentum is entirely dominated by the last few orbits prior to merger; see e.g. Fig. 9 in Brügmann et al. (2008). Astrophysical studies involving BHs, however, make predictions (often in statistical form for an ensemble of BHs) for the parameters at a time when the BHs are still much farther apart; see e.g. Gerosa et al. (2013). The ensuing inspiral from such large scales to the late inspiral regime of numerical relativity covers hundreds of thousands of orbits if not more. The question then is how this long inspiral phase will modify the BH parameters and how we can generate legitimate input for kick formulae valid for the last few orbits. This question is largely of statistical nature and in practice we will not be concerned too much with how the parameters of one particular binary are affected, but rather how the inspiral gradually influences and distorts a parameter distribution. This question has first been explored by Bogdanović et al. (2007) who use PN calculations (Blanchet 2006) to evolve an ensemble of BHs with mass ratio $q = 9/11$, maximal spin magnitudes and isotropic distribution of the spin directions from a separation $r = 1{,}000\ M$ to $r = 10\ M$ where M is the total BH binary

mass. They find this ensemble to remain isotropic, i.e. statistically unchanged, during the inspiral. As we shall see in the next section this does not remain the case when we start with initially anisotropic ensembles. In applications of kick formulae, this necessitates a prescription for how the statistical properties of a given distribution change under the inspiral.

(3) ***Astrophysical observations of SMBHs in galaxies:*** The third question is of astrophysical nature. As mentioned above, astrophysical observations suggest that large galaxies ubiquitously harbor SMBHs. On the other hand, the magnitude of superkicks or hangup kicks comfortably exceeds the escape velocity from even the most massive galaxies and thus would suffice to eject BHs from their hosts. Why then do we not observe more galaxies without a BH? Possibly, we simply need a larger statistical sample of observations. Schnittman and Buonanno (2007) have estimated the statistical distribution of kicks generated in BH mergers using effective-one-body calculations for an ensemble of BH binaries with $q \in [0.1, 1]$, $\chi_i = 0.9$ and find that about 12 % of the mergers result in $v_{\text{kick}} > 500$ km/s and about 3 % in $v_{\text{kick}} >1{,}000$ km/s. Recent work by Gerosa and Sesana (2014) models the consequences of superkicks in the merger history of brightest-cluster galaxies and finds that the BH occupation fraction f of these galaxies is $0.9 < f < 0.99$ in the local universe. A statistically robust determination of the frequency of BH ejection therefore seems to require hundreds of observations which will be made possible by future thirty-meter-class telescopes. An alternative explanation of ubiquitous presence on SMBHs in galaxies may be that superkicks, while theoretically possible, are statistically suppressed by some mechanism. This could be achieved, for example, by the alignment of BH spins with the orbital angular momentum through torques from accreting gas in gas-rich mergers (Bogdanović et al. 2007). The degree of alignment is likely to depend on the properties of the gas disks and may reduce the angle between the spins and $\boldsymbol{L}$ to $10°$ ($30°$) for cold (hot) gas (Dotti et al. 2009). We emphasize here, that these angles are parameters valid at large separation and their validity as input parameters for kick formulae is still subject to the concerns raised in question 2. In the next section we will see that the inclusion of the long inspiral stage up to the last few orbits indeed has a profound statistical effect that may disfavor those spin configurations leading to superkicks.

4 Spin-Orbit Resonances

A BH binary system containing two spinning holes can be characterized by ten parameters: two BH masses m_1 and m_2, six parameters for the spins $\boldsymbol{S}_1$ and $\boldsymbol{S}_2$ and two parameters for the direction $\hat{\boldsymbol{L}}$ of the orbital angular momentum. The magnitude L is merely a measure of the separation of the BHs and does not characterize the actual system. The inspiral phase of a BH binary from large separations up to the last $\sim$10 orbits is rather well modeled by PN theory (Blanchet 2006) which in particular determines the time evolution of the above parameters including the BH separation and thus L. The physical description is simplified significantly by eliminating seven of these parameters as follows.

- BH binaries are invariant under a rescaling with the total mass $M = m_1 + m_2$ which leaves only one parameter for the mass ratio: $q = m_2/m_1$.
- At the PN orders considered, the BH masses, and hence the mass ratio q, and the individual BH spin magnitudes S_1, S_2 are conserved.
- We choose the z axis of our coordinate system such that it points in the direction of the orbital angular momentum.
- We choose the x axis such that it points in the direction of the projection of $\boldsymbol{S}_1$ onto the orbital plane.

The time evolution of the system is therefore described by three remaining parameters which we choose to be the angles θ_1 and θ_2 between the individual BH spins and $\boldsymbol{L}$ and the angle $\Delta\Phi$ between the projections of $\boldsymbol{S}_1$ and $\boldsymbol{S}_2$ onto the orbital plane; cf. Fig. 3. These angles can be directly obtained from the individual spins and the orbital angular momentum whose time evolution is determined by the PN equations; see e.g. Appendix A in Gerosa et al. (2013).

In general, the BH spins and $\boldsymbol{L}$ precess in a complicated way around the total angular momentum vector $\boldsymbol{J} \equiv \boldsymbol{L} + \boldsymbol{S}$, where $\boldsymbol{S} \equiv \boldsymbol{S}_1 + \boldsymbol{S}_2$. Schnittman (2004), however, has found a subset of configurations where all three vectors $\boldsymbol{L}$, $\boldsymbol{S}_1$ and $\boldsymbol{S}_2$ are locked in a plane as they jointly precess around $\boldsymbol{J}$. These configurations are often referred to as "spin-orbit resonances" and are a consequence of the fact that the three time scales involved in a BH binary inspiral, the orbital time $t_{\rm orb}$, the precession time $t_{\rm pr}$ and the radiation reaction time scale $t_{\rm GW}$ obey a clear hierarchy $t_{\rm orb} \ll t_{\rm pr} \ll t_{\rm GW}$. Evidently, for these resonance configurations $\Delta\Phi = 0°$ or $\Delta\Phi = \pm 180°$ but Schnittman also showed that for each of the two resonances there exist a one-parameter family of values (θ_1, θ_2) which remain constant on the precession time

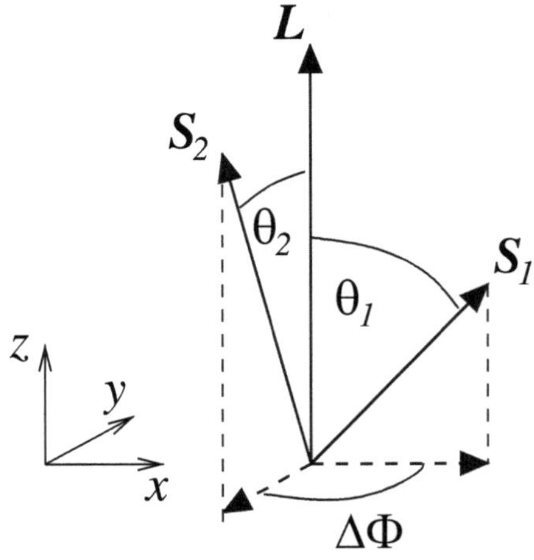

Fig. 3 The orbital angular momentum vector $\boldsymbol{L}$ and the individual BH spins $\boldsymbol{S}_1$, $\boldsymbol{S}_2$ define the angles θ_1, θ_2 and $\Delta\Phi$. x, y and z denote our specific choice of coordinates. Note that $\Delta\Phi$ is negative for the example shown

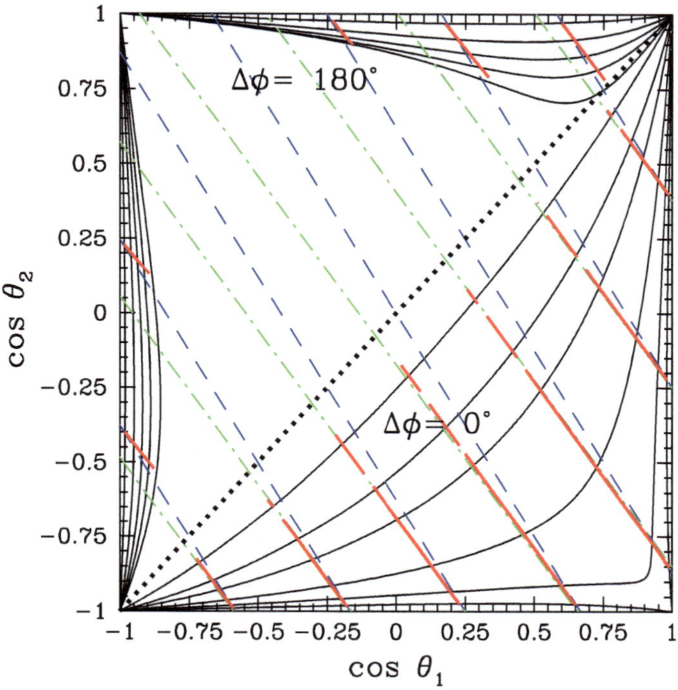

Fig. 4 Resonance families (*black solid curves* corresponding to $r = 1{,}000\ M$, $500\ M$, $250\ M$, $100\ M$, $50\ M$ and $10\ M$ from the square's edges inwards) are shown in the θ_1-θ_2 plane. As resonant binaries inspiral, they move towards the diagonal $\theta_1 = \theta_2$ along the *red dashed lines*. It can be shown that freely precessing binaries move along the same lines, although not in monotonic fashion towards the diagonal but back and forth. The *blue short-dashed curves* correspond to constant $\boldsymbol{S} \cdot \hat{\boldsymbol{L}}$ and *green dash-dotted curves* to constant $\boldsymbol{S}_0 \cdot \hat{\boldsymbol{L}}$. Figure taken from Kesden et al. (2010a) (Color figure online)

scale. On the much longer radiation reaction timescale, however, θ_1 and θ_2 slowly evolve while $\Delta\Phi$ remains constant at 0° or ±180°. At a given moment in time we therefore have two curves in the θ_1-θ_2 plane, one curve for the $\Delta\Phi = 0°$ resonance and one for the $\Delta\Phi = \pm 180°$ resonance. As the BHs inspiral on the t_{GW} time scale, these two curves gradually sweep through the θ_1-θ_2 plane. Moreover, the resonances act as an attractor and capture freely precessing binaries into resonance or near-resonance configurations; cf. Figs. 6 and 7 in Schnittman (2004). The resonances are illustrated in Fig. 4 for BH parameters $\chi_1 = \chi_2 = 1$ and $q = 9/11$. The resonant families are displayed as the solid (black) curves in the bottom right ($\Delta\Phi = 0°$) and the upper left triangle ($\Delta\Phi = \pm 180°$ resonance). The diagonal separating the two regions corresponds to $\theta_1 = \theta_2$. The figure demonstrates that $\theta_1 < \theta_2$ for the $\Delta\Phi = 0$ resonance solutions and $\theta_1 > \theta_2$ for those with $\Delta\Phi = \pm 180°$. This result is in fact general and applies to any choice of χ_1, χ_2 and q (Schnittman 2004). In the limit of infinite BH separation, the resonance curves coincide with the edges of the square. As the BHs inspiral, the resonance curves gradually approach the diagonal, but this motion is more pronounced for the $\Delta\Phi = 0°$ family. For each family, the figure shows six curves corresponding to a BH separation $r = 1{,}000\ M$, $500\ M$, $250\ M$, $100\ M$, $50\ M$ and $10\ M$. The dashed (red curves) represent the curves along which individual resonant binary systems evolve. The short-dashed (blue) lines are curves of constant $\boldsymbol{S} \cdot \hat{\boldsymbol{L}}$ and the dash-dotted (green) lines are curves of constant $\boldsymbol{S}_0 \cdot \hat{\boldsymbol{L}}$ where $\boldsymbol{S}_0 \equiv (1+q)\boldsymbol{S}_1 + (1+q^{-1})\boldsymbol{S}_2$. From the figure, and bearing in mind the attractive character of the resonance families, we can arrive at the following conclusions.

- Initially non-isotropic ensembles of spinning BH binaries can dramatically change their characteristics. In particular, binaries starting with $\theta_1 < \theta_2$ (i.e. the more massive BH's spin is more aligned with $\boldsymbol{L}$) but with isotropic distribution in $\Delta\Phi$ are gradually captured by $\Delta\Phi = 0°$ resonances and therefore cluster near $\Delta\Phi = 0°$. Likewise, ensembles starting with $\theta_1 > \theta_2$ (the more massive BH's spin is more misaligned with $\boldsymbol{L}$) preferentially cluster near $\Delta\Phi = \pm 180°$. This is important since superkick configurations have $\Delta\Phi = 180°$.
- As binary systems move towards the diagonal, θ_1 and θ_2 approach each other. For $\Delta\Phi = 0°$ resonances this means $\boldsymbol{S}_1$ and $\boldsymbol{S}_2$ align. For the $\Delta\Phi = \pm 180°$ resonances, instead, the angle θ_{12} between $\boldsymbol{S}_1$ and $\boldsymbol{S}_2$ approaches $\theta_1 + \theta_2$.
- The dashed (red) and dash-dotted (green) lines in Fig. 4 are coincident (within numerical accuracy) in the figure. As the binaries inspiral, the projection of the weighted spin $\boldsymbol{S}_0$ onto the orbital angular momentum is therefore conserved; an analytic calculation at the used PN order confirms this conclusion (Kesden et al. 2010a). The short-dashed (blue) curves corresponding to constant $\boldsymbol{S} \cdot \hat{\boldsymbol{L}}$ are steeper than the dashed (red) curves. The total spin $\boldsymbol{S}$ therefore becomes gradually more misaligned (aligned) with the orbital angular momentum for $\Delta\Phi = 0°$ ($\Delta\Phi = \pm 180°$) resonance configurations.

It can furthermore be shown that $\boldsymbol{S}_0 \cdot \hat{\boldsymbol{L}}$ is also conserved for non-resonant binaries and the corresponding conclusions therefore apply for generic binaries. The most important conclusion for our discussion is the first item in the above list: binaries are attracted towards $\Delta\Phi = 0°$ ($\Delta\Phi = \pm 180°$) if the more massive (the less massive) BH's spin is more aligned with the orbital angular momentum. The intuitive conclusion is that preferential alignment (misalignment) of the more massive BH with $\boldsymbol{L}$ leads to a statistical suppression (enhancement) of superkicks. We shall see that this is indeed the case in the next section.

5 Suppression of Superkicks

The impact of the spin-orbit resonances on the recoil velocities is of statistical nature. We have seen that for a given ensemble of BH binaries, the distribution of the BH spins, characterized by the angles θ_1, θ_2 and $\Delta\Phi$, can change substantially as the binaries inspiral from $r = 1{,}000\ M$ to $r = 10\ M$. In order to quantify the resulting effect on the expected kick distribution, Kesden et al. (2010b) consider two types of ensembles with fixed values $\chi_1 = \chi_2 = 1$, $q = 9/11$. Ensemble 1 contains

$10 \times 10 \times 10$ binaries equally spaced in $\Delta\Phi = [-180°, 180°)$ and $\cos\theta_1$, $\cos\theta_2 \in [0, 1]$, i.e. with isotropic distribution in the spin directions. Ensembles of type 2 consist of 30×30 binaries equally spaced in $\cos\theta_2$ and $\Delta\Phi$ but with a fixed value of θ_1. We consider six ensembles of type 2 for the specific values $\theta_1 = 10°$, $20°$, $30°$, $150°$, $160°$ and $170°$. Ensembles with a small θ_1 represent binaries where the more massive BH is preferentially more aligned with $\boldsymbol{L}$ whereas those ensembles with large θ_1 have the less massive BH preferentially aligned with $\boldsymbol{L}$. From the results of the previous section, we expect the former to be attracted by the $\Delta\Phi = 0°$ resonances and therefore to result in smaller kicks. For large θ_1 we expect the converse.

This expectation is borne out in Fig. 5 which displays histograms of the recoil velocities for the different ensembles. Let us first consider the upper panel of

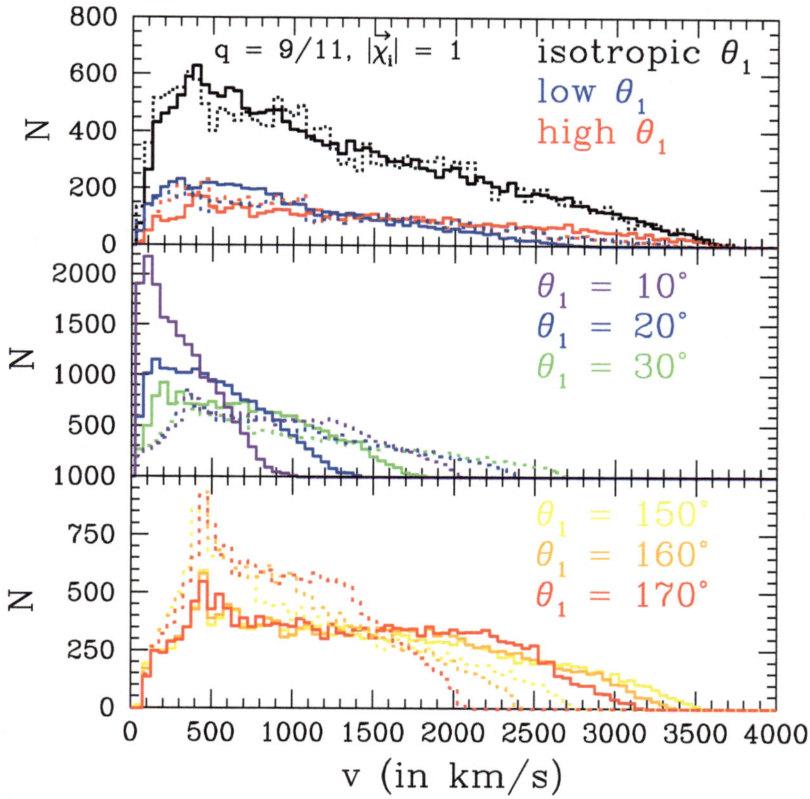

Fig. 5 Histograms of the kick distribution obtained for Ensemble 1 (*top panel*) and six ensembles of type 2 (*center and bottom panel*). *Dashed curves* are obtained for the BH binary parameters as initialized at large separation $r = 1{,}000\ M$ and *solid curves* are obtained for the parameters at the end of the inspiral $r = 10\ M$. Changes in the histograms due to the inspiral depend on whether the more massive BH is initially more aligned (small values of θ_1) or more misaligned (large values of θ_1) with the orbital angular momentum $\boldsymbol{L}$. Figure taken from Kesden et al. (2010b)

the figure. The dashed black curve shows the kick distribution for ensemble 1 as predicted by Eq. (3) for the input parameters of the binaries at the start of the inspiral, i.e. at $r = 1{,}000\ M$. Likewise, the solid black curve shows the distribution when using the BH binary parameters at the end of the inspiral, i.e. at $r = 10\ M$. Ensemble 1 is an isotropic ensemble and as we have already discussed in the previous section, initially isotropic ensembles remain isotropic under the inspiral and the corresponding kick distributions are identical up to statistical noise. The blue (red) curves in the same panel correspond to subsets of ensemble 1 containing only those 30 % of the binaries with the lowest (highest) values of θ_1. For these non-isotropic subsets we observe the expected change in the kick distribution. For ensembles where the more massive BH is more (less) aligned with the orbital angular momentum, that is for the "blue" ("red") subsets, the kick distribution obtained for the parameters at the end of the inspiral is shifted towards smaller (larger) kick velocities. This is displayed in more detail in the center and bottom panel of the figure where the six ensembles of type 2 are displayed, again in the form of dashed curves for the initial values of θ_1, θ_2 and $\Delta\Phi$ at $r = 1{,}000\ M$ and solid curves for the values at the end of the inspiral at $r = 10\ M$. The central panels contains ensembles with $\theta_1 = 10°$ (purple), $\theta_1 = 20°$ (blue) and $\theta_1 = 30°$ (green). The kick distribution after inspiral is notably shifted towards smaller $v_{\rm kick}$, the effect being stronger the smaller θ_1. The opposite is observed in the bottom panel where $\theta_1 = 150°$ (yellow), $\theta_1 = 160°$ (orange) and $\theta_1 = 170°$ (red curves). The larger θ_1 is fixed initially, the more strongly the distributions are shifted towards larger $v_{\rm kick}$ after the inspiral.

We have focused here on maximal spin magnitude $\chi_1 = \chi_2 = 1$ and mass ratio $q = 9/11$. The complete study performed by Kesden et al. (2010a, b) also considers smaller spin magnitudes 0.75, 0.5 and 0.25 as well as other mass ratios $q = 2/3$ and $1/3$. The kick suppression or enhancement due to the spin-orbit resonances becomes less pronounced for smaller values of the spin magnitude and mass ratio. In these cases, however, the kick velocities are significantly smaller anyway (whether including resonance effects or not), so that alterations to the distribution due to capture by resonances during the inspiral are less important in the context of BH ejection or displacement from the centers of galaxies. See Figs. 3, 4 and Table 1 in Kesden et al. (2010b) for more quantitative details.

We finally note that the kick formula used for this investigation is given by Eq. (3) and therefore does not include the hang-up effect in the recoil velocities. A similar study to that described above has been done for the hang-up kicks in Berti et al. (2012) and observed very similar suppression (enhancement) of kick distributions when using a more recent formula by Lousto and Zlochower (2011); see also Lousto et al. (2012). The effective suppression or enhancement of hang-up kicks due to resonant capture may appear surprising at first glance since partial alignment of the individual BH spins with $\boldsymbol{L}$ is a vital ingredient in the hang-up kicks. The key angle affected by the resonances, however, is $\Delta\Phi$. And BH configurations leading to superkicks or hang-up kicks always require $\Delta\Phi \approx 180°$. As we have seen, the resonances tend to push BH binaries either close to this value of $\Delta\Phi$ (when the more massive BH is preferentially misaligned with $\boldsymbol{L}$, i.e. large θ_1) or to the opposite extreme $\Delta\Phi = 0°$ (when the more massive BH is preferentially aligned with $\boldsymbol{L}$, i.e. small θ_1).

Given these results, there remains one important outstanding question: is the more massive BH preferentially aligned with the orbital angular momentum or not? The answer to this question ultimately has to come from astrophysics and, in particular, detailed studies of the interaction of the individual BHs in a binary with surrounding accretion disks from the point of formation of the binary system up to the point where the inspiral is driven by GW emission, i.e. up to separations $r \sim 1{,}000\ M$. The seemingly ubiquitous presence of SMBHs at the center of galaxies may indicate that superkicks are indeed suppressed and therefore suggest stronger alignment of the more massive BH but the answer to this question remains at present unknown.

Conclusions

Gravitational recoil generated by the emission of anisotropic gravitational radiation may manifest itself in a variety of observational features of galaxies and is likely to play an important role in the formation of galaxies and the assembly of the observed SMBHs. Observational signatures in the electromagnetic spectrum include double AGNs, X ray flares and Doppler shifts between the broad and narrow-line regions of quasars.

Following the breakthroughs in numerical relativity, it has been possible to obtain precision estimates for the magnitude of the kick velocities as functions of the BH parameters. For non-spinning BH binaries the maximum kick is 175 km/s realized for a mass ratio $\eta = 0.195$. The most astounding result of the numerical studies has been the discovery of *superkicks* of several thousand km/s generated for BH binaries with comparable mass and spin components in the orbital plane that are equal in magnitude and opposite in direction. These kicks are comfortably large enough to eject BHs from even the most massive host galaxies and may explain some of the observed features mentioned above although kicks are not the only explanations for these observations.

The large magnitude of the superkicks naturally prompts the question why SMBHs do not appear to be efficiently kicked out of their host galaxies. The answer to this question may simply be of statistical nature and BH ejection may feature more prominently in larger future surveys. There are indications, however, that superkick configurations may not be the favored arrangements in astrophysical binaries. Accretion torques tend to align the BH spins with the orbital angular momentum. During the GW driven inspiral all the way to merger, spin-orbit resonances are likely to populate preferentially specific portions of the parameter space. This effect depends on whether the more massive BH is initially more or less aligned with the orbital angular momentum than its lighter companion. In the former case, superkicks are

(continued)

suppressed whereas in the latter case they become more likely. It is at present unknown which of these scenarios is more common. Future observations of the BH occupation fraction of galaxies should provide valuable insight into this question. Either way, the spin-orbit resonances demonstrate the importance of including the long GW driven inspiral in the usage of BH parameters in formulae predicting kick velocities. Further calibration of these kick formulae is also required, in particular to obtain accurate determination of the hang-up or superkicks' dependency on the mass ratio.

Acknowledgements The author thanks E. Berti, D. Gerosa, M. Kesden, R. O'Shaughnessy for many fruitful discussions. This work was supported by NSF-XSEDE Grant No. PHY-090003, FP7-PEOPLE-2011-CIG Grant No. 293412 "CBHEO", FP7-PEOPLE-2011-IRSES Grant No.295189 "NRHEP", STFC GR Roller Grant No. ST/L000636/1, the Cosmos system, part of DiRAC, funded by STFC and BIS under Grant Nos. ST/K00333X/1, ST/H008586/1 and ST/J005673/1, ERC-2010-StG Grant No. DyBHo, and CESGA Grant No. ICTS-2013-249. Computations were performed on the Cambridge Cosmos system, the SDSC Trestles and NICS Kraken clusters, and CESGA's Finis Terra.

References

J.G. Baker, J. Centrella, D.I. Choi, M. Koppitz, J. van Meter, Phys. Rev. Lett. **96**, 111102 (2006a). doi:10.1103/PhysRevLett.96.111102. gr-qc/0511103

J.G. Baker, J. Centrella, D.I. Choi, M. Koppitz, J. van Meter, M.C. Miller, Astrophys. J **653**, L93 (2006b). doi:10.1086/510448. gr-qc/0603204

J.G. Baker et al., Astrophys. J. **668**, 1140 (2008a). doi:10.1086/521330. astro-ph/0702390

J.G. Baker et al., Astrophys. J. **682**, L29 (2008b). doi:10.1086/590927. arXiv:0802.0416 [astro-ph]

J.D. Bekenstein, Astrophys. J. **183**, 657 (1973). doi:10.1086/152255

E. Berti, M. Kesden, U. Sperhake, Phys. Rev. D **85**, 124049 (2012). doi:10.1103/PhysRevD.85.124049. arXiv:1203.2920 [astro-ph]

L. Blanchet, Living Rev. Relat. **9**(4) (2006). http://www.livingreviews.org/lrr-2006-4

L. Blecha, F. Civano, M. Elvis, A. Loeb, (2012). arXiv:1205.6202 [astro-ph]

J.S. Bloom et al., Science **333**, 203 (2011). doi:10.1126/science.1207150. arXiv:1104.3257 [astro-ph]

T. Bogdanović, C.S. Reynolds, M.C. Miller, Astrophys. J. **661**, L147 (2007). doi:10.1086/518769. astro-ph/0703054

T. Bogdanović, M. Eracleous, S. Sigurdsson, Astrophys. J. **697**, 288 (2009). doi:10.1088/0004-637X/697/1/288. arXiv:0809.3262 [astro-ph]

W.B. Bonnor, M.A. Rotenberg, Proc. R. Soc. Lond. A. **265**, 109 (1961)

L. Boyle, M. Kesden, S. Nissanke, Phys. Rev. Lett. **100**, 151101 (2008). doi:10.1103/PhysRevLett.100.151101. arXiv:0709.0299 [gr-qc]

L. Boyle, M. Kesden, Phys. Rev. D **78**, 024017 (2008). doi:10.1103/PhysRevD.78.024017. arXiv:0712.2819 [astro-ph]

B. Brügmann, J.A. González, M.D. Hannam, S. Husa, U. Sperhake, Phys. Rev. D **77**, 124047 (2008). doi:10.1103/PhysRevD.77.124047. arXiv:0707.0135 [gr-qc]

M. Campanelli, C.O. Lousto, P. Marronetti, Y. Zlochower, Phys. Rev. Lett. **96**, 111101 (2006a). doi:10.1103/PhysRevLett.96.111101. gr-qc/0511048

M. Campanelli, C.O. Lousto, Y. Zlochower, Phys. Rev. D **74**, 041501 (2006b). doi:10.1103/PhysRevD.74.041501. gr-qc/0604012

M. Campanelli, C.O. Lousto, Y. Zlochower, D. Merritt, Astrophys. J. **659**, L5 (2007a). doi:10.1086/516712. gr-qc/0701164

M. Campanelli, C.O. Lousto, Y. Zlochower, D. Merritt, Phys. Rev. Lett. **98**, 231102 (2007b). doi:10.1103/PhysRevLett.98.231102. gr-qc/0702133

M. Dotti, M. Volonteri, A. Perego, M. Colpi, M. Ruszkowski, F. Haardt, Mon. Not. R. Astron. Soc. **402**, 682 (2009). arXiv:0910.5729 [astro-ph]

L. Ferrares, D. Merritt, Astrophys. J. **539**, L9 (2000). doi:10.1086/312838. astro-ph/0006053

M.J. Fitchett, Mon. Not. R. Astron. Soc. **203**, 1049 (1983)

K. Gebhardt, R. Bender, G. Bower, A. Dressler, S.M. Faber, A.V. Filippenko, R. Green, C. Grillmair, L.C. Ho, J. Kormendy, T.R. Lauer, J. Magorrian, J. Pinkney, S. Richstone, D. Tremaine, Astrophys. J **539**, L13 (2000). doi:10.1086/312840. astro-ph/0006289

D. Gerosa, M. Kesden, E. Berti, R. O'Shaughnessy, U. Sperhake, Phys. Rev. D **87**, 104028 (2013). doi:10.1103/PhysRevD.87.104028. arXiv:1302.4442 [gr-qc]

D. Gerosa, A. Sesana, (2014). arXiv:1405.2072 [astro-ph]

J.A. González, U. Sperhake, B. Brügmann, M.D. Hannam, S. Husa, Phys. Rev. Lett. **98**, 091101 (2007a). doi:10.1103/PhysRevLett.98.091101. gr-qc/0610154

J.A. González, M.D. Hannam, U. Sperhake, B. Brügmann, S. Husa, Phys. Rev. Lett. **98**, 231101 (2007b). doi:10.1103/PhysRevLett.98.231101. gr-qc/0702052

J.A. Gonzalez, U. Sperhake, B. Brügmann, Phys. Rev. D **79**, 124006 (2009). doi:10.1103/PhysRevD.79.124006. arXiv:0811.3952 [gr-qc]

Google Dictionary (2011). http://google-dictionary.so8848.com/

Z. Haiman, Astrophys. J **613**, 36 (2004). doi:10.1086/422910. astro-ph/0404196

F. Herrmann, I. Hinder, D. Shoemaker, P. Laguna, Class. Quantum Grav. **24**, S33 (2007a). doi:10.1088/0264-9381/24/12/S04. gr-qc/0601026

F. Herrmann, I. Hinder, D. Shoemaker, P. Laguna, R.A. Matzner, Astrophys. J. **661**, 430 (2007b). doi:10.1086/513603. gr-qc/0701143

K. Holley-Bockelmann, K. Gultekin, D. Shoemaker, N. Yunes, Astrophys. J. **686**, 829 (2008). doi:10.1086/591218. arXiv:0707.1334 [astro-ph]

M. Kesden, U. Sperhake, E. Berti, Phys. Rev. D **81**, 084054 (2010a). doi:10.1103/PhysRevD.81.084054. arXiv:1002.2643 [astro-ph]

M. Kesden, U. Sperhake, E. Berti, Astrophys. J. **715**, 1006 (2010b). doi:10.1088/0004-637X/715/2/1006. arXiv:1003.4993 [astro-ph]

L.E. Kidder, Phys. Rev. D **52**, 821 (1995). doi:10.1103/PhysRevD.52.821. gr-qc/9506022

S. Komossa, Adv. Astron. **2012**, 364973 (2012). doi:10.1155/2012/364973. arXiv:1202.1977 [astro-ph]

S. Komossa, N. Bade, Astron. Astrophys. **343**, 775 (1999). astro-ph/9901141

S. Komossa, D. Merritt, Astrophys. J. **683**, L21 (2008a). doi:10.1086/591420. arXiv:0807.0223 [astro-ph]

S. Komossa, D. Merritt, Astrophys. J. **689**, L89 (2008b). doi:10.1086/595883. arXiv:0811.1037 [astro-ph]

S. Komossa, H. Zhou, H. Lu, Astrophys. J. **678**, L81 (2008). doi:10.1086/588656. arXiv:0804.4585 [astro-ph]

M. Koppitz, D. Pollney, C. Reisswig, L. Rezzolla, J. Thornburg, P. Diener, E. Schnetter, Phys. Rev. Lett. **99**, 041102 (2007). doi:10.1103/PhysRevLett.99.041102. gr-qc/0701163

J. Kormendy, D. Richstone, Ann. Rev. Astron. Astrophys. **33**, 581 (1995). doi:10.1146/annurev.aa.33.090195.003053

F.D. Lora-Clavijo, F.S. Guzmán, Mon. Not. R. Astron. Soc. **429**, 3144 (2013). doi:10.1093/mnras/sts573. arXiv:1212.2139 [astro-ph]

C.O. Lousto, Y. Zlochower, Phys. Rev. D **77**, 044028 (2008). doi:10.1103/PhysRevD.77.044028. arXiv:0708.4048 [gr-qc]

C.O. Lousto, Y. Zlochower, Phys. Rev. D **79**, 064018 (2009). doi:10.1103/PhysRevD.79.064018. arXiv:0805.0159 [gr-qc]

C.O. Lousto, Y. Zlochower, Phys. Rev. Lett. **107**, 231102 (2011). doi:10.1103/PhysRevLett.107.231102. arXiv:1108.2009 [gr-qc]
C.O. Lousto, Y. Zlochower, Phys. Rev. D **87**(8), 084027 (2013). doi:10.1103/PhysRevD.87.084027. arXiv:1211.7099 [gr-qc]
C.O. Lousto, Y. Zlochower, M. Dotti, M. Volonteri, Phys. Rev. D **85**, 084015 (2012). doi:10.1103/PhysRevD.85.084015. arXiv:1201.1923 [gr-qc]
G. Lovelace, M. Boyle, M.A. Scheel, B. Szilágyi, Class. Quantum Grav. **29**, 045003 (2012). doi:10.1088/0264-9381/29/4/045003. arXiv:1110.2229 [gr-qc]
J. Magorrian, S. Tremaine, D. Richstone, R. Bender, G. Bower, A. Dressler, S.M. Faber, K. Gebhardt, R. Green, C. Grillmair, J. Kormendy, T. Lauer, Astron. J **115**, 2285 (1998). doi:10.1086/300353. astro-ph/9708072
I. Mandel, D.A. Brown, J.R. Gair, M.C. Miller, Astrophys. J. **681**, 1431 (2008). doi:10.1086/588246. arXiv:0705.0285 [astro-ph]
D. Merritt, M. Milosavljević, M. Favata, S. Hughes, D. Holz, Astrophys. J. **607**, L9 (2004). doi:10.1086/421551. astro-ph/0402057
M. Micic, T. Abel, S. Sigurdsson, Mon. Not. R. Astron. Soc. **372**, 1540 (2006). doi:10.1111/j.1365-2966.2006.11013.x. astro-ph/0512123
M.C. Miller, V.M. Lauburg, Astrophys. J. **692**, 917 (2009). doi:10.1088/0004-637X/692/1/917. arXiv:0804.2783 [astro-ph]
A. Peres, Phys. Rev. **128**, 2471 (1962)
P.C. Peters, Phys. Rev. **136**, B1224 (1964). doi:10.1103/PhysRev.136.B1224
D. Pollney et al., Phys. Rev. D **76**, 124002 (2007). doi:10.1103/PhysRevD.76.124002. arXiv:0707.2559 [gr-qc]
F. Pretorius, Phys. Rev. Lett. **95**, 121101 (2005). doi:10.1103/PhysRevLett.95.121101. gr-qc/0507014
F. Pretorius, in *Physics of Relativistic Objects in Compact Binaries: From Birth to Coalescence*, ed. by M. Colpi et al. (Springer, New York, 2009). arXiv:0710.1338 [gr-qc]
J.D. Schnittman, Phys. Rev. D **70**, 124020 (2004). doi:10.1103/PhysRevD.70.124020. astro-ph/0409174
J.D. Schnittman, A. Buonanno, Astrophys. J. **662**, L63 (2007). doi:10.1086/519309. astro-ph/0702641
G. Shields, E. Bonning, S. Salviander, Astrophys. J. **696**, 1367 (2009). doi:10.1088/0004-637X/696/2/1367. arXiv:0810.2563 [astro-ph]
M. Volonteri, F. Haardt, P. Madau, Astrophys. J **582**, 559 (2003). doi:10.1086/344675. astro-ph/0207276
A. Wongwathanarat, H.T. Janka, E. Mueller, Astron. Astrophys. **552**, A126 (2013). doi:10.1051/0004-6361/201220636. arXiv:1210.8148 [astro-ph]
J. Wrobel, J. Comerford, E. Middelberg, Astrophys. J. **782**, 116 (2014). doi:10.1088/0004-637X/782/2/116. arXiv:1401.4756 [astro-ph]

Fully Covariant and Conformal Formulation of the Z4 System Compared to the BSSN Formulation in Spherical Symmetry

Nicolas Sanchis-Gual, Pedro J. Montero, José A. Font, Ewald Müller, and Thomas W. Baumgarte

Abstract We have generalized a covariant and conformal version of the Z4 system of the Einstein equations by adopting a reference metric approach, that we denote as fCCZ4, well suited for curvilinear as well as Cartesian coordinates. We implement this formalism in spherical polar coordinates under the assumption of spherical symmetry using a partially-implicit Runge-Kutta (PIRK) method, without using any regularization scheme, and show that our code can evolve both vacuum and non-vacuum spacetimes without encountering instabilities. We have performed several tests and compared the Hamiltonian constraint violations of the fCCZ4 system, for different choices of certain free parameters, with these of BSSN. For an optimal choice of these parameters, and for neutron star spacetimes, the violations of the Hamiltonian constraint can be between 1 and 3 orders of magnitude smaller in the fCCZ4 system than in the BSSN formulation. For black hole spacetimes, on the other hand, any advantages of fCCZ4 over BSSN are less evident.

1 Introduction

The Einstein equations are among the most difficult partial differential equations to solve in mathematical physics. The numerical evolution must produce solutions that satisfy the constraints of the system, namely the Hamiltonian and the momentum constraint equations, which are a set of four elliptic equations. At the continuum

N. Sanchis-Gual (✉) • J.A. Font
Departamento de Astronomía y Astrofísica, Universitat de València, Dr. Moliner 50, 46100 Burjassot, València, Spain
e-mail: nicolas.sanchis@uv.es; j.antonio.font@uv.es

P.J. Montero • E. Müller
Max-Planck-Institute für Astrophysik, Karl-Schwarzschild-Str. 1, 85748 Garching bei München, Germany
e-mail: montero@mpa-garching.mpg.de; ewald@mpa-garching.mpg.de

T.W. Baumgarte
Department of Physics and Astronomy, Bowdoin College, Brunswick, ME 04011, USA
e-mail: tbaumgar@bowdoin.edu

C.F. Sopuerta (ed.), *Gravitational Wave Astrophysics*, Astrophysics and Space Science Proceedings, DOI 10.1007/978-3-319-10488-1_17

level, if they are satisfied at the initial time slice, they will remain so during the time evolution. At the discrete level that is however not the case. One common approach is to perform a free evolution, in which the constraints are only explicitly solved for the initial data. As a result, the constraints will be violated due to the non-vanishing numerical error, but the violation should converge away with resolution. In this case, the constraints are used to monitor the accuracy of the solution.

Most current numerical relativity codes use the 3 + 1 decomposition of the Einstein equations proposed by Baumgarte, Shapiro, Shibata and Nakamura, known as the BSSN formulation (Nakamura et al. 1987; Shibata and Nakamura 1995; Baumgarte and Shapiro 1998), which is a free evolution scheme. A major topic of research has been the development of techniques and formulations to minimize the constraint violation. One promising approach is the Z4 formulation proposed by Bona et al. (2003).

The Z4 formulation is an extension of the Einstein equations that adds a four vector ${}^{(4)}Z_\mu$ to the original Einstein equations. Given a set of initial data for the original Einstein equations will give the same solution for the extended equations by taking at the initial time slice ${}^{(4)}Z_\mu = 0$. Any deviation from the original Einstein equations solution will then propagate away.

Two conformal and traceless decomposition of the Z4 system have been recently proposed: the Z4c formulation (Bernuzzi and Hilditch 2010; Ruiz et al. 2011; Weyhausen et al. 2012; Hilditch et al. 2013) and the CCZ4 formulation (Alic et al. 2012, 2013). Both of them incorporate the constraint damping scheme developed by Gundlach et al. (2005), that allows for the dynamical control of the constraint violations by means of constraint damping terms, that are parametrized by two constants κ_1 and κ_2. They have been extensively tested and results show a reduction of the Hamiltonian constraint violations for some spacetimes, from one to three orders of magnitude with respect to BSSN.

However, all the aforementioned formulations, BSSN, Z4c and CCZ4, in their original form are particularly tuned for Cartesian coordinates. A covariant formulation of the evolution equations well-suited for curvilinear coordinate systems is desirable as many astrophysical phenomena are symmetric with respect to the rotation axis or spherical coordinates adapt better to their geometry. Obtaining such a formulation and assess it numerically is the purpose of this work (see also Sanchis-Gual et al. 2014 for further details).

2 The Fully Covariant and Conformal Z4 Formulation

In the Z4 system, the Hamiltonian and momentum constraints result in equations for the four-vector ${}^{(4)}Z_\mu$. In a 3 + 1 decomposition, these equations can be written as evolution equations for the projection of the ${}^{(4)}Z_\mu$ along the normal n^μ, which, following convention, we define as $\Theta \equiv -n_\mu {}^{(4)}Z^\mu = \alpha {}^{(4)}Z^0$, and the spatial projection of ${}^{(4)}Z_\mu$, $Z_i \equiv \gamma_i{}^{\mu} {}^{(4)}Z_\mu$. Here Z_i now denotes a spatial vector whose index can be raised with the (inverse) spatial metric, $Z^i = \gamma^{ij} Z_j$.

Defining $\partial_\perp \equiv \partial_t - \mathscr{L}_\beta$ where $\mathscr{L}_\beta$ denotes the Lie derivative along the shift vector β^i, the fully covariant and conformal Z4 system in a reference-metric approach (Brown 2009) (fCCZ4) is then given by the following set of evolution equations:

$$\partial_\perp \bar{\gamma}_{ij} = -\frac{2}{3}\bar{\gamma}_{ij}\bar{D}_k\beta^k - 2\alpha\bar{A}_{ij}, \tag{1}$$

$$\begin{aligned}\partial_\perp \bar{A}_{ij} = &-\frac{2}{3}\bar{A}_{ij}\bar{D}_k\beta^k - 2\alpha\bar{A}_{ik}\bar{A}^k_j + \alpha\bar{A}_{ij}(K-2\Theta) + e^{-4\phi}\big[-2\alpha\bar{D}_i\bar{D}_j\phi - \bar{D}_i\bar{D}_j\alpha \\ &+4\alpha\bar{D}_i\phi\bar{D}_j\phi + 4\bar{D}_{(i}\alpha\bar{D}_{j)}\phi + \alpha(\bar{R}_{ij} + D_iZ_j + D_jZ_i - 8\pi S_{ij})\big]^{\mathrm{TF}},\end{aligned} \tag{2}$$

$$\partial_\perp \phi = \frac{1}{6}\bar{D}_i\beta^i - \frac{1}{6}\alpha K, \tag{3}$$

$$\begin{aligned}\partial_\perp K = &\, e^{-4\phi}\big[\alpha\big(\bar{R} - 8\bar{D}^i\phi\bar{D}_i\phi - 8\bar{D}^2\phi\big) - \big(2\bar{D}^i\alpha\bar{D}_i\phi + \bar{D}^2\alpha\big)\big] + \alpha(K^2 - 2\Theta K) \\ &+2\alpha D_iZ^i - 3\alpha\kappa_1(1+\kappa_2)\Theta + 4\pi\alpha(S-3E),\end{aligned} \tag{4}$$

$$\begin{aligned}\partial_\perp \Theta = &\, \frac{1}{2}\alpha\big[e^{-4\phi}\big(\bar{R} - 8\bar{D}^i\phi\bar{D}_i\phi - 8\bar{D}^2\phi\big) - \bar{A}^{ij}\bar{A}_{ij} + \frac{2}{3}K^2 - 2\Theta K + 2D_iZ^i\big] \\ &-Z^i\partial_i\alpha - \alpha\kappa_1(2+\kappa_2)\Theta - 8\pi\alpha E,\end{aligned} \tag{5}$$

$$\begin{aligned}\partial_\perp \tilde{\Lambda}^i = &\, \bar{\gamma}^{jk}\hat{D}_j\hat{D}_k\beta^i + \frac{2}{3}\Delta\Gamma^i\bar{D}_j\beta^j - 2\bar{A}^{jk}(\delta^i{}_j\partial_k\alpha - 6\alpha\delta^i{}_j\partial_k\phi - \alpha\Delta\Gamma^i_{jk}) \\ &+\frac{1}{3}\bar{D}^i\bar{D}_j\beta^j - \frac{4}{3}\alpha\bar{\gamma}^{ij}\partial_jK + 2\bar{\gamma}^{ki}\big(\alpha\partial_k\Theta - \Theta\partial_k\alpha - \frac{2}{3}\alpha KZ_k\big) - 2\alpha\kappa_1\bar{\gamma}^{ij}Z_j \\ &-16\pi\alpha\bar{\gamma}^{ij}S_j.\end{aligned} \tag{6}$$

The superscript TF denotes the trace-free part of a tensor, κ_1 and κ_2 are the damping coefficients introduced by Gundlach et al. (2005), and $\hat{D}_i$, D_i and $\bar{D}_i$ denote the covariant derivatives built from the connection associated with the reference metric $\hat{\gamma}_{ij}$, the physical metric γ_{ij} and the conformal metric $\bar{\gamma}_{ij}$, respectively. The matter sources E, S_i, S_{ij} and S denote the density, momentum density, stress, and the trace of the stress as observed by a normal observer, respectively. We have also defined $\tilde{\Lambda}^i \equiv \bar{\Lambda}^i + 2\bar{\gamma}^{ij}Z_j$, where $\bar{\Lambda}^i \equiv \Delta\Gamma^i = \bar{\gamma}^{jk}\Delta\Gamma^i_{jk}$. The vector $\tilde{\Lambda}^i$ plays the role of the "conformal connection functions" in the original CCZ4 system; its evolution equation (6) involves the evolution equation for the variables Z_i.

When $\Theta = Z_i = 0$, the evolution equations (1)–(6) imply that the Hamiltonian constraint hold in the form

$$\mathscr{H} \equiv \frac{2}{3}K^2 - \bar{A}_{ij}\bar{A}^{ij} + e^{-4\phi}\big(\bar{R} - 8\bar{D}^i\phi\bar{D}_i\phi - 8\bar{D}^2\phi\big) - 16\pi E = 0, \tag{7}$$

where $\bar{R}$ is the trace of the Ricci tensor $\bar{R}_{ij}$.

3 Numerical Results

The Einstein equations coupled to the general relativistic hydrodynamics equations are solved using either the BSSN or the fCCZ4 formalisms in a 1D-code using spherical coordinates without any regularization scheme at the origin. We present here two tests and refer to Sanchis-Gual et al. (2014) for further details and additional tests.

3.1 *Schwarzschild Black Hole*

We evolve a single Schwarzschild black hole given by wormhole initial data and follow the coordinate evolution to the trumpet geometry. We fix the gauge freedom by imposing the so called "non-advective 1+log" condition for the lapse (Bona et al. 1997): $\partial_t \alpha = -2\alpha(K - 2\Theta)$, and a variation of the "Gamma-driver" condition for the shift vector (Alcubierre et al. 2003).

In the left panel of Fig. 1, we plot the L2-norm of the Hamiltonian constraint computed in the whole computational domain. The larger violation of the Hamiltonian constraint takes place due to the finite differencing close to the puncture. The right panel of Fig. 1, shows that the L2-norm of the Hamiltonian constraint violation computed outside the apparent horizon (AH) presents some differences between the two formulations which also depends on the values for the damping coefficients. We observe that the numerical evolutions develop instabilities for $\kappa_2 = 0$ and $\kappa_1 M = (0, 0.02)$. Selecting $\kappa_1 M = 0.07$ and $\kappa_2 = 0.5$ leads to an over-damped behavior that is responsible for an exponential growth of the constraint violation at late times. We find that $\kappa_2 = 0$ with $\kappa_1 M = 0.07$ or $\kappa_1 M = 0.2$ give the best results, leading to constraint violations that are comparable to those achieved with BSSN.

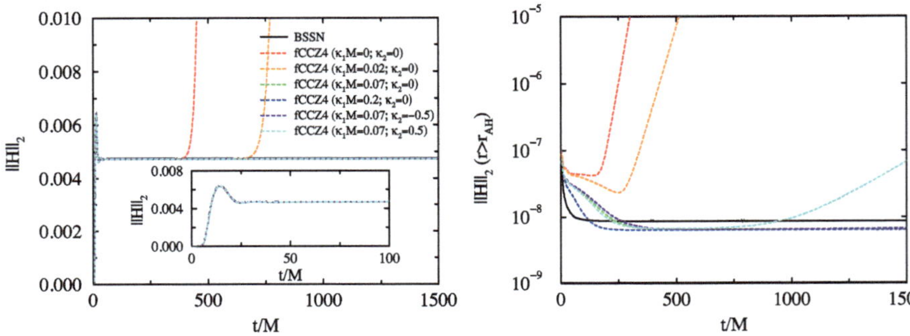

Fig. 1 *Left panel:* L2-norm of the Hamiltonian constraint in the single puncture black hole simulation. The *inset* shows a magnified view of the initial 100 M in the evolution. *Right panel:* Same quantity but computed outside the apparent horizon

The AH mass is defined as $M_{\text{AH}} = \sqrt{A/16\pi}$, where A is the proper area of the horizon. For stable black hole evolutions the difference between the AH mass for BSSN and fCCZ4 is less than 0.005 % at the end of the simulation ($t = 1,875\,M$), while the error with respect to the initial ADM mass is $\sim 0.7\,\%$. We note, however, that the black hole mass continues to drift for the fCCZ4 formulation, while it remains constant after an initial transition for the BSSN formulation.

3.2 Stable Spherical Relativistic Star

We have constructed spherically symmetric initial data for a relativistic star by solving the Tolman-Oppenheimer-Volkoff (TOV) equations for a polytropic equation of state $P = K\rho^{1+1/N}$, where K is the polytropic constant and N the polytropic index.

We investigate the effect of the damping parameter κ_1 during the time evolution of the TOV solution and choosing $\kappa_1 M = \{0, 0.098, 0.28\}$. In the left panel of Fig. 2, we show the time evolution of the normalized central rest-mass density of the star. This figure shows the distinctive periodic radial oscillations which are triggered by finite-difference errors. The amplitude of the oscillations is reduced when the damping parameter is increased (compare the red dashed line and the green dashed line). Choosing $\kappa_1 M = 0.28$, causes overdamping effects which lead to a drift in the central rest-mass density and a growth in the L2-norm of the Hamiltonian constraint (see right panel of Fig. 2). For smaller values of $\kappa_1 M$ (i.e. $\kappa_1 M = 0$ or 0.098) the drift in the central density at late times is very similar for fCCZ4 and BSSN. We observe that the amplitude of the oscillations decreases slightly faster for the fCCZ4 system than for BSSN, indicating that BSSN has a slightly smaller numerical viscosity.

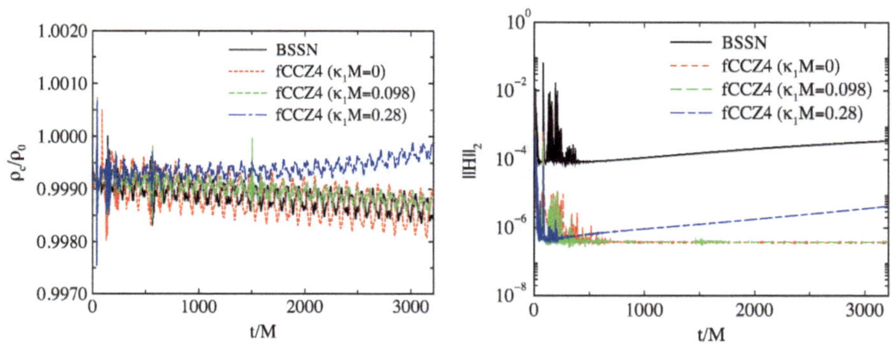

Fig. 2 *Left panel:* Time evolution of the normalized central density with fCCZ4 for different values of κ_1 and BSSN. *Right panel:* Comparison of the time evolution of the L2-norm of the Hamiltonian constraint for the stable spherical relativistic star for BSSN (*solid line*) and fCCZ4 (*dashed lines*) with $\kappa_1 M = \{0, 0.098, 0.28\}$

4 Summary

The fCCZ4 formulation, which is a generalization of the CCZ4 formulation, allows us to write the evolution equations in a fully covariant form suitable for curvilinear coordinate systems. We have implemented the fCCZ4 system in spherical coordinates under the assumption of spherical symmetry adopting a PIRK scheme for the time evolution and we have obtained stable evolutions—without regularization of the equations—for both vacuum and non-vacuum spacetimes.

We have tested the fCCZ4 formulation in a 1D code: for optimal choices of the damping parameters, we obtain 1–3 orders of magnitude reduction in the violations of the Hamiltonian constraint with respect to BSSN in non-vacuum spacetimes. For black hole spacetimes, on the other hand, any advantages of fCCZ4 over BSSN are less evident.

Acknowledgements Nicolas Sanchis-Gual thanks the Max-Planck-Institut für Astrophysik for its hospitality during the development of part of this project. Pedro J. Montero thanks Sebastiano Bernuzzi and David Hilditch for helpful discussions. Thomas W. Baumgarte gratefully acknowledges support from the Alexander-von-Humboldt Foundation. This work was supported in part by the Spanish MICINN (AYA 2010-21097-C03-01), by the Generalitat Valenciana (PROMETEO-2009-103), by the Deutsche Forschungsgesellschaft (DFG) through its Transregional Center SFB/TR 7 "Gravitational Wave Astronomy", and by NSF grant PHY-1063240 to Bowdoin College.

References

M. Alcubierre, B. Brügmann, P. Diener, M. Koppitz, D. Pollney, E. Seidel, R. Takahashi, Phys. Rev. D **67**, 084023 (2003)
D. Alic, C. Bona-Casas, C. Bona, L. Rezzolla, C. Palenzuela, Phys. Rev. D **85**, 064040 (2012)
D. Alic, W. Kastaun, L. Rezzolla, Phys. Rev. D **88**, 064049 (2013)
T.W. Baumgarte, S.L. Shapiro, Phys. Rev. D **59**, 024007 (1998)
S. Bernuzzi, D. Hilditch, Phys. Rev. D **81**, 084003 (2010)
C. Bona, J. Massó, E. Seidel, J. Stela, Phys. Rev. D **56**, 3405 (1997)
C. Bona, T. Ledvinka, C. Palenzuela, M. Zacek, Phys. Rev. D **67**, 104005 (2003)
J.D. Brown, Phys. Rev. D **79**, 104029 (2009)
C. Gundlach, J.M. Martin-Garcia, G. Calabrese, I. Hinder, Class. Quantum Grav. **22**, 3767 (2005)
D. Hilditch, S. Bernuzzi, M. Thierfelder, Z. Cao, W. Tichy, B. Brgmann, Phys. Rev. D **88**, 084057 (2013)
T. Nakamura, K. Oohara, Y. Kojima, Prog. Theor. Phys. Suppl. **90**, 1 (1987)
M. Ruiz, D. Hilditch, S. Bernuzzi, Phys. Rev. D **83**, 024025 (2011)
N. Sanchis-Gual, P. Montero, J.A. Font, E. Müller, T.W. Baumgarte, Phys. Rev. D **89**, 104033 (2014)
M. Shibata, T. Nakamura, Phys. Rev. D **52**, 5428 (1995)
A. Weyhausen, S. Bernuzzi, D. Hilditch, Phys. Rev. D **85**, 024038 (2012)

Extraction of GWs from a Numerical Simulation

Nigel T. Bishop and Luciano Rezzolla

Abstract The methods commonly used for gravitational wave (GW) extraction from a numerical simulation are the quadrupole formula, perturbative matching, ψ_4, and characteristic extraction. These methods are reviewed, and the strengths and weaknesses of each method are discussed. The accuracy of the methods is compared for astrophysical scenarios that are commonly simulated numerically.

1 Introduction

This paper discusses the estimation of Gravitational Wave (GW) emission within the context of a numerical simulation of Einstein's equations. It is assumed that the spacetime is asymptotically flat, and that ADM-like "3+1" coordinates are being used. It is well known that GWs are invariantly defined only in the limit as future null infinity ($\mathscr{I}^+$) is approached, but the extent of a "3+1" simulation is necessarily finite. Thus a GW extraction method is needed to estimate the gravitational field near $\mathscr{I}^+$ based on data in a finite region. The following GW extraction methods are described (a) Quadrupole formula (QF), (b) Perturbative matching (PM), (c) ψ_4 extraction, and (d) Characteristic extraction (CE). Finally, in Sect. 6, some quantitative comparisons of the accuracy of the various methods are presented.

Apart from QF, all the methods require the existence of a worldtube in the wavezone, that is where the spacetime can be regarded as Schwarzschild plus a small perturbation. Further, the intersection of each time slice with the worldtube is a spherical 2-surface which should approximately be of constant curvature; in practice, the shift drivers in modern numerical relativity codes do achieve this condition. The difference between the methods is the effects that are taken into account in transporting data from the worldtube to $\mathscr{I}^+$.

N.T. Bishop (✉)
Department of Mathematics (Pure & Applied), Rhodes University, Grahamstown 6140, South Africa
e-mail: n.bishop@ru.ac.za

L. Rezzolla
Institut fur Theoretische Physik, 6038 Frankfurt am Main, Germany
e-mail: luciano.rezzolla@aei.mpg.de

C.F. Sopuerta (ed.), *Gravitational Wave Astrophysics*, Astrophysics and Space Science Proceedings, DOI 10.1007/978-3-319-10488-1_18

GWs are most commonly described in terms of the wave strain $h_+, h_\times$ in the transverse-traceless (*TT*) gauge, or by means of the Newman-Penrose quantity ψ_4. Another descriptor that will be used later is the Bondi news $\mathcal{N}$, which is defined only at $\mathcal{J}^+$. Both the wave strain and ψ_4 fall off as r^{-1} where r is a surface-area radial coordinate, and the relation between the various descriptors is

$$\lim_{r\to\infty} r\psi_4 = 2\partial_t \bar{\mathcal{N}} = \partial_t^2 \lim_{r\to\infty} r(h_+ - ih_\times) \tag{1}$$

where $i = \sqrt{-1}$ and $\bar{\mathcal{N}}$ is the complex conjugate of $\mathcal{N}$.

2 Quadrupole Formula (QF)

The well-known QF applies in the approximation of weak field and low velocity. The reduced quadrupole moment is

$$\text{I}\!\!\!-_{jk} = \int \rho \left(x_j x_k - \frac{1}{3}\delta_{jk} x_i x^i \right) dV \,, \tag{2}$$

where ρ is the matter density, and then the GW field is

$$h_{jk}^{TT}(t, x^i) = \frac{2}{r} T_{jk}^{mn} \ddot{\text{I}\!\!\!-}_{mn} \tag{3}$$

with T_{jk}^{mn} the projection operator into the *TT*-gauge. The QF is used extensively in obtaining order of magnitude astrophysical estimates of GWs. In numerical relativity, the key advantages of the QF are that it is simple to compute, and that the domain that needs to be computed is limited to the region containing matter; this latter point is a key factor in simulations in which the domain size is restricted and does not extend into the wave zone. The key disadvantages are that the most powerful GW sources are those for which the assumptions behind the QF break down, but note that surprisingly accurate results have been obtained with the QF (Sect. 6); and that it does not give higher order multipoles ($\ell > 2$) of the GW field.

There have been a number of adaptations of the QF. Firstly, it has been found that numerical noise is reduced by using the equations of motion for the fluid to replace time derivatives by spatial derivatives (Finn 1989). Secondly, including the Lorentz factor, and other factors, in evaluating the density and velocity fields has led to several versions of the formula with the correct Newtonian limit (Nagar et al. 2005), but it is not clear if any version, in some formal sense, is a better approximation than the QF.

3 Perturbative Methods (PM)

The basic idea is that in the far field region spacetime is regarded as a linear perturbation of the Schwarzschild geometry. Then the linearized Einstein equations are decomposed into spherical harmonics, and the gravitational field may be described by the linear wave equation (Chandrasekhar 1983)

$$\partial_t^2 \Psi - \partial_{r*}^2 \Psi + V_\ell(r_*)\Psi = 0\,, \tag{4}$$

where the master function $\Psi = \Psi(t, r_*)$, t is the Schwarzschild time coordinate, and r_* is the Schwarzschild radial tortoise coordinate. The potential function $V_\ell(r_*)$ depends on the spherical harmonic index ℓ, and also on whether the perturbation is of even or odd parity, but these details are not pursued here.

Data on a worldtube from a Cauchy simulation is used to construct Ψ, thus providing the necessary boundary data for Eq. (4) to be solved. In this way Ψ and hence $h_+, h_\times$ (decomposed into spherical harmonics) are found at large r_*.

While the implementation of the above seems quite straightforward, the theory behind it has taken some time to develop (Regge and Wheeler 1957; Zerilli 1970; Moncrief 1974) because of the issue of gauge freedom: is a non-zero perturbation from the Schwarzschild metric a GW or can it be transformed away by a change of gauge? More recently Martel and Poisson (2005) has formulated the perturbation equations independently of the background coordinate choice.

The PM is relatively straightforward to implement numerically, and in principle the only source of error is that the linearized Einstein equations, rather than the full nonlinear equations, are used between the worldtube and null infinity. In practice, there are problems for which it does not perform well (see Sect. 6), but these tests are very limited in scope and should not be regarded as conclusive.

4 ψ_4 (Fixed Radius) and ψ_4 (Extrapolated)

Newman-Penrose scalars (Newman and Penrose 1962) are defined in terms of the Weyl tensor $C_{\alpha\beta\gamma\delta}$ and a null tetrad $(n^\alpha, \ell^\alpha, m^\alpha, \bar{m}^\alpha)$ on a surface S^2. The scalar that describes outgoing gravitational radiation is $\psi_4 = C_{\alpha\beta\gamma\delta} n^\alpha \bar{m}^\beta n^\gamma \bar{m}^\delta$, but it is usually more convenient to compute ψ_4 directly from ADM quantities (Gunnarsen et al. 1995)

$$\psi_4 = (-R_{ij} - KK_{ij} + K_{ik}K^k_j + i\epsilon_i^{k\ell}\nabla_k K_{\ell j})\bar{m}^i\bar{m}^j\,, \tag{5}$$

and decomposed into spherical harmonics $\psi_4^{\ell m}$.

The method can be implemented very simply. The surface S^2 is constructed as $x^2 + y^2 + z^2 = w^2$ (constant), where (x, y, z) are the spatial coordinates in the code. Then angular coordinates ϕ^A are defined in the standard way, so that the null tetrad and then $\psi_4^{\ell m}$ may be calculated. It has been found that considerable improvement in accuracy has been obtained (Baker et al. 2006; Pollney et al. 2009; Boyle and Mroué 2009; Taylor et al. 2013) by evaluating $\psi_4^{\ell m}$ on shells of different radii and then extrapolating to $\mathscr{I}^+$ on (approximately) outgoing **null** slices. In order to do so, $\psi_4^{\ell m}(t,w)$ must be re-expressed as $\psi_4^{\ell m}(t_*,r)$ where r is a surface area coordinate, and t_* is (approximately) constant on outgoing null slices. In practice, t_* is not exactly null, and the real and imaginary parts of $\psi_4^{\ell m}$ can vary rapidly with r. Better performance has been obtained by writing $\psi_4^{\ell m} = |\psi_4^{\ell m}| \exp(i \arg(\psi_4^{\ell m}))$, and two fitting problems are solved

$$|\psi_4^{\ell m}(t_*,r_k)| \text{ to } \sum_{n=1}^{N} \frac{A_n(t_*)}{r^n}, \qquad \arg(\psi_4^{\ell m}(t_*,r_k)) \text{ to } \sum_{n=0}^{N} \frac{\phi_n(t_*)}{r^n}, \tag{6}$$

leading to the estimate

$$\lim_{r\to\infty} r\psi_4^{\ell m} = A_1 \exp(i\phi_0). \tag{7}$$

The method is straightforward to implement, and has become the most popular choice for extracting GWs from a worldtube in the wave zone. In the ψ_4 (fixed radius) version, the estimate for GWs at $\mathscr{I}^+$ is identical to that on the worldtube, and clearly this limits the accuracy of the method. The ψ_4 (extrapolated) version has produced excellent results (see Sect. 6). It should also be noted that if the method is used to calculate a waveform (i.e. $h_+, h_\times$), then a double time integration is required, which has led to aliasing and other inaccuracies (Reisswig and Pollney 2011).

5 Characteristic Extraction (CE)

In CE, "3+1" data on a worldtube is used to construct boundary data for the full Einstein equations expressed in characteristic form using the Bondi-Sachs metric (Bondi et al. 1962)

$$\begin{aligned} ds^2 = &-\left(e^{2\beta}(1+W_c r) - r^2 h_{AB}U^A U^B\right) du^2 \\ &-2e^{2\beta} du dr - 2r^2 h_{AB} U^B du d\phi^A + r^2 h_{AB} d\phi^A d\phi^B, \end{aligned} \tag{8}$$

where u is constant on outgoing null cones, r is a surface area coordinate, and ϕ^A are angular coordinates. Thus h_{AB} has only two independent components and so can be represented by one complex number, J, with $J = 0$ representing spherical symmetry (Bishop et al. 1997). The Einstein equations remain regular under compactification, e.g. $r \to \rho = 1/r$, so $\mathscr{I}^+$ can be included in a finite grid. Given data on an inner worldtube as well as J at $u = 0$, the Einstein equations may be solved numerically to give the gravitational field at $\mathscr{I}^+$ (Bishop et al. 1997).

Coordinates which naturally exhibit asymptotic flatness always exist, now known as the Bondi gauge. In this gauge, the various GW descriptors are

$$\frac{\psi_4}{\rho} = \partial_u^2 \partial_\rho \bar{J}, \; \mathscr{N} = \frac{\partial_u \partial_\rho J}{2}, \; \frac{h_+ + ih_\times}{\rho} = \partial_\rho J, \tag{9}$$

where $\rho = 1/r$ and evaluated in the limit as $\rho \to 0$. In practice, the gauge is not Bondi, and a coordinate transformation, usually implicit, has to be made to the Bondi gauge. This is a complicated process, and in the end the GW descriptors are given in terms of (general gauge) metric variables (Bishop et al. 1997; Babiuc et al. 2009; Bishop and Reisswig 2014), but these details are not discussed here.

Construction of characteristic boundary data (Bishop et al. 1998; Babiuc et al. 2005; Reisswig et al. 2009, 2010; Babiuc et al. 2011) from "3+1" worldtube data is a matter of implementing a coordinate transformation, but this needs to be done in two steps since the surface area coordinate r cannot be defined explicitly in terms "3+1" coordinates (t, x, y, z). The first step of the transformation is to null affine coordinates (u, λ, ϕ^A), and then r can be defined in terms of the angular part of the null affine metric.

The key drawback of CE is that, even though code is publicly available in the Einstein toolkit, the successful implementation of CE does require skill; further, the computation requirements of CE are not negligible, and it introduces an overhead relative to the "3+1" computation of typically about 1 %. However, the key advantage is that it is free of systematic error, that is only the following errors are expected: (a) Truncation and round-off error, (b) Unphysical initial "3+1" data, (c) Unphysical "3+1"outer boundary condition, unless the outer boundary is far away and is not causally connected to the worldtube during GW extraction, (d) Unphysical initial characteristic data. On this last point, it should be noted that methods for setting initial data free of incoming GWs are known (Babiuc et al. 2011; Bishop et al. 2011).

6 Comparisons

It would be desirable to test extraction results against exact solutions. However, in astrophysical scenarios, GW wavefronts are spherical and vary in both amplitude and frequency. Exact solutions with these properties are not known, so extraction methods are tested against each other.

Comparisons between ψ_4 (fixed radius), ψ_4 (extrapolated) and CE have been made for binary black hole inspiral and merger (Taylor et al. 2013; Reisswig et al. 2010). It was shown that (1) there are differences for ψ_4 (fixed radius), but that ψ_4 (extrapolated) can be as accurate as characteristic extraction; (2) CE is not needed for constructing templates for event searches in detector data. Recently (Bishop and Reisswig 2014), CE has been adapted to produce a waveform directly without double time integration, and there are small differences to the waveform from

ψ_4 methods. It is not known whether, with this development, the conclusion that ψ_4 (extrapolated) can be as accurate as CE remains valid for the computation of waveforms.

In the case of stellar core collapse, a comparison (Reisswig et al. 2011) of CE, ψ_4 (fixed radius), PM and QF, found noticeable differences between all methods. Relative to CE, (1) ψ_4 (fixed radius) performed best; (2) QF gave good results for the phase with some error in the amplitude; (3) The PM performed worst, containing spurious high frequency components.

In the case of oscillating accretion tori, various modifications of QF were compared to PM (Nagar et al. 2005). The results were very similar when backscattering is negligible, otherwise noticeable differences in amplitude do occur.

Pollney et al. (2007) compared PM to ψ_4 (fixed radius) extraction for the recoil resulting from a binary black hole merger. It was found that results for the recoil velocities are consistent between the two methods.

The methods PM, modified QF and ψ_4 (fixed radius) were compared for a perturbed neutron star (Baiotti et al. 2009). While the results are generally consistent, each method experienced some drawback. The PM method has a spurious initial junk component that gets larger as the worldtube radius is increased. In ψ_4 (fixed radius), fixing the constants of integration that arise in obtaining the wave strain can be a delicate issue, although such problems did not arise in this case. The generalized QF led to good predictions of the phase, but to noticeable error in the signal amplitude.

Acknowledgements NTB thanks the National Research Foundation, South Africa, for financial support.

References

M. Babiuc, B. Szilágyi, I. Hawke, Y. Zlochower, Class. Quantum Grav. **22**, 5089 (2005)
M.C. Babiuc, N.T. Bishop, B. Szilágyi, J. Winicour, Phys. Rev. D **79**, 084011 (2009)
M.C. Babiuc, B. Szilágyi, J. Winicour, Y. Zlochower, Phys. Rev. D **84**, 044057 (2011)
L. Baiotti, S. Bernuzzi, G. Corvino, R. De Pietri, A. Nagar, Phys. Rev. D **79**, 024002 (2009). doi:10.1103/PhysRevD.79.024002
J.G. Baker, J. Centrella, D.I. Choi, M. Koppitz, J. van Meter, Phys. Rev. Lett. **96**, 111102 (2006)
N.T. Bishop, C. Reisswig, Gen. Rel. Grav. **46**, 1643 (2014)
N.T. Bishop, R. Gómez, L. Lehner, M. Maharaj, J. Winicour, Phys. Rev. D **56**(10), 6298 (1997)
N.T. Bishop, R. Isaacson, R. Gómez, L. Lehner, B. Szilágyi, J. Winicour, in *On the Black Hole Trail*, ed. by B. Iyer, B. Bhawal (Kluwer, Dordrecht, 1998)
N.T. Bishop, D. Pollney, C. Reisswig, Class. Quantum Grav. **28**, 155019 (2011)
H. Bondi, M.G.J. van der Burg, A.W.K. Metzner, Proc. R. Soc. Lond. **A269**, 21 (1962)
M. Boyle, A.H. Mroué, Phys. Rev. D **80**(12), 124045 (2009). doi:10.1103/PhysRevD.80.124045
S. Chandrasekhar, *The Mathematical Theory of Black Holes* (Oxford University Press, Oxford, 1983)
L.S. Finn, in *Frontiers in Numerical Relativity*, ed. by C. Evans, L. Finn, D. Hobill (Cambridge University Press, Cambridge, 1989), pp. 126–145
L. Gunnarsen, H. Shinkai, K. Maeda, Class. Quantum Grav. **12**, 133 (1995)

K. Martel, E. Poisson, Phys. Rev. D **71**, 104003 (2005)
V. Moncrief, Ann. Phys. **88**, 323 (1974)
A. Nagar, J.A. Font, O. Zanotti, R. de Pietri, Phys. Rev. D **72**(2), 024007 (2005). doi:10.1103/PhysRevD.72.024007
E.T. Newman, R. Penrose, J. Math. Phys. **3**(3), 566 (1962). doi:10.1063/1.1724257. Erratum in J. Math. Phys. **4**, 998 (1963)
D. Pollney, C. Reisswig, L. Rezzolla, B. Szilágyi, M. Ansorg, B. Deris, P. Diener, E.N. Dorband, M. Koppitz, A. Nagar, E. Schnetter, Phys. Rev. D **76**, 124002 (2007)
D. Pollney, C. Reisswig, N. Dorband, E. Schnetter, P. Diener, Phys. Rev. D **80**, 121502 (2009). doi:10.1103/PhysRevD.80.121502
T. Regge, J. Wheeler, Phys. Rev. **108**(4), 1063 (1957)
C. Reisswig, D. Pollney, Class. Quantum Grav. **28**, 195015 (2011)
C. Reisswig, N.T. Bishop, D. Pollney, B. Szilagyi, Phys. Rev. Lett. **103**, 221101 (2009)
C. Reisswig, N.T. Bishop, D. Pollney, B. Szilagyi, Class. Quantum Grav. **27**, 075014 (2010)
C. Reisswig, C.D. Ott, U. Sperhake, E. Schnetter, Phys. Rev. D **83**, 064008 (2011)
N.W. Taylor, M. Boyle, C. Reisswig, M.A. Scheel, T. Chu, L.E. Kidder, B. Szilagyi, Phys. Rev. **D88**, 124010 (2013). doi:10.1103/PhysRevD.88.124010
F.J. Zerilli, J. Math. Phys. **11**, 2203 (1970)

Quasinormal Modes Beyond Kerr

Aaron Zimmerman, Huan Yang, Zachary Mark, Yanbei Chen, and Luis Lehner

Abstract The quasinormal modes (QNMs) of a black hole spacetime are the free, decaying oscillations of the spacetime, and are well understood in the case of Kerr black holes. We discuss a method for computing the QNMs of spacetimes which are slightly deformed from Kerr. We mention two example applications: the parametric, turbulent instability of scalar fields on a background which includes a gravitational QNM, and the shifts to the QNM frequencies of Kerr when the black hole is weakly charged. This method may be of use in studies of black holes which are deformed by external fields or are solutions to alternative theories of gravity.

1 Introduction

The quasinormal modes (QNMs) of a black hole are the characteristic oscillations of the spacetime. They decay in time as energy flows through the horizon and disperses to infinity as waves. They are present for all types of perturbations (scalar, electromagnetic, and gravitational), and are excited whenever the spacetime is transiently perturbed. In particular, the gravitational wave QNMs make up the

A. Zimmerman (✉)
Canadian Institute for Theoretical Astrophysics, 60 St. George Street, Toronto, ON, Canada
e-mail: azimmer@cita.utoronto.ca

H. Yang
Perimeter Institute for Theoretical Physics, Waterloo, Ontario, N2L2Y5, Canada

Institute for Quantum Computing, University of Waterloo, Waterloo, Ontario N2L3G1, Canada

Z. Mark
Theoretical Astrophysics 350-17, California Institute of Technology, Pasadena, California 91125, USA

Department of Physics and Astronomy, Oberlin College, Oberlin, Ohio, 44074, USA

Y. Chen
Theoretical Astrophysics 350-17, California Institute of Technology, Pasadena, California 91125, USA

L. Lehner
Perimeter Institute for Theoretical Physics, Waterloo, Ontario, N2L2Y5, Canada

C.F. Sopuerta (ed.), *Gravitational Wave Astrophysics*, Astrophysics and Space Science Proceedings, DOI 10.1007/978-3-319-10488-1_19

ringdown phase following a binary black hole merger or the formation of a black hole, and so they play an important role in gravitational wave astrophysics. Berti et al. (2009) provides a detailed review of black hole QNMs.

The QNM frequencies are the solutions to an eigenvalue problem, analogous to the problem of computing the energy spectrum of the hydrogen atom in quantum mechanics. Efficient methods are available for computing the QNM frequencies and wavefunctions for Kerr black holes (e.g. Leaver's method (Leaver 1985)), as well as many spherically symmetric generalizations of the Schwarzschild black hole. It is difficult, however, to solve for the frequency spectrum for generalizations of Kerr black holes, because the fundamental equations may no longer separate, or even decouple (as in the case of Kerr-Newman black holes).

Here we describe a new method for computing small shifts to the QNM frequencies for spacetimes which are slightly deformed from Kerr. This method is analogous to the problem of solving for the shifts in the energy levels of a quantum system that is perturbed by a small change in the Hamiltonian. We have applied this method to explore a parametric instability among the oscillation modes of rapidly rotating black holes (Yang et al. 2014), and are currently using the method to calculate the QNM frequencies of weakly charged black holes with generic spin for the first time.

2 Quasinormal Modes of Deformed Spacetimes

Perturbations of rotating black holes are well understood, and can be treated using the Teukolsky formalism (Teukolsky 1973) which unifies scalar, electromagnetic, and gravitational perturbations into a single master equation. This equation can be solved for spin-weighted scalar quantities ${}_s\psi$, where $s = 0, \pm 1, \pm 2$ correspond to scalar, electromagnetic, and gravitational perturbations respectively. We write the master equation schematically as

$$L_T\left[{}_s\psi(x^\mu)\right] = \mathscr{T}\,, \tag{1}$$

where L_T is a second order differential operator constructed using the Newman-Penrose formulation of the field equations (Newman and Penrose 1962), and $\mathscr{T}$ represents sources of stress energy which generate the perturbations. Since we are interested in the free oscillation frequencies of the spacetime, we set $\mathscr{T} = 0$. All relevant gravitational and electromagnetic fields can be reconstructed from the spin-weighted scalars ${}_s\psi$. Remarkably, Eq. (1) separates when we expand the scalars in the frequency domain,

$${}_s\psi = \sum_{lm} \int d\omega\, \mathrm{e}^{-\mathrm{i}\omega t + \mathrm{i}m\phi}\, {}_sS_{lm\omega}(\theta)\, {}_sR_{lm\omega}(r)\,, \tag{2}$$

where ${}_sS_{lm\omega}$ are the spin-weighted spheroidal harmonics, and ${}_sR_{lm\omega}$ are the radial wavefunctions (Teukolsky 1973). Inserting this expansion into Eq. (1) results in a pair of coupled eigenvalue equations for ${}_sS_{lm\omega}$ and ${}_sR_{lm\omega}$, which give the angular eigenvalues ${}_sA_{lm}(\omega)$ and the QNM frequencies ω_{lm}.

We are interested in how this situation changes when we introduce a small deformation to the black hole spacetime, $g_{\mu\nu} = g^{(0)}_{\mu\nu} + \eta h^{(1)}_{\mu\nu}$ with $\eta \ll 1$. We call this a "deformation" to distinguish $h^{(1)}_{\mu\nu}$ from the further QNM perturbations of the spacetime. The deformation can represent any small modification to the spacetime, for example the impact of additional multipole moments if the central object is a "bumpy" black hole, or the addition of a small amount of charge to the hole (which carries additional complications). As we proceed, we only keep terms linear in η.

The small deformation should only change the QNM spectrum slightly, introducing frequency shifts $\omega \to \omega^{(0)} + \eta\omega^{(1)}$. To compute the frequency shifts $\omega^{(1)}$, we first write $\psi_{lm\omega}(r,\theta) = {}_sS_{lm\omega}\,{}_sR_{lm\omega}$ and expand the master equation in frequency and azimuthal harmonics,

$$\tilde{L}_T\left[\psi_{lm\omega}(r,\theta)\right] = 0\,. \tag{3}$$

Introducing the deformation and the shifts to the QNM frequencies gives

$$\left(\tilde{L}^{(0)}_T + \eta\omega^{(1)}\frac{\partial\tilde{L}^{(0)}_T}{\partial\omega} + \eta\tilde{L}^{(1)}_T\right)\left[\psi^{(0)}_{lm\omega} + \eta\psi^{(1)}_{lm\omega}\right] \approx 0\,. \tag{4}$$

Equation (4) is a perturbed eigenvalue equation, and is reminiscent of the problem of solving for the eigenvalues of a perturbed Hamiltonian in quantum mechanics. In that case, we have a Hamiltonian $H = H^{(0)} + \eta H^{(1)}$, perturbed energies $E_n \approx E^{(0)}_n + \eta E^{(1)}_n$, and perturbed eigenstates $|n\rangle \approx |n^{(0)}\rangle + \eta|n^{(1)}\rangle$. In order to solve for $E^{(1)}_n$, we left multiply the equation $H|n\rangle = E_n|n\rangle$ with background energy eigenstate $\langle n^{(0)}|$, expand to leading order in η, and arrive at

$$E^{(1)}_n = \frac{\langle n^{(0)}|H^{(1)}|n^{(0)}\rangle}{\langle n^{(0)}|n^{(0)}\rangle}\,. \tag{5}$$

Here we have written $E^{(1)}_n$ in a manner that emphasizes that this expression does not require a normalized (or even orthogonal) basis. We only require that $\langle n^{(0)}|(H^{(0)} - E^{(0)}_n)|n^{(1)}\rangle = 0$, by acting the Hamiltonian on the left. In order to solve for the shifted energies, we need a self-adjoint, finite inner product $\langle|\rangle$.

We can isolate the shifted frequencies of Eq. (4) by defining such a product. However, at spatial infinity the outgoing boundary conditions for the radial functions mean that ${}_sR_{lm\omega} \sim r^{-2s-1}e^{i\omega_R r_* + \omega_I r_*}$ where $\omega = \omega_R - i\omega_I$. In order for the QNMs to decay in time $\omega_I > 0$, so that the radial wavefunctions diverge at infinity. By analytically continuing ${}_sR_{lm\omega}$ into the complex-r plane, these same solutions are exponentially decaying as we move into the upper half plane, $r \to +i\infty$. Thus we define the finite inner product between two wavefunctions with the asymptotic behavior of QNMs,

$$\langle \psi | \chi \rangle = \int_{\mathscr{C}} (r - r_+)^s (r - r_-)^s dr \int \sin\theta d\theta \psi(r,\theta)\chi(r,\theta), \tag{6}$$

where the contour $\mathscr{C}$ begins at $+i\infty$, encircles a branch cut in ${}_sR_{lm\omega}$ running from the horizon r_+ parallel to the imaginary axis, and returns to $+i\infty$ on the other side of the cut. The weights of the integral guarantee that the product is self-adjoint, so that $\langle \psi^{(0)}_{lm\omega} | \tilde{L}^{(0)}_T [\psi^{(1)}_{lm\omega}] \rangle = 0$. Using this inner product, the frequency shifts are

$$\omega^{(1)} = -\frac{\langle \psi^{(0)}_{lm\omega} | \tilde{L}^{(1)}_T | \psi^{(0)}_{lm\omega} \rangle}{\langle \psi^{(0)}_{lm\omega} | \partial_\omega \tilde{L}^{(0)}_T | \psi^{(0)}_{lm\omega} \rangle}. \tag{7}$$

Note that this procedure works even when $\tilde{L}^{(1)}_T$ cannot be separated in r and θ. We have tested that this method recovers the corrections to the QNM frequencies of Schwarzschild when slow rotation is added.

3 Parametric Instability of Rapidly Rotating Kerr Black Holes

As a first application of the method described above, we have investigated the possibility of parametric resonances around rapidly rotating Kerr black holes. For rapidly rotating black holes, with $\epsilon = 1 - a \ll 1$, the QNM spectrum is weakly damped and nearly evenly spaced in the azimuthal mode number m, $\omega = m/2 + O(\sqrt{\epsilon})$. This spacing leads to the possibility of resonant mode interactions beyond linear order in perturbations about Kerr. In particular, parametric resonance is a possibility, since parametric resonance occurs when an oscillator is driven at a frequency ω approximately twice its natural frequency, $\omega \approx 2\omega'$. This happens when a QNM of mode number m drives a mode with half its mode number $m' = m/2$. The strength of the driving depends on the small amplitude of the driving mode, and competes with the slow $O(\sqrt{\epsilon})$ decay of the driven mode.

To investigate the possibility that an initially excited QNM can parametrically drive one or more modes into growth, we must use second order perturbation theory. Since this is quite challenging, we instead deal with the simpler but conceptually similar situation of a scalar field evolving in a dynamic background consisting of the black hole plus a weak gravitational QNM perturbation. The QNM then serves as the driving mode, and the scalar field as a test oscillator that can be driven.

The scalar field obeys

$$\Box_{g^{(0)}+h^{(1)}} \psi = \left(\Box_{g^{(0)}} + \mathscr{H}[h^{(1)}]\right) [\psi] = 0, \tag{8}$$

where $\mathscr{H}$ is the linearization of the wave operator with respect to $h^{(1)}_{\mu\nu}$. Using the ansatz that the scalar field solution is a background QNM with an additional time dependence,

$$\psi \sim e^{\int \alpha(t)dt - \omega'_I t} e^{-i(\omega_R/2)t + i(m/2)\phi} \left(\psi^{(0)}_{lm\omega'} + \psi^{(1)}_{lm\omega'}\right) + \text{c.c.}\,, \tag{9}$$

we can use the method described above to solve for the growth parameter $\alpha(t)$, which is time-dependent because the background QNM decays in time. In the above equation, c.c. indicates the complex conjugate of the preceding terms. Our self-adjoint inner product allows us to eliminate the unknown correction to the wavefunctions $\psi^{(1)}_{lm\omega}$, and $\alpha(t)$ can be determined solely from integrals involving the unperturbed QNM wavefunctions and the metric. Whether or not growth occurs for some time is a sensitive function of the initial amplitude of the gravitational perturbation, but may be possible for a moderately small mass ratio merger, provided the final black hole is rapidly spinning (Yang et al. 2014). The growth is always transient, because the driving mode decays in time.

An intriguing aspect of this instability is the fact that a single gravitational mode with azimuthal number m can drive modes with many different angular numbers l, provided they have azimuthal number $m' = m/2$. In addition, modes of lower frequencies are driven by modes with higher frequencies. This situation is analogous to the recently discovered turbulent behavior of perturbations of four-dimensional AdS black hole spacetimes (Adams et al. 2014) and their three-dimensional conformal fluid duals (Green et al. 2014), which feature an inverse cascade of mode energy to lower frequencies. In fact, it is the connection to the fluid-gravity correspondence that first motivated the study of the parametric instability, which may represent the onset of turbulence around a rapidly rotating black hole (Yang et al. 2014). This would be the first example of turbulent gravitational behavior in an asymptotically flat spacetime.

4 The Quasinormal Modes of Kerr-Newman Black Holes

A second application of this method is to weakly charged, rotating black hole spacetimes. These are the small-charge limit of Kerr-Newman (KN) black holes, parameterized by both a and the charge Q in geometric units. In this case, the background spacetime contains electromagnetic (EM) fields. A perturbation to the gravitational fields couples to the background EM fields and vice versa, so that these types of perturbations are coupled. Because of this, the problem of perturbations of KN black holes has only been treated in the cases $a = 0$ (Reissner Nordstrom black holes) (Leaver 1990) and $a \ll 1$ (Pani et al. 2013a, b). The methods discussed here allow for the computation of the QNM frequencies in small charge limit, where $q = Q^2 \ll 1$. In this case the KN solution is simply a small deformation of Kerr.

The perturbation equations were derived by Chandrasekhar in the phantom gauge for the coupled $s = 2$ and $s = 1$ scalars (Chandrasekhar 1983), and are schematically given by

$$\left[\tilde{L}_T^{(0)} + q\tilde{L}_T^{(1)} + q\omega^{(1)}\partial_\omega \tilde{L}_T^{(0)}\right][{}_2\psi_{lm\omega}] = q\tilde{G}^{(1)}[{}_1\psi_{lm\omega}], \tag{10}$$

$$\left[\tilde{L}_T^{(0)} + q\tilde{L}_T^{(1)} + q\omega^{(1)}\partial_\omega \tilde{L}_T^{(0)}\right][{}_1\psi_{lm\omega}] = q\tilde{F}^{(1)}[{}_2\psi_{lm\omega}]. \tag{11}$$

However, we are interested in how the background gravitational ($s = 2$) modes change with the addition of a small charge, and in this case the electromagnetic ($s = 1$) modes must be $O(q)$,

$$ {}_2\psi_{lm\omega} = {}_2\psi^{(0)}_{lm\omega} + q\ {}_2\psi^{(1)}_{lm\omega} + O(q^2), \tag{12}$$

$$ {}_1\psi_{lm\omega} = q\ {}_1\psi^{(1)}_{lm\omega} + O(q^2). \tag{13}$$

Inserting this expansion into the pair of Eqs. (10) and (11), we recover Eq. (4) for the shift of the gravitational QNM frequencies. Similarly, we can recover the leading correction to the electromagnetic QNM frequencies.

5 Outlook

The eigenvalue perturbation method described here can be applied to a variety of situations of theoretical and astrophysical interest. It can be applied directly when the deformation of the black hole spacetime is stationary and axisymmetric, and it can also be generalized to dynamical or non-axisymmetric deformations when these are decomposed into modes, as the case of the parametric instability illustrates. In addition, the analysis of the Kerr-Newman black hole indicates that the method applies even when coupling exists between perturbations and the sources of the deformed metric, such as stress-energy in the spacetime. Such couplings shift the QNM frequencies beyond $O(\eta)$ and can be neglected at that order. The method is quite general, and can be applied to deformations of other systems whose QNM spectrum is known. In particular, application to higher dimensional black holes may provide new insights about these objects.

A major technical challenge is to compute the correction to the Teukolsky equation, $\tilde{L}_T^{(1)}$ given the deformation $h^{(1)}_{\mu\nu}$ to the background metric, but in principle this is straightforward. The case of scalar fields is especially simple. A general procedure for generating $\tilde{L}_T^{(1)}$ for all spin-weights is left to future work, along with additional applications of the eigenvalue perturbation method.

Acknowledgements We thank Stephen Green and Rob Owen for useful conversations during the course of this work. AZ, HY, and YC were supported by NSF Grant PHY-1068881, CAREER

Grant 0956189, and the David and Barbara Groce Startup Fund at Caltech. ZM was supported by the LIGO SURF program at Caltech. LL was supported by NSERC through Discovery Grants and CIFAR. This research was supported in part by Perimeter Institute for Theoretical Physics. Research at Perimeter Institute is supported by the Government of Canada through Industry Canada and by the Province of Ontario through the Ministry of Research and Innovation.

References

A. Adams, P.M. Chesler, H. Liu, Phys. Rev. Lett. **112**, 151602 (2014). doi:10.1103/PhysRevLett.112.151602

E. Berti, V. Cardoso, A.O. Starinets, Class. Quantum Grav. **26**(16), 163001 (2009). doi:10.1088/0264-9381/26/16/163001

S. Chandrasekhar, *The Mathematical Theory of Black Holes* (Clarendon Press, Oxford, 1983)

S.R. Green, F. Carrasco, L. Lehner, Phys. Rev. **X4**, 011001 (2014). doi:10.1103/PhysRevX.4.011001

E. Leaver, Proc. R. Soc. Lond. **A402**, 285 (1985)

E.W. Leaver, Phys. Rev. **D41**, 2986 (1990). doi:10.1103/PhysRevD.41.2986

E. Newman, R. Penrose, J. Math. Phys. **3**, 566 (1962). doi:10.1063/1.1724257

P. Pani, E. Berti, L. Gualtieri, Phys. Rev. Lett. **110**(24), 241103 (2013a). doi:10.1103/PhysRevLett.110.241103

P. Pani, E. Berti, L. Gualtieri, Phys. Rev. **D88**, 064048 (2013b). doi:10.1103/PhysRevD.88.064048

S.A. Teukolsky, Astrophys. J. **185**, 635 (1973). doi:10.1086/152444

H. Yang, A. Zimmerman, L. Lehner, Turbulent Black Holes (2014). arXiv:1402.4859

The Scientific Potential of Space-Based Gravitational Wave Detectors

Jonathan R. Gair

Abstract The millihertz gravitational wave band can only be accessed with a space-based interferometer, but it is one of the richest in potential sources. Observations in this band have amazing scientific potential. The mergers between massive black holes with mass in the range 10^4–$10^7 M_{\odot}$, which are expected to occur following the mergers of their host galaxies, produce strong millihertz gravitational radiation. Observations of these systems will trace the hierarchical assembly of structure in the Universe in a mass range that is very difficult to probe electromagnetically. Stellar mass compact objects falling into such black holes in the centres of galaxies generate detectable gravitational radiation for several years prior to the final plunge and merger with the central black hole. Measurements of these systems offer an unprecedented opportunity to probe the predictions of general relativity in the strong-field and dynamical regime. Millihertz gravitational waves are also generated by millions of ultra-compact binaries in the Milky Way, providing a new way to probe galactic stellar populations. ESA has recognised this great scientific potential by selecting *The Gravitational Universe* as its theme for the L3 large satellite mission, scheduled for launch in ~2034. In this article we will review the likely sources for millihertz gravitational wave detectors and describe the wide applications that observations of these sources could have for astrophysics, cosmology and fundamental physics.

1 Introduction

Gravitational waves (GWs) are expected to be generated in all frequency bands by a wide range of different sources. Over the next two decades several different frequency bands will be opened up observationally by a variety of different instruments. Extremely-low frequency gravitational waves ($\sim 10^{-16}$ Hz) generated by quantum fluctuations during inflation imprint polarisation on the cosmic microwave background (CMB) which could be detected by ongoing experiments.

J.R. Gair (✉)
Institute of Astronomy, University of Cambridge, Madingley Road, Cambridge CB3 0HA, UK
e-mail: jgair@ast.cam.ac.uk

C.F. Sopuerta (ed.), *Gravitational Wave Astrophysics*, Astrophysics and Space Science Proceedings, DOI 10.1007/978-3-319-10488-1_20

In March 2014 the BICEP2 collaboration announced a detection of B-mode polarisation based on measurements of one part of the sky which were consistent with being produced by primordial gravitational radiation (Ade et al. 2014). These results are not yet confirmed and could be explained by foreground contamination, but they suggest that the signature of inflationary gravitational waves in the CMB could soon be detected. In the nanohertz frequency band, gravitational waves can be detected by the accurate timing of millisecond pulsars. These objects rotate very stably and so the time of arrival of pulses can be predicted very accurately. A gravitational wave propagating between the observer and the pulsar will cause the pulse to arrive earlier or later than expected and so gravitational waves can be detected by looking for periodic fluctuations in the pulse arrival times that are correlated between different pulsars on the sky. Three major pulsar timing array (PTA) efforts are currently underway—the European PTA, NANOGrav, and the Parkes PTA—and agreements exist to combine the data to create an international PTA (Hobbs et al. 2010). GWs in the 10–1,000 Hz band can be detected using kilometre-scale ground-based laser interferometers. A network of such detectors has been constructed and taken data over the last two decades (LIGO, Virgo, GEO, Kagra). No detections have yet been made, but scientifically interesting upper limits have been set and these detectors are currently undergoing upgrades to advanced configurations, with a factor of $\sim$10 improvement in sensitivity, which will gradually start to take data from 2015 (Aasi et al. 2013). These instruments could detect gravitational waves from coalescing compact binaries, from deformed rotating neutron stars, from transient events such as supernova or cosmic string cusps and perhaps from a stochastic background of radiation generated in the early Universe.

A frequency band missing from this list is the millihertz band from $\sim 10^{-4}$ to 10^{-1} Hz. This frequency range is inaccessible from the ground due to the large seismic noise background that exists below $\sim$10 Hz and therefore to open it up we must put a detector in space. Proposals for a space-based interferometer have been discussed for over 20 years and it will be another 20 years before such an instrument is realised. However, this prospect has become more certain since ESA selected *The Gravitational Universe* (The eLISA Consortium 2013) as the science theme to be addressed by the L3 large science mission due to be launched in 2034. In this article we will describe the scientific potential of such an instrument. This article is organised as follows. In the remaining sections of this introduction we will describe millihertz gravitational wave detectors (Sect. 1.1) and potential sources of millihertz GW radiation (Sect. 1.2). The remainder of the article will describe the scientific applications of millihertz GW observations to astrophysics (Sect. 2), cosmology (Sect. 3) and fundamental physics (Sect. 4). We finish with a summary in Sect. 5.

1.1 Detectors

The canonical concept for a space-based millihertz gravitational wave detector is the Laser Interferometer Space Antenna (LISA) (Danzmann et al. 1996). This design calls for a constellation of three spacecraft, arranged at the corners of an equilateral triangle, 5 million km apart in a heliocentric Earth-trailing orbit, 20° behind the Earth. Two laser beams would pass along each arm of the triangle, one in each direction, and these would be used to make one-way measurements of the distance between the satellites from which the presence of gravitational waves can be inferred. LISA was developed as a joint NASA-ESA project until 2011 when NASA withdrew from the mission due to funding constraints. A rescoped concept, NGO, was proposed as a candidate for the ESA-only L1 mission opportunity (Amaro-Seoane et al. 2012, 2013), but was not selected. NGO differed from LISA in the arm length—1 million km; in the number of arms—only two of the sides of the constellation would have laser links instead of three; in the orbit—9° behind the Earth and drifting away; in the mission duration—2 years instead of 5; and in specifics such as telescope diameter (20 cm instead of 40 cm) and acceleration noise requirements (3 $\mathrm{fm\,s^{-2}\,Hz^{-\frac{1}{2}}}$, about a factor of 5 less stringent than the LISA baseline). The primary difference between a two-arm and three-arm configuration is that the three-arm configuration allows the construction of two independent data streams, while only one can be constructed in the two-arm case, which has important consequences for parameter estimation. These differences allowed NGO to fit within the L1 mission budget cap, but lead to significant changes in sensitivity (see Fig. 1). *The Gravitational Universe* (The eLISA Consortium 2013) science theme selected by ESA for the L3 mission used essentially the same NGO concept to illustrate the

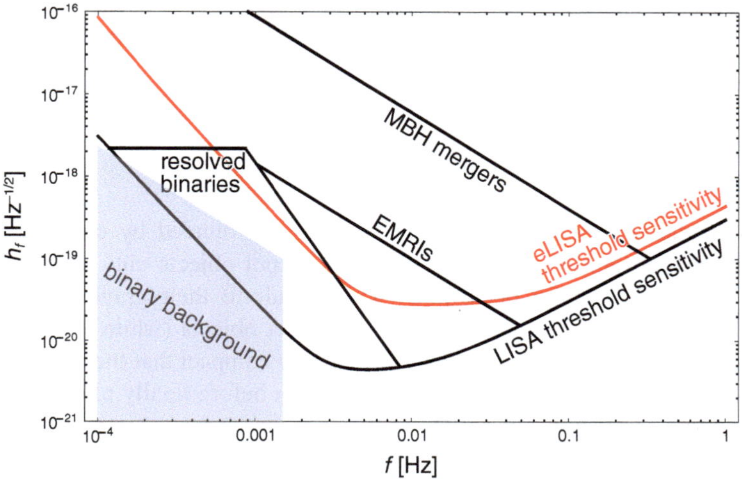

Fig. 1 Sensitivity curves for LISA and eLISA, with the approximate frequency range and amplitude of expected sources indicated. Reproduced from Gair et al. (2013)

scientific potential. This mission is now referred to as eLISA (evolved-LISA) to reflect the fact that the design has not yet been fixed. We will refer to both LISA and eLISA in this article, but in all cases scientific results quoted for eLISA have been calculated using the NGO concept.

1.2 Sources

We describe here the four major types of millihertz gravitational wave source.

1.2.1 Massive Black Hole Mergers

Massive black holes (MBHs) are ubiquitous in the centres of galaxies (Croton et al. 2006). When galaxies merge it is expected that their black holes will sink to the centre of the merged galaxy through dynamical friction and eventually merge via GW emission. If the MBH has mass in the range 10^4–$10^7 M_\odot$ the emitted radiation will be in the millihertz band. eLISA would be able to observe these mergers with very high signal to noise ratio (S/N) anywhere in the observable Universe, for instance a merger of two $10^5 M_\odot$ black holes would be observable with S/N of 50 at a redshift of $z = 20$ (Amaro-Seoane et al. 2012, 2013). LISA would have been expected to observe a few tens of MBH merger events per year in 'heavy seed' models and about twice as many events under 'light seed' models (Arun et al. 2009). The expected event rate for eLISA is comparable for 'heavy seed' models and a factor of $\sim$1.5 lower for 'light seed' models, as eLISA will not be able to observe the mergers between lighter black holes at higher redshift that make up a significant fraction of the LISA events in that case (Amaro-Seoane et al. 2012, 2013). The ability of a space-based GW detector to distinguish these and other models will be discussed in detail in Sect. 2.

1.2.2 Extreme-Mass-Ratio Inspirals

MBHs in the centres of galaxies are typically surrounded by clusters of stars. Interactions between stars in these clusters can put objects onto orbits that pass very close to the central black hole, which leads to their gravitational capture by and eventual inspiral into the MBH. Compact objects (white dwarfs, neutron stars and stellar mass black holes) are sufficiently compact that they are not tidally disrupted but gradually inspiral via GW emission before finally plunging into and merging with the MBH. For MBHs in the appropriate mass range these extreme-mass-ratio inspirals (EMRIs) will generate millihertz GWs (Amaro-Seoane et al. 2007). LISA would have been able to observe these systems out to redshifts $z \gtrsim 1$, with an expected event rate of a few hundreds to several thousand per year (Gair 2009). eLISA would be able to observe EMRIs out to redshift $z \sim 0.7$ with an expected event rate of several tens per year (Amaro-Seoane et al. 2012; The eLISA Consortium 2013). In both cases, the event rate is dominated by inspirals of black holes, partially due to the enhancement of the intrinsic rate for these sources due to mass segregation and partially because they are detectable to greater distance. There are considerable uncertainties in these event rates due to the poorly understood physics of dense stellar clusters, but even a single EMRI observation could have a profound impact on our understanding of fundamental physics (see Sect. 4).

1.2.3 Ultra-Compact Binaries in the Milky Way

The majority of stars are formed in systems containing two or more stellar objects. Roughly half of the stellar binaries that form are sufficiently compact that they remain bound through the whole process of stellar evolution and evolve into compact systems containing white dwarfs, neutron stars or black holes. The shortest period ($\sim$1 h) systems, known as ultra-compact binaries, generate millihertz gravitational waves (Nelemans et al. 2001). eLISA would be able to resolve several thousand individual systems with S/N greater than 7 and measure the first time derivative of the frequency for many systems (Nissanke et al. 2012). LISA would be able to individually resolve as many as 20,000 individual systems, measuring first derivatives of the orbital frequency for many and second derivatives for a handful (Nissanke et al. 2012). At frequencies below about 2 mHz, there are so many ultra-compact binaries that the population creates an unresolvable stochastic foreground. This was expected to dominate over instrumental noise for LISA, but will most likely lie below the eLISA noise, although eLISA could possibly still detect it from its annual modulation (The eLISA Consortium 2013).

1.2.4 Cosmological Sources

Processes occurring on high energy scales in the very early Universe can generate GWs at all frequencies. The typical frequency scale is determined by the horizon size at the epoch when the GWs were created and for the millihertz band this corresponds to the TeV energy scale. TeV GWs could be generated by phase transitions or by inflationary reheating in certain braneworld scenarios, allowing a millihertz GW detector to probe Higgs self-couplings, the presence of supersymmetry or conformal dynamics at TeV scales and the dynamics of compact extra dimensions (Binetruy et al. 2012). eLISA will be able to place bounds on the energy density of relic TeV gravitational radiation at the level $\Omega_{\rm GW} \sim 10^{-5}$ (Amaro-Seoane et al. 2012, 2013; The eLISA Consortium 2013). For LISA this bound would be a factor of a few better. Phase transitions often create one-dimensional topological defects, cosmic strings, which can generate GWs from cusps (points on the string which are travelling at nearly the speed of light formed by the interaction of waves propagating along string loops in different directions) or kinks. Millihertz GW detectors offer the best prospects for detecting and constraining the properties of these cosmic string networks (The eLISA Consortium 2013).

2 Science Applications: Astrophysics

From observations of millihertz gravitational wave sources it will be possible to determine the systems parameters very precisely. This offers great potential to constrain the astrophysics of the sources. eLISA measurements of MBH mergers should be able to determine the masses of the two black holes to a fractional precision of $\sim 10^{-3}$–10^{-2}, the spin of the primary/secondary to a fractional precision of $\sim 10^{-2}/10^{-1}$, the sky location of the source to approximately $100\,\text{deg}^2$ and the luminosity distance to a few tens of percent (Arun et al. 2009). Measurements with LISA would be moderately (a factor of 2–3) better for the intrinsic parameters, but considerably better (a factor of 10) for the sky position and distance, thanks primarily to the addition of a second independent data stream due to the third arm. Extremely precise measurements of the parameters of EMRIs are also possible. For a source observed with a S/N of 30, LISA observations would be expected to determine the masses of the two objects to a part in 10^4, the central black hole spin magnitude to a few parts in 10^4, the sky position to a few square degrees and the luminosity distance to $\sim 10\,\%$ (Barack and Cutler 2004). This parameter precision comes from being able to track the EMRI waveform phase over many hundreds of thousands of waveform cycles and so equally precise measurements are possible with eLISA, for a source at the same S/N. Typical S/N's for eLISA will be lower than for LISA, but by less than a factor of 2 and since the expected parameter errors scale like the inverse of S/N, the order of magnitude expected for eLISA errors is comparable.

Very precise measurements for individual sources will be interesting but do not encode much astrophysical information. Astrophysical results will come from looking at the statistics of the set of sources of a particular type that are observed. We will discuss some specific questions that millihertz gravitational wave observations will be able to address in the following sections.

2.1 MBH Constraints on Growth of Structure

There is general consensus about the overall process of hierarchical structure formation. Galaxies were formed very early in the Universe (the most distant known galaxy is at a redshift $z \sim 10.7$ (Coe et al. 2013)) and from shortly after their formation many galaxies already contained massive black holes in their centres. The largest of these black holes are seen as QSOs at redshifts as high as $z \sim 7.08$ (Mortlock et al. 2011). Galaxies evolve through a sequence of mergers over cosmic time, while the black holes are expected to merge following mergers of their hosts and also grow through accretion. Several tight correlations exist between the properties of galaxies and the massive black holes they contain, in particular the M-sigma relation between the central black hole mass and the velocity dispersion of stars in the centre of the galaxy (Tremaine et al. 2002). These correlations are indicative of the close co-evolution of a MBH and its host.

While this overall picture is widely accepted, there are some uncertainties, for instance in the initial mass distribution of the seed black holes from which the MBHs have grown, in the metallicity of the material from which the MBHs formed, in the efficiency of accretion and in the accretion geometry. Several models exist which can reproduce current observational constraints, but these constraints are mostly at the high-mass and low-redshift end of the black hole distribution, which is the regime accessible to current electromagnetic observations. The electromagnetically accessible regime is unlikely to change significantly over the next 20 years. Different models make different predictions for the high-redshift and low-mass regime of the black hole distribution, which is precisely the regime that will be explored by millihertz gravitational wave observations. GW observations with eLISA could therefore have a significant impact on our understanding of structure formation.

The constraints that LISA observations could place on models of MBH growth were explored in detail in Sesana et al. (2011). They considered two different seeding mechanisms—light 'POPIII' seeds, formed at high redshift with typical mass $\sim 100M_{\odot}$, or heavy 'quasistar' seeds, formed from direct collapse of dust at lower redshift and with mass $\sim 10^5M_{\odot}$—two different metallicities for the dust from which the MBHs form—$Z = 0$ or a distribution across all Z—two different accretion models—accretion always at the Eddington limit or using the more complex Merloni-Heinz prescription (Merloni and Heinz 2008)—and two different accretion geometries—'chaotic' accretion, in which small amounts of mass are added sequentially from random orientations, or 'efficient' accretion, in which all mass is added in a single coherent accretion episode. The analysis considered LISA measurements of MBH mass and redshift only and considered two different detection criteria (S/N detection threshold of 8 or 20) and two different detector configurations (1 data stream or 2). All possible model comparisons were considered and it was found that for most pairs of models 3 months of LISA observations would be sufficient to distinguish the two models with 95 % confidence. Models that differed only in the accretion efficiency could not be distinguished with a 3 month observation, although the predictions of such model pairs differ primarily in the predicted spins of the black holes, which were not included in the analysis. With only 3 months of observations it was also not possible to distinguish between the zero-metallicity/light-seed model and the metallicity-distribution/heavy-seed model. However, after 1 year of observation we would expect that all model pairs could be confidently distinguished.

The distinguishability of a subset of these models using eLISA observations has also been considered and is summarised in Table 1. Using only mass and redshift measurements, eLISA can confidently distinguish the seed mass properties, but not the accretion efficiency. Once spin measurements are included, however, it should be possible to confidently distinguish all four models.

It is very unlikely that the Universe will be precisely described by one of these existing models, but instead all the processes will be taking place in some unknown proportion and the resulting evolution will represent some amalgam of possible scenarios. This was also explored in Sesana et al. (2011), in two different ways. In the first approach, mixed catalogues of mergers were constructed by linearly

Table 1 Probability of choosing the row model over the column model with better than 95 % confidence when the row model is true (upper half table) or when the column model is true (lower half table)

	Without spins				With spins			
	LE	LC	HE	HC	LE	LC	HE	HC
LE	×	0.48	0.99	0.99	×	0.96	0.99	0.99
LC	0.53	×	1.00	1.00	0.13	×	1.00	1.00
HE	0.01	0.01	×	0.79	0.01	0.01	×	0.97
HC	0.02	0.02	0.22	×	0.02	0.02	0.06	×

We consider models that differ in black hole seed prescriptions—either light (L) or heavy (H) seeds—and in the accretion efficiency—either efficient accretion (E) or chaotic accretion (C). The results in the left-hand portion of the table labelled 'without spins' are based on measurements of the black hole mass and redshift only, while those in the right-hand portion labelled 'with spins' are based on measurements of the spin as well as mass and redshift

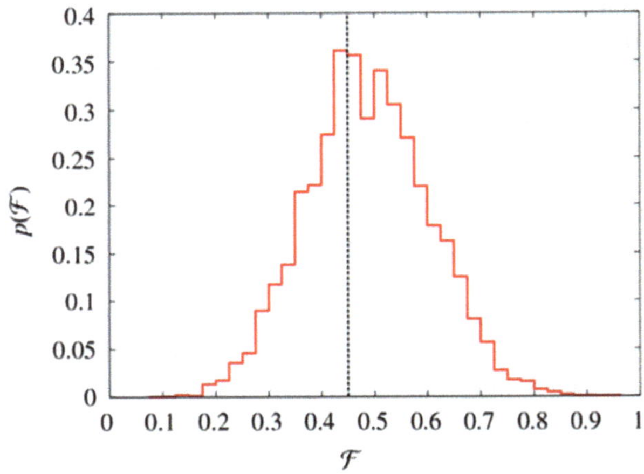

Fig. 2 Posterior distribution recovered by eLISA for the mixing fraction of a combination of the LE and HE models used in Table 1. The injected model, a 0.45:0.55 mix of LE:HE, is indicated by the *vertical line*. Figure reproduced from Amaro-Seoane et al. (2012)

combining the catalogues from two or four different models in relative proportions $f_i \in [0, 1]$, with $\sum f_i = 1$. This was done in two different ways—either including the relative number of events predicted by the model when combining the two catalogues, or combining the parameter distributions renormalised to one event and then marginalising over the number of observed events in the analysis. In each case, using both open and blind analyses, it was found that LISA observations would be able to derive posterior distributions on the mixing fractions that were consistent with the injected values and had typical widths of $\sim \pm 0.1$. eLISA will also be able to distinguish these mixed models, albeit with slightly less precision, $\sim \pm 0.15$ (see Fig. 2). This linear mixing of catalogues is artificial since it does not include cross-mergers of black holes from the two different models. The second model mixing analysis in Sesana et al. (2011) used numerical simulations to directly compute mergers for a hybrid model that included both light and heavy black hole seeds. Analysis of the hybrid models demonstrated that the linear-mixing model could successfully identify that the hybrid models comprised a mixture of light and heavy seed models and correctly indicated that the hybrid model with more efficient quasistar seeding had a greater fraction of heavy seeds.

More work is needed to fully understand the implications of eLISA observations for characterising the physics of structure formation. For instance, rather than merely determining the relative fractions of some distinct discrete models that are consistent with the observed data, it would be interesting to see how well eLISA could measure physical parameters that characterise the growth of structure, for instance the efficiency of quasistar formation, accretion efficiency etc. Nonetheless, it is already clear that eLISA will have a profound impact on our understanding of these processes, as it is extremely difficult to constrain these lower mass merging black holes using any other techniques.

2.2 *EMRI Measurements of the MBH Mass Function*

EMRI observations will be primarily restricted to the low redshift Universe, $z \lesssim 0.7$. The 'cosmic high noon' (as described by the CANDELS team, Grogin et al. 2011, Koekemoer et al. 2011) of peak star formation rate and maximal QSO activity occurred at redshifts $z \sim 3$–1.5 and EMRI observations will therefore probe low mass black holes during the period of decline in AGN activity. Quiescent black holes in the relevant mass range are again poorly constrained observationally. Black hole masses in the regime $M < 10^7 M_\odot$ can be inferred from observed galaxy luminosity functions using correlations between galaxy luminosity and stellar velocity dispersion and between velocity dispersion and central black hole mass (Greene and Ho 2007). It is also possible to directly measure velocity dispersions for some galaxies using galaxy surveys, e.g., SDSS (Sheth et al. 2003). However, the luminosity function is incomplete at lower luminosities and the sensitivity limit of direct velocity dispersion measurements approximately corresponds to a black hole mass of $10^7 M_\odot$. EMRI observations will therefore provide the first direct measurement of the low-mass end of the black hole mass function in the local Universe.

In Gair et al. (2010) the ability of LISA EMRI observations to recover the parameters of a simple power-law black hole mass function, $dn/d \log M = AM^\alpha$, was considered. The analysis assumed that a threshold S/N of 30 would be required for EMRI detection, used a circular-equatorial EMRI waveform model for fixed values of the central black hole spin of $a = 0$ or $a = 0.9$ and considered both a 'pessimistic' (1 data channel, 2 year observation) and 'optimistic' (2 data channels, 5 year observation) configuration of LISA. The precision with which LISA would be able to determine the parameters of the mass function was found to scale simply with one over the square root of the number of detected events, $N_{\rm obs}$. In the worst case (MBH spin $a = 0$ and pessimistic detector assumptions) typical errors were $\Delta \ln A \approx 0.8\sqrt{10/N_{\rm obs}}$ and $\Delta\alpha \approx 0.3\sqrt{10/N_{\rm obs}}$, while for the best case (MBH spin $a = 0.9$ and optimistic detector assumptions) the typical errors were $\Delta \ln A \approx 0.5\sqrt{10/N_{\text{obs}}}$, $\Delta\alpha \approx 0.2\sqrt{10/N_{\text{obs}}}$. The corresponding results for eLISA are similar to the pessimistic detector case, $\Delta \ln A \approx 1.1\sqrt{10/N_{\text{obs}}}$ and $\Delta\alpha \approx 0.35\sqrt{10/N_{\text{obs}}}$. Current electromagnetic constraints suggest that the black hole mass function is flat in log mass in the eLISA range, with an uncertainty in slope of ± 0.3. Therefore, with just ten EMRI observations we will constrain the black hole mass function to a precision better than it is currently known. We expect eLISA to observe several tens of EMRI events per year, so it is likely that the derived constraint will be a considerable improvement on our current knowledge.

eLISA is actually sensitive to the product of the number density of black holes with the EMRI rate per black hole and the preceding analysis assumed that the latter took a known form and scaled with black hole mass to the power of -0.15, which is implied by numerical simulations. This is a source of uncertainty in the result, although it is hoped that over the next two decades this scaling will be constrained with sufficient precision that the degeneracy between the mass function and EMRI rate scaling can be broken. EMRI constraints on a redshift-dependent mass function of the form $dn/d \log M = A_0(1+z)^{A_1} M^{\alpha_0 - \alpha_1 z}$ were also considered in Gair et al. (2010), but the conclusion was that LISA would only be able to place very weak constraints on A_1 and α_1. eLISA observations will be restricted to a smaller range of redshifts than those for LISA and so it is unlikely that eLISA EMRI observations will place interesting constraints on the evolution of the mass function.

2.3 Other Astrophysical Applications

Millihertz GW observations will also provide a rich variety of astrophysical information in other areas that have so far not been studied in detail. The analyses described above have considered the impact of MBH mergers or EMRIs alone, but the MBHs into which EMRIs are falling are the quiescent, low-redshift remnants of the same population that eLISA will observe merging at higher redshift. By using observed MBH merger and EMRI events in combination it will be possible to learn more about these systems. Combined observations might, for example, break the degeneracy between the EMRI rate and black hole mass function and provide insight into the evolution of the MBH mass function. In addition, MBH merger observations will provide constraints on the magnitudes and directions of the spins of the component black holes, which will offer indirect constraints on the processes that drive spin evolution, in particular accretion, and on the efficiency of mechanisms of spin alignment. EMRI observations will also provide spin measurements, which will provide insights into the spin distribution and accretion history of quiescent low-mass black holes in the local Universe. The characteristics of the orbits (eccentricity and inclination) of observed EMRIs encode information about the relative importance of different EMRI formation mechanisms—EMRIs started through dynamical capture onto a highly eccentric orbit will have moderate eccentricity and random inclination in band; EMRIs started by the tidal splitting of a binary star or through tidal stripping of the envelope of a massive star will be circular but with non-zero inclination; and EMRIs that occur after the formation of stars in a disc around the central MBH will tend to be both circular and of zero inclination (Amaro-Seoane et al. 2007). EMRI measurements of compact object masses will reveal the properties of stellar populations in galactic central clusters and the efficiency of mass segregation in bringing heavier compact objects to the centre of the cluster. Observations of ultra-compact galactic binaries will provide a comprehensive survey of the population of these objects in the Milky Way, offering new constraints on stellar evolution. The binaries for which frequency derivatives are measured will provide information on the physics of mass transfer and of tidal interactions. Measuring the frequency derivative will also allow the distance to the source to be determined and we expect to have about 100 such sources concentrated in the inner Galaxy, allowing a direct measurement of the distance to the Galactic centre to be made (The eLISA Consortium 2013). Finally, although eLISA will not directly observe coalescing compact binaries, the statistics of the observed population will significantly improve estimates of the binary merger rate.

3 Science Applications: Cosmology

As described in Sect. 1.2.4, a millihertz GW detector could detect a stochastic background of radiation generated in the very early Universe, which would place strong constraints on TeV physics. We will not consider this further here but refer the reader to Binetruy et al. (2012) for a comprehensive discussion of the subject. However, as well as directly detecting cosmological GWs, millihertz GW sources can be used indirectly as "standard sirens" to probe the expansion history of the Universe. This basic idea was first proposed in Schutz (1986). The observed strain of a GW source scales as $h \sim (1+z)\mathcal{M}/D_L(z)$, where $\mathcal{M}$ is the chirp mass, z the redshift and $D_L(z)$ the luminosity distance to the source. A GW detector can determine $\mathcal{M}$ very accurately by tracking the phase of the waveform, and so, if the mass-redshift degeneracy can be broken, the observed amplitude of the GW source can be used to determine $D_L(z)$ and hence probe the expansion history. The primary complication is to determine the redshift. This can be done if an electromagnetic counterpart to an event is observed. MBH mergers were thought to be promising candidates, in particular so-called 'golden binaries' observed with high S/N at relatively low redshift (Hughes and Holz 2003). However, no robust counterpart mechanism is currently known and weak lensing dominates the distance measurement error for sources at $z \gtrsim 1$. The weak lensing error can be mitigated to some extent by mass reconstruction, but not sufficiently to reach useful precision (Hilbert et al. 2011).

An alternative and more promising approach to GW cosmography is to combine measurements of multiple events statistically. This was the approach advocated by Schutz in his original paper (Schutz 1986) and it was first explored in the context of space-based detectors by MacLeod and Hogan (2008). They considered measurements of the Hubble constant, H_0, using LISA observations of EMRIs. Given an observed LISA EMRI, a galaxy catalogue is used to identify the redshifts of all galaxies consistent with the LISA error box for that EMRI. For each galaxy, this redshift distribution is used to construct a distribution of distance estimates and hence cosmological parameters. These distributions are then combined for all the events observed. Although employing galaxy catalogues, the method does not rely on the fact that the true host galaxy is in the catalogue, but will work as long as the distribution of EMRI host galaxies is similar to the distribution of observed galaxies. McLeod and Hogan found that LISA observations of $\sim$20 EMRI events at redshifts $z < 0.5$ would be sufficient to determine the Hubble constant to $\sim$1 %. This calculation has not been repeated for eLISA. We expect eLISA to detect the required $\sim$20 events at redshift $z < 0.5$, but eLISA's typical distance errors will be larger. Assuming, conservatively, that the eLISA error would be a factor of 2 times larger, we would expect to determine H_0 to $\sim$2 % with 20 EMRI events or to $\sim$1 % with 80 events. We expect to observe a few tens of EMRI events per year with eLISA, so this number would only be reached with an extended mission. In Petiteau et al. (2011) a similar analysis was carried out for the constraints that LISA observations of MBHs could place on cosmological parameters. They found that LISA would be able to place constraints on the equation of state of dark energy that are a factor of 2–8 better than current constraints from electromagnetic observations. The analysis has not been repeated for eLISA, but eLISA determination of MBH merger distances is somewhat poorer so it is likely that eLISA constraints will be no better than those based on current electromagnetic data.

As eLISA is not scheduled for launch until 2034 it is likely that any constraints on cosmological parameters derived from eLISA observations will not be competitive with the best constraints obtained from other methods by that time. However, these gravitational wave constraints will be completely independent of all other measurements, which will provide important confirmation of existing results and could, in principle, reveal the presence of previously unappreciated systematics in the electromagnetic results.

4 Science Applications: Fundamental Physics

Gravitational wave observations offer a unique way to probe the strong-field, dynamical regime of gravity that has so far not been constrained experimentally. All GW observations can be used to test the theory of general relativity (GR), but the strongest constraints will come from millihertz sources. MBH mergers are the loudest GW events that we expect to observe, while EMRIs should be detectable for hundreds of thousands of waveform cycles over the last several years of inspiral. A one cycle phase difference over the observation should be detectable in principle, allowing very small fractional deviations from the model, $\lesssim 10^{-5}$, to be identified. The potential impact of a space-based gravitational wave detector on our understanding of fundamental physics has been discussed extensively in the literature. We will briefly summarise some of the key ideas in the following, but refer the reader to Gair et al. (2013) for a thorough review of this topic.

4.1 Tests of Gravitational Physics

GW observations will provide tests of several different aspects of gravitational physics.

GW polarisation: In GR, there are only two possible polarisation states for gravitational waves—the transverse tensor 'plus' and 'cross' modes (TT)—but up to four additional polarisation modes—scalar-transverse (ST), scalar-longitudinal (SL) and two vector-longitudinal (VL) modes—may exist in alternative metric theories of gravity (Will 2006). At high frequencies, LISA would be ten times more sensitive to SL and VL modes than to ST and TT modes (Tinto and da Silva Alves 2010), although whether these would be detected in practice depends on the relative amplitude at which the different modes are generated by a given source in a particular alternative theory.

GW propagation: In GR, GWs are predicted to travel at the speed of light. GW observations can measure the speed of GWs and hence constrain the effective 'graviton mass' either by comparing the GW arrival time to that of an electromagnetic counterpart or by detecting the dispersion of gravitational wave chirps (different frequency components travel at different speeds). Bounds on the Compton wavelength of the graviton of $\lambda_g = h/m_g \gtrsim$ a few $\times\ 10^{16}$ km should be possible using the population of observed eLISA MBH merger events (Berti et al. 2011).

Inspiral rate: If energy is lost from a system to other forms of radiation in addition to the TT modes of GR this will manifest itself as a difference in the observed rate of inspiral of a system. In scalar-tensor gravity dipole radiation is generated in addition to the TT modes. The best GW bounds on this dipole radiation will come from observations of neutron star-MBH inspirals observed with space-based detectors. The derivable bound on the coupling constant of Brans-Dicke gravity scales as (Will and Yunes 2004)

$$\omega_{\mathrm{BD}} > 2 \times 10^4 \left(\frac{\mathscr{S}}{0.3}\right)^2 \left(\frac{100}{\Delta\Phi_D}\right) \left(\frac{T}{1\mathrm{yr}}\right)^{\frac{7}{8}} \left(\frac{10^4 M_\odot}{M_\bullet}\right)^{\frac{3}{4}} \quad (1)$$

where $\mathscr{S}$ is the 'sensitivity', the difference between the self-gravitational binding energy per unit rest mass of the neutron star and the MBH, $\Delta\Phi_D$ is the dipole contribution to the GW phase over the observation, T is the observation time and $M_\bullet$ is the MBH mass. This is comparable to the current bound obtained from the Cassini spacecraft, but including the effects of spin-orbit coupling and eccentricity, ignored in the preceding equation, can reduce the bound by as much as a factor of ten (Yagi and Tanaka 2010).

Phase evolution: A general gravitational wave can be written in the frequency domain in the form $\tilde{h}(f) = \mathscr{A}(f)\exp[i\Psi(f)]$. GR makes definite predictions for the coefficients in an expansion of the phase function in powers of frequency. A natural thing to do in order to provide sensitivity to un-modelled departures from GR is to try to measure these coefficients directly, or include extra terms in the expansion. Two different approaches to this problem have been explored. The first is to consider the coefficients, $\{\psi_k\}$, of the expansion of the phase in powers of $f^{(k-5)/3}$ to be free parameters to be measured from the data (Arun et al. 2006a). For an MBH merger source at a distance of 3 Gpc, LISA would be able to measure ψ_0 to $\sim$0.1 % and ψ_2 and ψ_3 to $\sim$10 %, but no other parameters. Alternatively, if ψ_0 and ψ_2 are used to determine the masses of the two objects in the binary and then the other coefficients are just checked for consistency with GR, all of the parameters can be constrained to $\sim$ a few % (Arun et al. 2006b). The second approach, the so-called 'parameterised post-Einsteinian' formalism (ppE) (Yunes and Pretorius 2009), in its simplest form, involves multiplying the GR waveform by a factor $(1 + \alpha(\pi\mathscr{M}f)^a)\exp[i\beta(\pi\mathscr{M}f)^b]$ and attempting to constrain the parameters α, a, β, b that represent the leading order departures in the amplitude and phase from GR. All currently known alternative theories of gravity can be represented in this ppE formalism. It was shown in Cornish et al. (2011) that for $b \gtrsim -1.75$, LISA would be able to place more stringent constraints on the amplitude of the deviation parameter β than is possible with either binary-pulsar observations or the post-Newtonian coefficient modification approach described above. The ppE formalism has now been generalised to alternative polarisation states, higher harmonics and eccentricity (see description and references in Gair et al. 2013).

4.2 Tests of the Nature and Structure of Black Holes

In GR, with certain additional assumptions made on physical grounds (the space-time is stationary, vacuum, asymptotically flat and contains an event horizon, but no closed time-like curves exterior to the horizon), the unique end state of gravitational collapse is a Kerr black hole (Hawking and Ellis 1973). The Kerr metric depends on just two free parameters—the mass of the black hole, M, and its (dimensionless) angular momentum, a. All higher multipole moments of the Kerr metric are determined by these two quantities, $M_l + iS_l = M(ia)^l$, so this uniqueness result is sometimes referred to as the 'no-hair' theorem. EMRIs are expected to be observed on eccentric and inclined orbits for hundreds of thousands of GW cycles while the inspiralling object is in the strong field region of the space-time. The small body thus explores the whole strong-field regime and the emitted GWs encode a detailed map of the space-time structure which can be used to make precise tests of the consistency of the space-time with the Kerr metric. This idea was first discussed in depth in Ryan (1995), who showed that the multipole moments of the space-time were encoded in GW observables—the precession frequencies of the orbit and the number of waveform cycles generated at a given frequency. Different multipole moments enter at different orders in the expansion of the precession frequencies as functions of the orbital frequency and so by observing the evolution of the frequencies over an inspiral a map of the space-time can be constructed. Many aspects of space-time mapping have been explored over the subsequent two decades and we summarise some of the key results in the following. Most of these results have been computed in the context of observations with LISA. However, the key thing that they rely on is the ability to track EMRI phase over the inspiral, which will be equally possible with eLISA. The conclusions are therefore unlikely to be changed by modifications of the mission design.

The nature of the central object: In Ryan (1997) it was shown that LISA could measure the first three multipole moments of the space-time with reasonable precision, which is enough for a test of the no-hair property. However, as more multipole moments were added in the model, the expected precision with which the lower multipole moments could be determined was degraded. The multipole moment expansion is an inefficient way to carry out space-time mapping, as an infinite number of multipoles are required to characterise the expected Kerr solution. Subsequent studies, starting with Collins and Hughes (2004), focussed on 'bumpy black holes', space times which are Kerr plus a small perturbation. Perturbative modifications of Schwarzschild (Collins and Hughes 2004) and Kerr (Glampedakis and Babak 2006) have been considered as well as exact solutions to the field equations of GR that violate one of the physicality assumptions of the uniqueness theorem (Gair et al. 2008) and the linear spin black hole space time predicted in dynamical Chern Simons modified gravity (a parity-violating modification to GR inspired by string theory) (Sopuerta and Yunes 2009). Fisher matrix analyses of the precision with which LISA observations would be able to determine the size of the deviations from GR have been carried out for quadrupole-deformed space-times (Barack and Cutler 2007) and the Chern Simons deformation (Canizares et al. 2012). These studies provided a consistent conclusion that LISA EMRI observations would be able to measure deviations from the Kerr metric at the level of $\sim$0.1 %, while simultaneously measuring the mass and spin to $\sim$0.01 %. LISA could also place a bound on the parameter characterising the Chern-Simons deviation at the level of $\xi^{\frac{1}{4}} < 10^4$ km, four orders of magnitude better than the current Solar System bound (Canizares et al. 2012).

Differences in the central object from a Kerr black hole are also encoded in qualitative features of the emitted GWs. If the object does not have a horizon, e.g., if it were a compact boson star, GW emission could continue after the object 'plunges' into the horizon (Kesden et al. 2005). Energy is lost into the horizon through tidal coupling and so information about this tidal coupling can be inferred from comparing the observed inspiral rate to the flux of energy carried to infinity in the gravitational waves. LISA would be able to distinguish an $\sim O(1)$ proportional change in the tidal coupling efficiency (Li and Lovelace 2008). Finally, differences in the nature of the horizon also manifest themselves in the quasi-normal mode (QNM) structure, e.g., 'gravastars' in which the horizon is replaced by a thin shell (Pani et al. 2009). These QNMs can be excited by an EMRI, leading to resonances in the observed inspiral evolution. The excitement of QNMs following a black hole merger can also be used to constrain deviations from GR, which will be discussed further below.

The effect of matter: Departures in our observations from the model predictions could also come about from astrophysical perturbations. Various possibilities have been considered. In Barausse et al. (2007) the gravitational influence of a torus of material on an EMRI was investigated and it was found to be significant. However, making small adjustments to the mass and spin of the space-time (which are unknown and have to be measured from the data) made the signals indistinguishable. The effect of hydrodynamic drag from a disc around an MBH was considered in Barausse and Rezzolla (2008) and estimated to be detectable, but only for very compact discs around low-mass MBHs. The signature of the drag is a decrease in orbital inclination, which is opposite to the evolution under GW emission. EMRIs inspiraling within a disc were considered in Yunes et al. (2011a) and it was found that a $\sim$1 radian dephasing relative to a vacuum EMRI could accumulate over an observation in some circumstances. The effect of a distant massive perturber (another MBH within the same galaxy in a hierarchical triple) was considered in Yunes et al. (2011b) and it was found that a perturber within $\sim$0.1 pc could leave a measurable imprint. A second EMRI occurring in the same system could also leave an imprint in $\sim$1 % of EMRIs by causing nominally chaotic evolution of the orbit (Amaro-Seoane et al. 2012). Finally, the presence of exotic material, e.g., axion (Arvanitaki and Dubovsky 2011) or boson (Macedo et al. 2013) clouds, could leave measurable imprints on the signal. While each of these processes could leave an imprint, they should not occur in much more than a few per cent of observed EMRIs. Moreover, the qualitative signature of an astrophysical perturbation should be different from an intrinsic deviation in the central object, as the former should become weaker as the inspiral progresses, while the latter should become stronger. Further study may be required, but it is likely that astrophysical perturbations will not adversely impact tests of fundamental physics. For a comprehensive discussion of environmental impacts on GW tests of relativity we refer the interested reader to Barausse et al. (2014).

Strong field dynamics: Departures from GR could lead to large qualitative differences in the behaviour of the orbits and emitted waves in the strong-field regime. Kerr has a complete set of integrals of the motion, which means the equations of motion are separable and the orbits are tri-periodic. A generic deviation from Kerr will break this property in general. Chaotic orbits have been found in several non-Kerr space times, for instance the Manko-Novikov solutions (Gair et al. 2008). Observing the signs of chaos would be a clear smoking-gun for a deviation from Kerr, but it is not obvious how to detect chaos in practice, since LISA data analysis will use EMRI templates based on Kerr, which will be for tri-periodic and predictable orbits. However, if an inspiral passed from a regime where it was Kerr-like and tri-periodic and then transitioned into a chaotic regime, this would be observable as an EMRI 'disappearing' from the data prematurely (Gair et al. 2008). Another qualitative signature of a departure from Kerr would be the behaviour of the EMRI as it passes through a resonance on which the EMRI frequencies become commensurate. Resonances persist generically when integrable systems are perturbed (Apostolatos et al. 2009), so the resonance would last much longer than expected.

Generic deviations from GR: In the spirit of the ppE formalism described in Sect. 4.1, it is interesting to consider the imprint of generic deviations from GR in the GWs generated during EMRIs. In Gair and Yunes (2011) EMRI waveforms were constructed for a set of modified gravity space-times described in Vigeland et al. (2011) that were built to be close to Kerr and maintain a complete set of integrals of the motion. These were thought to be the most relevant types of deviation as separable inspirals will be most easily detected using a Kerr EMRI analysis for the reasons discussed in the preceding section. This family includes the linear-spin metric of Chern Simons modified gravity mentioned earlier (Sopuerta and Yunes 2009). These metrics are characterised by the scaling of the perturbation with radius. LISA should be able to constrain the size of the deviation to a precision of $\sim 10^{-6}$ for an r^{-2} departure from GR, with the precision of the bound degrading by a factor of 10 for each additional order of r in the perturbation.

Quasi-normal ring down radiation: A perturbed black hole settles down to a stationary state by emitting gravitational radiation that is a superposition of quasi-normal modes. Each QNM is a damped sinusoid, characterised by a frequency and damping time which, due to the no-hair property, depend on the mass and spin of the black hole only. A measurement of two QNMs will therefore provide a consistency check that could reveal departures from the Kerr solution. EMRIs do not excite significant QNMs due to the large mass-ratio, but QNMs excited by MBH mergers can be used for no-hair tests. In Berti et al. (2006) it was concluded that it should be possible for LISA to resolve one QNM and either the frequency or damping time of a second QNM, provided that the second mode radiates at least $\sim 10^{-2}$ of the total ringdown energy. Numerical relativity simulations suggest that this will be the case (Bert et al. 2007). Using these simulations it was estimated that a ringdown S/N of ~ 30 would be needed to carry out a no-hair test (Berti et al. 2007) and therefore that a 1 % departure from the QNMs of GR could be identified for a $10^8 M_\odot$ black hole at 50 Gpc and a 10 % departure for a $10^6 M_\odot$ black hole at 6 Gpc (Gossan et al. 2012). A full review of studies of no-hair tests using LISA observations of QNMs can be found in Sect. 6.3 of Gair et al. (2013).

5 Summary

The millihertz GW band contains a rich variety of sources which have tremendous potential for science. Observations in this band could transform our understanding of astrophysics, cosmology and fundamental physics. They will provide constraints on the properties of MBHs and their evolution over cosmic time and on stellar populations in galactic centres and throughout the Milky Way. Millihertz GW sources can be used to probe the expansion history of the Universe in a manner that is completely independent of all current constraints from electromagnetic observations. Finally, these observations will push tests of GR into a regime of strong-fields and dynamical gravity that has never been probed. We will test all aspects of gravitational physics and probe the nature and structure of black hole space times to an unprecedented precision. *The Gravitational Universe* (The eLISA Consortium 2013) has been selected by ESA as the science theme for the L3 large satellite mission to launch in ∼2034. This mission will open up the millihertz gravitational wave window for the first time and enable us to realise all of this outstanding scientific potential.

Acknowledgements JG's work is supported by the Royal Society.

References

J. Aasi et al., Preprint. arxiv:1304.0670 (2013)
P.A.R. Ade et al., Phys. Rev. Lett. **112**, 241101 (2014)
P. Amaro-Seoane et al., Class. Quantum Grav. **24**, R113 (2007)
P. Amaro-Seoane et al., Class. Quantum Grav. **29**, 124016 (2012)
P. Amaro-Seoane et al., Astrophys. J. Lett. **744**, L20 (2012)
P. Amaro-Seoane et al., GW Notes **6**, 4 (2013)
T.A. Apostolatos et al., Phys. Rev. Lett. **103**, 111101 (2009)
K.G. Arun et al., Class. Quantum Grav. **23**, L37 (2006a)
K.G. Arun et al., Phys. Rev. D **74**, 024006 (2006b)
K.G. Arun et al., Class. Quantum Grav. **26**, 4027 (2009)
A. Arvanitaki, S. Dubovsky, Phys. Rev. D **873**, 044026 (2011)
L. Barack, C. Cutler, Phys. Rev. D **69**, 082005 (2004)
L. Barack, C. Cutler, Phys. Rev. D **75**, 042003 (2007)
E. Barausse, L. Rezzolla, Phys. Rev. D **77**, 104027 (2008)
E. Barausse et al., Phys. Rev. D **75**, 064026 (2007)
E. Barausse et al., Phys. Rev. D **89**, 104059 (2014)
E. Berti et al., Phys. Rev. D **73**, 064030 (2006)
E. Bert et al., Phys. Rev. D **76**, 064034 (2007)
E. Berti et al., Phys. Rev. D **76**, 104044 (2007)
E. Berti et al., Phys. Rev. D **84**, 101501 (2011)
P. Binetruy et al., J. Cosmol. Astropart. Phys. **6**, 27 (2012)
P. Canizares et al., Phys. Rev. D **86**, 044010 (2012)
D. Coe et al., Astrophys. J. **762**, 32 (2013)
N.A. Collins, S.A. Hughes, Phys. Rev. D **69**, 124022 (2004)

N. Cornish et al., Phys. Rev. D **84**, 062003 (2011)
D.J. Croton et al., Mon. Not. R. Astron. Soc. **365**, 11 (2006)
K. Danzmann et al., Class. Quantum Grav. **13**, 247 (1996)
J.R. Gair, Class. Quantum Grav. **26**, 094034 (2009)
J.R. Gair, N. Yunes, Phys. Rev. D **84**, 064016 (2011)
J.R. Gair et al., Phys. Rev. D **77**, 024035 (2008)
J.R. Gair et al., Phys. Rev. D **81**, 104014 (2010)
J.R. Gair et al., Liv. Rev. Rel. **16**, 7 (2013) (cited on July 2nd 2014)
K. Glampedakis, S. Babak, Class. Quantum Grav. **23**, 4167 (2006)
S. Gossan et al., Phys. Rev. D **85**, 124056 (2012)
J.E. Greene, L.C. Ho, Astrophys. J. **667**, 131 (2007)
N.A. Grogin et al., Astrophys. J. Suppl. **197**, 35 (2011)
S.W. Hawking, G.F.R. Ellis, *The Large-Scale Structure of Space-Time* (Cambridge University Press, Cambridge, 1973)
S. Hilbert et al., Mon. Not. R. Astron. Soc. **412**, 1023 (2011)
G. Hobbs et al., Class. Quantum Grav. **27**, 084013 (2010)
S.A. Hughes, D.E. Holz, Class. Quantum Grav. **20**, 65 (2003)
M. Kesden et al., Phys. Rev. D **71**, 044015 (2005)
A.M. Koekemoer et al., Astrophys. J. Supp. **197**, 36 (2011)
C. Li, G. Lovelace, Phys. Rev. D **77**, 064022 (2008)
C.F.B. Macedo et al., Astrophys. J. **774**, 48 (2013)
C.L. MacLeod, C.J. Hogan, Phys. Rev. D **77**, 3512 (2008)
A. Merloni, S. Heinz, Mon. Not. R. Astron. Soc. **388**, 1011 (2008)
D.J. Mortlock et al., Nature **474**, 616 (2011)
G. Nelemans et al., Astron. Astrophys. **365**, 491 (2001)
S. Nissanke et al., Astrophys. J. **758**, 131 (2012)
P. Pani et al., Phys. Rev. D **80**, 124047 (2009)
A. Petiteau et al., Astrophys. J. **732**, 82 (2011)
F.D. Ryan, Phys. Rev. D **52**, 5707 (1995)
F.D. Ryan, Phys. Rev. D **56**, 1845 (1997)
B.F. Schutz, Nature **323**, 310 (1986)
A. Sesana et al., Phys. Rev. D **83**, 044036 (2011)
R.K. Sheth et al., Astrophys. J. **594**, 225 (2003)
C.F. Sopuerta, N. Yunes, Phys. Rev. D **80**, 064006 (2009)
The eLISA Consortium, *The Gravitational Universe*. arxiv:1375.5720 (2013)
M. Tinto, M.E. da Silva Alves, Phys. Rev. D **82**, 122003 (2010)
S. Tremaine et al., Astrophys. J. **574**, 740 (2002)
S. Vigeland et al., Phys. Rev. D **83**, 104027 (2011)
C.M. Will, Liv. Rev. Rel. **9**, 3 (2006) (cited on July 2nd 2014)
C.M. Will, N. Yunes, Class. Quantum Grav. **21**, 4367 (2004)
K. Yagi, T. Tanaka, Phys. Rev. D **81**, 064008 (2010)
N. Yunes, F. Pretorius, Phys. Rev. D **80**, 122003 (2009)
N. Yunes et al., Phys. Rev. Lett. **107**, 171103 (2011a)
N. Yunes et al., Phys. Rev. D **83**, 044030 (2011b)

Connecting Numerical Relativity and Data Analysis of Gravitational Wave Detectors

Deirdre Shoemaker, Karan Jani, Lionel London, and Larne Pekowsky

Abstract Gravitational waves deliver information in exquisite detail about astrophysical phenomena, among them the collision of two black holes, a system completely invisible to the eyes of electromagnetic telescopes. Models that predict gravitational wave signals from likely sources are crucial for the success of this endeavor. Modeling binary black hole sources of gravitational radiation requires solving the Einstein equations of General Relativity using powerful computer hardware and sophisticated numerical algorithms. This proceeding presents where we are in understanding ground-based gravitational waves resulting from the merger of black holes and the implications of these sources for the advent of gravitational-wave astronomy.

1 Introduction

During the GR 1 conference held on January 1957 at the University of North Carolina, Chapel Hill, Charlie Misner famously said, "... we assume that we have a computer machine better than anything we have now, and many programmers and a lot of money, and you want to look at a nice pretty solution of the Einstein equations. ... Now, if you don't watch out when you specify these initial conditions, then either the programmer will shoot himself or the machine will blow up (Misner 1957)." From that date in 1957 to Frans Pretorius' landmark paper in 2005 (Pretorius 2005) that successfully followed an orbit and merger of two BHs, many programmers and money tackled the problem that Misner so eloquently points out. With two other groups following the Pretorius paper within a year (Baker et al. 2006; Campanelli et al. 2006), a new age of numerical relativity (NR) was begun. Progress was being made over the intervening decades (Brandt and Bruegmann 1997; Bruegmann 1999; Abrahams et al. 1998; Cook et al. 1998; Gomez et al. 1998;

D. Shoemaker (✉) • K. Jani • L. London • L. Pekowsky
Center for Relativistic Astrophysics, School of Physics, Georgia Institute of Technology, Atlanta, GA 30332, USA
e-mail: deirdre@gatech.edu; kjani3@gatech.edu; london@gatech.edu; larne.pekowsky@physics.gatech.edu

C.F. Sopuerta (ed.), *Gravitational Wave Astrophysics*, Astrophysics and Space Science Proceedings, DOI 10.1007/978-3-319-10488-1_21

Brandt et al. 2000; Alcubierre et al. 2001, 2003; Bruegmann et al. 2004; Gundlach et al. 2005) to reference just a few of the many pivotal papers during that time.

Arguably the driving force behind the decades of work on solving Einstein's equations for the coalescence of compact binary systems has been and continues to be the imminent detection of gravitational waves (GWs). The world-wide network of detectors is coming online with Advanced LIGO (Harry 2010) and Advanced Virgo (Acernese et al. 2009) leading the way and KAGRA (Somiya 2012) on the way. The era of gravitational wave astronomy is upon us (Aasi et al. 2013a) and NR has an important role to play. Just some of the important issues currently being addressed by NR for GW astronomy are

- determining where approximations to full solutions to the Einstein equation break down as we approach merger (Ohme et al. 2011; MacDonald et al. 2013),
- detecting the difference between NS and BH sources (Foucart et al. 2013; Hannam et al. 2013),
- creating template banks that cover the complete parameter space (Kumar et al. 2014; Privitera et al. 2014),
- determining the impact of higher harmonics on detection and parameter estimation,
- settling down to final state of the black hole, and
- investigating mergers as GW transients.

This proceeding focuses on that last three items in the above list. First, we discuss a historical perspective of NR and GWs in Sect. 2 and a more detailed look into BBH mergers in NR Sect. 3 before discussing the role of harmonics Sect. 4, the approach to the final state Sect. 5 and mergers as transients Sect. 6. Conclusions are presented in the final section "Conclusions".

2 Historical Perspective on NR and GWs

While NR was struggling to capture the spacetime and dynamics of a merging compact binary on computers, the gravitational wave community was preparing for initial LIGO and VIRGO. Cutler et al. (1993) lays out the case for a deep understanding of the theoretical waveforms, called templates, from compact binary coalescence in preparation for the GW detectors. These theoretical templates are needed to unearth the signal within the noisy data using matched filtering, the optimal choice for detecting a signal in Gaussian noise where the template is known (Thorne 1987). Although at the time NR had not yet produced waveforms from BBH mergers, post-Newtonian theory was making progress in producing waveforms that work well for binary neutron stars and low mass BH binaries. Low-mass objects, a couple of solar masses to tens of solar masses, have long inspirals in the LIGO frequency band well modeled by post-Newtonian approximations. In fact, neutron star binaries may merge outside of the band making NR less useful in predicting these signals.[1]

BBH signals that are more massive could be the first detected (Brady et al. 1998) motivating several methods to bridge the gap between post-Newtonian inspiral and ringdown. Energy conditions were studied (Flanagan and Hughes 1998a, b) and approximations were developed to produce waveforms that would cover the full inspiral, merger, ringdown (Buonanno and Damour 1999; Khanna et al. 1999; Baker et al. 2002). Fortunately, NR was on the brink of success culminating in the series of papers in 2005–2006 (Pretorius 2005; Campanelli et al. 2006; Baker et al. 2006).

3 Binary Black Hole Mergers

Post 2005, NR groups world wide began an assault on the BBH parameter space, focusing primarily on gravitational recoil and the spin of the final black hole. A few turned their attention to the ramifications of these new NR waveforms on data analysis efforts to detect BBHs (Baumgarte et al. 2008; Buonanno et al. 2007; Baker et al. 2007; Ajith et al. 2007; Pan et al. 2008; Vaishnav et al. 2007). This culminated in three collaborations

1. NINJA (Aylott et al. 2009b, a; Ajith et al. 2012; Aasi et al. 2014): the Numerical INJection Analysis project whose goal is to use NR waveforms as signal and test the detection and parameter estimation pipelines for the ground-based GW detectors,
2. SAMURI (Hannam et al. 2009): this was a one-time project that did a consistency check of several NR codes; and,
3. NRAR (Hinder et al. 2014): Numerical Relativity and Analytical Relativity Collaboration whose goal is to use NR waveforms to produce the next generation of waveform models.

For more details on these collaborative efforts in the numerical relativity and gravitational wave community, see the paper by Sascha Husa in this collection.

Figure 1 is a figure from the NINJA 2 paper (Ajith et al. 2012) where a catalog of NINJA waveforms were publicly released. The figure depicts one hybrid waveform at three different total masses. The merger is at the extremum of the waveform. The triangles represent the region where we created hybrids by stitching together the post-Newtonian NR waveforms. As mass increases, the signal-to-noise ratio increases and the number of cycles in band decreases at a fixed distance for an optimally oriented binary with the parameters held fixed; however, at some point, the reduction in the number of cycles will result in a drop of signal-to-noise ratio. The figure illustrates the impact of merger on the BBH signals for masses above

[1] To understand matter's effect on neutron star waveforms see Read et al. (2013) for binary neutron stars and Andersson and Kokkotas (1998) for isolated neutron stars.

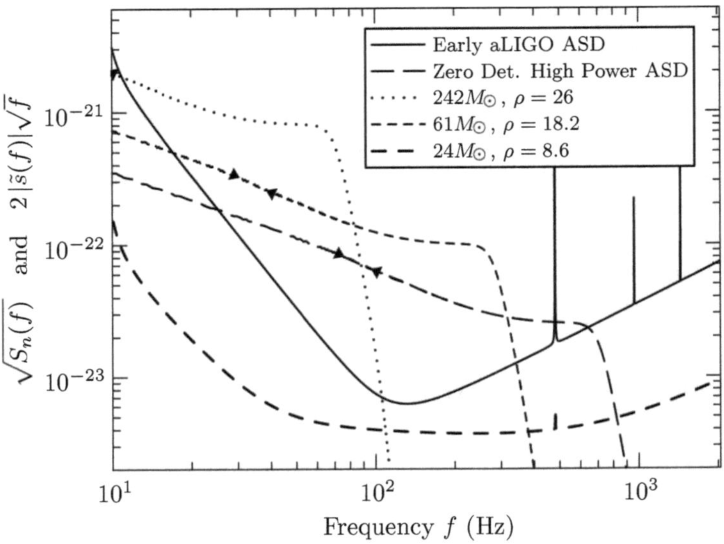

Fig. 1 A Ninja 2 Collaboration (Ajith et al. 2012) figure. Here we see the GW from a NINJA2 waveform versus frequency against two predictions of the advanced LIGO noise curve. A single NR waveform is represented at three different total masses. The waveform is a hybrid, non-spinning, 2 to 1 mass ratio waveform scaled to three different total masses. The *triangles* represent the starting and ending frequencies of the hybridization region where the NR and post-Newtonian waveforms were stitched together. The waveform is scaled such that it is optimally oriented at a distance of 1 Gpc from the detector. The signal-to-noise ratios, ρ, were computed at that distance with the early aLIGO sensitivity

$20M_\odot$ where the merger sits in the most sensitive region of the detector. As the masses grow higher than a couple of hundred solar masses, we begin to only detect the merger and ringdown addressed in Sect. 6.

One of the challenges in NR is covering the full parameter space of the binary black hole (BBH) system. There are no stringent constraints on the initial spins of the black hole, although there are some expectations (Gerosa et al. 2013; Kesden et al. 2010; Schnittman 2004); and, in fact, we expects GWs to be the main method for observing BH parameters. A similar scenario holds for the mass ratio, $q = m_1/m_2$ where m_1 and m_2 are the BH masses, and total mass, $M = m_1 + m_2$. Recent work points to the possibility of higher mass BHs than expected (Dominik et al. 2014) and see the proceedings by Thomas Bulik in this collection. Simulating extreme spins (Lovelace et al. 2012) and mass ratios (Lousto and Zlochower 2011) is still a challenge, especially in producing quality waveforms from such simulations. Despite these challenges, several groups are running generic, precessing BBH systems. The SXS collaboration has a published catalog of 174 BBH waveforms (Mroue et al. 2013) including spins of 0.99 and mass-ratios of 1:10 (Lovelace 2014), RIT has generic runs (Lousto et al. 2010) and GT (Pekowsky et al. 2013b) has several hundred shown here in Fig. 2.

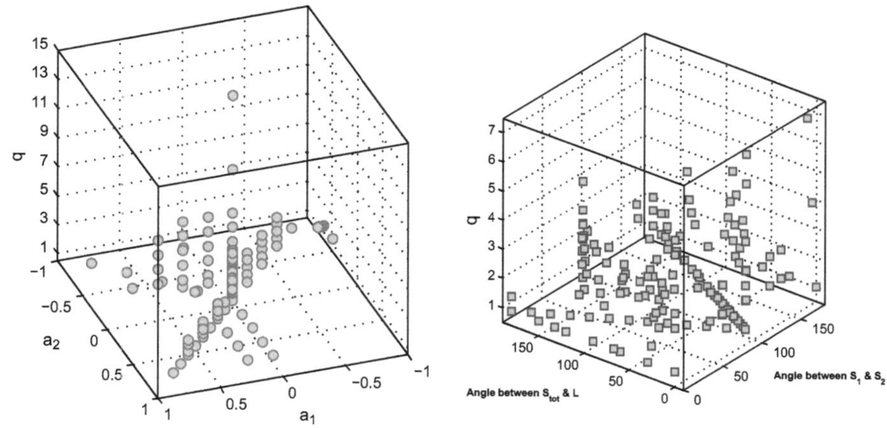

Fig. 2 The *left figure* is GT's non-precessing runs and the *right* the precessing runs, The *left plot* indicates our set of runs, one for each *circle*, corresponding to the mass ratio, $q = m_1/m_2$ where m_1 and m_2 are the BH masses, and the initial spin of each black hole, $a_1 = S_1/m_1^2$ and $a_2 = S_2/m_2^2$. The figure on the *right*, again depicts each run now by a square bot plots versus mass ratio on the vertical axis, q, and the angle between the total spins $S_{tot} = S_1 + S_2$ and L, the total momentum and the angle between S_1 and S_2

The open problem and important challenge for NR and GW data analysis is in preparing to detect and characterize the parameters from a precessing BBH system, see Hannam (2014) and references therein.

4 Role of Higher Harmonics on BBH Mergers and GWs

Gravitational waves encode the dynamics and history of the compact binary coalescence. When the GW reaches the detector, the radiation also encodes the position of the binary on the sky and its inclination with respect to the detector(s). In NR, we choose to represent the GWs extracted from our simulations via spin-weighted spherical harmonics that map the waveform from a function of angles to a function of modes, or harmonics, labeled by ℓ and m. This mode description can be very useful theoretically since inspiralling binaries will have the quadrupolar mode dominating, given by $\ell = |m| = 2$. However, as the black holes inspiral, merge and settle down to its final BH, the information about the system is radiated in those harmonics. In addition, the sky position also mixes in with these parameters as seen in Fig. 3 where we plot a BBH system at two different lines of site with respect to the detector. The binary in this figure has a mass ratio 1:10 and is non-spinning. To the left, you see that the binary was optimally oriented and the $\ell = |m| = 2$ mode, given by h_{22}, overlaps with the full waveform, h, containing all the non-zero modes well. The overlap is the scalar product between h and h_{22} normalized to one.

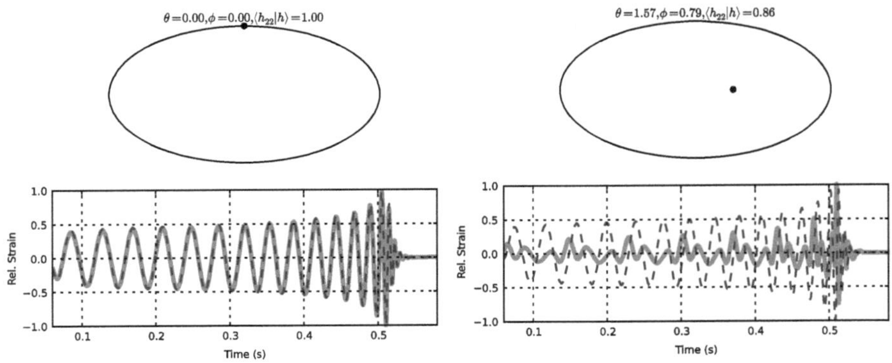

Fig. 3 The *upper figures* show the location of the $q = 10$ binary with respect to the detector. The *lower plots* are of the strain of the $\ell = |m| = 2$ waveform and the full waveform plotted on top of each other. In the *left*, the binary is optimally oriented and the *right*, randomly oriented. The strains are scaled to be one at the peak

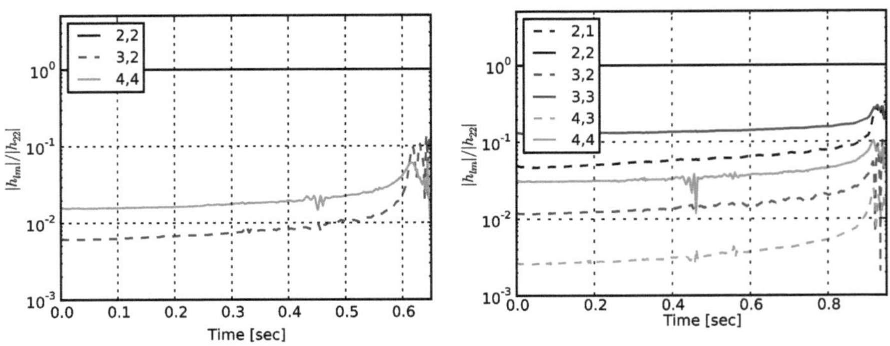

Fig. 4 Figure from Healy et al. (2013) shows the strain amplitude relative to the $\ell = |m| = 2$ mode. The *left panel* shows the case of an equal mass binary, and the *right panel* that of a 1:4 mass ratio system. In both figures the systems have been scaled to total mass $M = 100M_\odot$ and neither have spin

To the right, we place the binary at a non-optimal position and now see that h_{22} no longer as good of a representation of the full waveform along that line of site.

The question then is, for the current set of detectors, how important is this information? Can we detect binaries of generic initial states without accounting for the modes? If so, how does it bias the extracted parameters? This topic is addressed in a recent set of papers (Healy et al. 2013; Pekowsky et al. 2013a; Capano et al. 2014) that are reviewed here.

First we can look back at Fig. 3 and see which modes are needed to optimize our sky coverage. We know that for highly precessing binaries or binaries with moderately large mass ratios, the energy in the radiation is spread out among the modes. In Fig. 4 we see the relative amplitudes of higher modes in comparison with the $\ell = |m| = 2$ mode for both an equal mass (left) and unequal (right).

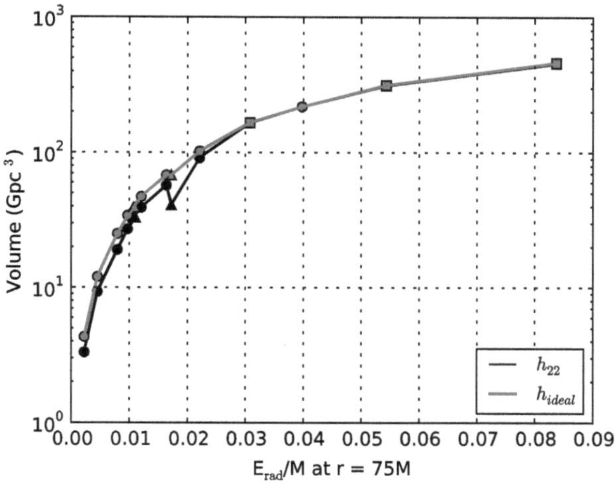

Fig. 5 Figure from Pekowsky et al. (2013a) shows the correlation between the total energy radiated and the detector reach measured in volume using just the $\ell = |m| = 2$ waveform or the full waveform labeled here as h_{ideal}. *Circles* label non-spinning systems, *squares* are spinning but non-precessing systems, and *triangles* are precessing systems

Clearly, as the mass ratio increases, the higher modes gain in amplitude. In fact, the relative importance of the modes also increases towards merger, which makes these modes and hence NR potentially more important for BBH systems. However, when we looked at sky coverage, defined as the percent of sky in a source-centric frame that a template will match with nature, which modes were important could deviate from a ranked list of amplitudes. To cover the entire sky to some measure, as mass ratio increases beyond $q = 7$, more than 4 modes are necessary. In the absence of precession, increasing spin magnitude reduces the number of needed modes. Precessing systems required up to 12 modes for the cases studies in Healy et al. (2013).

We then computed how the reach of a single detector would be impacted by not including all the relevant modes in Pekowsky et al. (2013a). Figure 5 is a plot from that paper that shows the energy radiated per run in both the $\ell = |m| = 2$ only and all the modes versus the volume of the detector for that system. The energy radiated in these systems corresponds to the amplitude of the waveform. The actual vales in Gpc are not to be used literally since these numbers do not incorporate event-loss and other important aspects of the real detector pipeline. We see that the fact the more modes are required for higher mass-ratio systems is diluted by the fact that these systems radiated less in total output; and, therefore, are reduced in volume. Likewise, the higher modes are less important as the spin increases, seen here as the points on the right that radiate the largest amount of energy. Precessing systems will depend on the parameters but the largest deviation between all the modes and just the quadrupolar is seen.

A further step was taken to understand the role of the sub-dominate modes for mass ratios of $q < 4$ in Capano et al. (2014). They found that the higher modes did not improve detection rates for advanced LIGO when using a re-weighed SNR (Abadie et al. 2012b; Aasi et al. 2013b) and in fact including the modes increases the false-alarm rate. We may conclude, that for unequal-mass binaries of $q < 4$, higher modes should not be included in the detection pipeline, although a definitive answer would require a all-mode injection into the real detector noise. Whether or not that will remain the case for generic runs with precessing spins is an open question, as is the potential for parameter bias in the unequal-mass case which we are addressing in a future work (Pekowsky et al. 2014).

5 End State of BBH Merger

We've studied how the initial data of the binaries is imprinted on the harmonics of the waveforms. In these final moments of the coalescence, the two BHs merge into a perturbed Kerr BH, whose gravitational radiation *rings down* like a struck bell. With the BH's mass and spin known, perturbation theory provides the quasi-normal modess (QNMs) that describe ringdown (Press and Teukolsky 1973; Leaver 1985; Berti et al. 2007); however, perturbation theory cannot inform us of the amplitudes of each mode. Recent studies have used NR waveforms to correlate the QNM excitation amplitudes with initial parameters (Berti et al. 2007; Kamaretsos et al. 2012a, b; Kelly and Baker 2013; London et al. 2014). Figure 6 depicts the end state of our GT runs with the final BH's spin versus its mass.[2]

In London et al. (2014), we studied the QNMs excited by a series of NR simulations that focused on non-spinning, unequal mass binaries. We found a

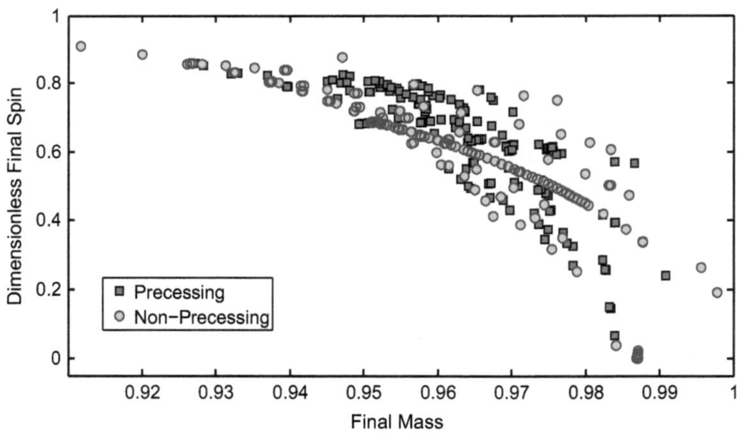

Fig. 6 Plotted are the spin and mass for the final, merged BH at the end of our NR simulations. Each point represents a different simulation from unique initial data. The *squares* are precessing runs and the *circles* are non-precessing

[2]Note that the final mass is constrained to be less than one. The mass and spin are computed using isolated horizons (Dreyer et al. 2003).

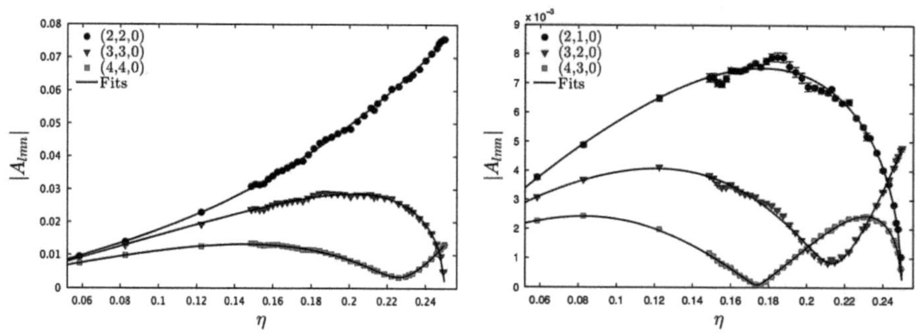

Fig. 7 Figure from London et al. (2014) of the amplitudes of three QNMs versus symmetric mass ratio and the corresponding fits

rich variety of fundamental QNMs, as well as overtones, and evidence for 2nd order QNMs. Rather than the spin-weighted spherical harmonics used in NR extraction, we represent the waveforms in terms of the spheroidal harmonics which are natural to perturbation theory. It has been an open question whether or not such a difference would make an impact on the modes. Inner-products between spherical and spheroidal harmonics suggest that for the $\ell = m$ modes, very little difference should be expected (Berti et al. 2006). For the $\ell \neq m$ modes we expect "beating" that arises in the QNMs when found using spherical harmonics which are combinations of the spheroidal harmonics used in perturbation theory (Kelly and Baker 2013). As the final spin magnitude increasing deviates from 0, the difference between spherical and spheroidal harmonics is more pronounced, as expected (London et al. 2014).

We find that the mass-ratio dependence of quasi-normal mode excitation is very well modeled by a Post-Newtonian like sum for most but not all QNMs. We presented new fitting formulas for the amplitudes versus the mass ratio in London et al. (2014). Figure 7 plots the amplitudes of the fundamental QNMs versus q and include our fits.

The modes would be difficult to detect for current GW detectors. However, there is the possibility that future detectors might be able to see them, as seen in Fig. 8 which depicts both Einstein Telescope and advanced LIGO noise predictions. It seems possible to do gravitational-wave astronomy with QNMs, if not as simply as first laid out in Kamaretsos et al. (2012a, b).

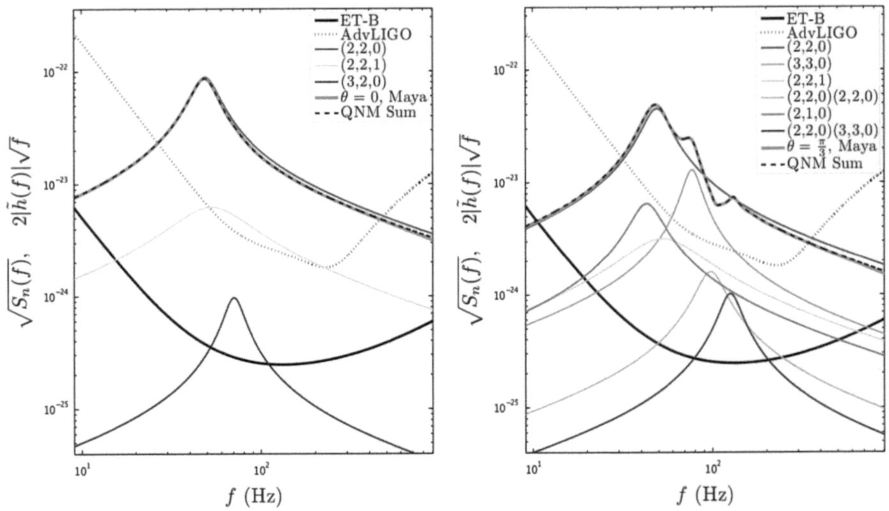

Fig. 8 This figure from London et al. (2014) plots the frequency domain envelopes of strain and fitted QNM amplitudes for a 2:1 mass ratio system of $350M_\odot$ at a distance of 100 Mpc. The *left figure* is for an optimally oriented binary and the *right* for a specific orientation. Noise curves for the Einstein Telescope and Adv. LIGO are shown for reference

6 Mergers as Transient GW Events

The LIGO and Virgo Collaborations search for gravitational wave transients, or *bursts*, without templates for sources like supernova or the unknown (Abadie et al. 2012a; Andersson et al. 2013). Recently, several groups have begun to explore the merger of BHs as a transient event (Fischetti et al. 2011; Clark et al. 2014; Klimenko 2014; Cornish 2014). This is a very interesting development that focuses on intermediate-mass BBH systems that may only have a few cycles in band. See Clark et al. (2014) in this volume for a discussion of one such approach to mergers as transients.

Conclusions

Undoubtably, numerical relativity, analytical relativity and gravitational wave data analysis have made great strides in the last few years. With the advanced LIGO scheduled to begin taking data next year and her sister detectors soon after, the theoretical efforts are focusing down on the crucial open questions for detecting GW sources and estimating their parameters. This proceeding highlighted just a few of those questions.

(continued)

Clearly, the parameter space of generic BH binaries is large and the NR community continues its efforts to provide waveforms for the full space and to turn these NR waveforms into viable GW templates. One avenue of research that is still open is using tools like principal component analysis (Clark et al. 2014) to predict where we need a new NR run that will make the most impact on the template bank.

Over the last couple of years, work has been done to measure the impact of higher modes on detection. We are optimistic that for lower mass-ratios and non-precessing spins, the dominate mode will be enough for detection. The problem is still open for a verification of this statement as mass ratio increases over the full BBH total mass range. As is the question of higher modes for highly precessing binaries where the $\ell = |m| = 2$ mode is not always dominate.

NR runs result in the BBH system ringing down to the final BH. We have found a rich and fascinating regime during this epoch, revealing a variety of QNMs. While it is still questionable whether these modes will be detectable by the current generation of GW detectors, it does hint at the tantalizing possibility of conduction GW astronomy with BH ringdown waveforms.

NR still has a story to tell GW astronomy. New and exciting work is being done in BH-neutron star and binary neutron stars. BBHs as transients are being seriously investigated. At the end of the day, it will be GW astronomy that will excite new work in NR as we learn about how general relativity is manifested in the Universe.

Acknowledgements The authors thank the organizers and participants of the Sant Cugat Forum on Astrophysics on Gravitational Wave Astrophysics. The work presented here was supported by NSF PHY-0955825.

References

J. Aasi et al., Prospects for localization of gravitational wave transients by the Advanced LIGO and Advanced Virgo observatories (2013a). Preprint. arXiv:1304.0670

J. Aasi et al., Search for gravitational waves from binary black hole inspiral, merger and ringdown in LIGO-Virgo data from 2009–2010. Phys. Rev. **D87**, 022002 (2013b)

J. Aasi et al., The NINJA-2 project: detecting and characterizing gravitational waveforms modelled using numerical binary black hole simulations. Class. Quantum Grav. **31**, 115004 (2014)

J. Abadie et al., All-sky search for gravitational-wave bursts in the second joint LIGO-Virgo run. Phys. Rev. D **85**, 122007 (2012a)

J. Abadie et al., Search for gravitational waves from low mass compact binary coalescence in LIGO's sixth science run and Virgo's science runs 2 and 3. Phys. Rev. **D85**, 082002 (2012b)

A.M. Abrahams et al., Gravitational wave extraction and outer boundary conditions by perturbative matching. Phys. Rev. Lett. **80**, 1812–1815 (1998)

F. Acernese et al., Virgo Technical Report VIR-0027A-09 (2009). https://tds.ego-gw.it/ql/?c=6589

P. Ajith, S. Babak, Y. Chen, M. Hewitson, B. Krishnan et al., Phenomenological template family for black-hole coalescence waveforms. Class. Quantum. Grav. **24**, S689–S700 (2007)
P. Ajith et al., The NINJA-2 catalog of hybrid post-Newtonian/numerical-relativity waveforms for non-precessing black-hole binaries. Class. Quantum Grav. **29**, 124001 (2012)
M. Alcubierre, W. Benger, B. Bruegmann, G. Lanfermann, L. Nerger et al., The 3-D grazing collision of two black holes. Phys. Rev. Lett. **87**, 271103 (2001)
M. Alcubierre, B. Bruegmann, P. Diener, M. Koppitz, D. Pollney et al. Gauge conditions for long term numerical black hole evolutions without excision. Phys. Rev. **D67**, 084023 (2003)
N. Andersson, K.D. Kokkotas, Towards gravitational wave asteroseismology. Mon. Not. R. Astron. Soc. **299**, 1059–1068 (1998)
N. Andersson et al., The transient gravitational-wave sky. Class. Quantum Grav. **30**, 193002 (2013)
B. Aylott, J.G. Baker, W.D. Boggs, M. Boyle, P.R. Brady et al., Status of NINJA: the Numerical INJection Analysis project. Class. Quantum Grav. **26**, 114008 (2009a)
B. Aylott, J.G. Baker, W.D. Boggs, M. Boyle, P.R. Brady et al., Testing gravitational-wave searches with numerical relativity waveforms: results from the first Numerical INJection Analysis (NINJA) project. Class. Quantum Grav. **26**, 165008 (2009b)
J.G. Baker, M. Campanelli, C.O. Lousto, The Lazarus project: a pragmatic approach to binary black hole evolutions. Phys. Rev. **D65**, 044001 (2002)
J.G. Baker, J. Centrella, D.-I. Choi, M. Koppitz, J. van Meter, Gravitational wave extraction from an inspiraling configuration of merging black holes. Phys. Rev. Lett. **96**, 111102 (2006)
J.G. Baker, S.T. McWilliams, J.R. van Meter, J. Centrella, D.-I. Choi et al., Binary black hole late inspiral: simulations for gravitational wave observations. Phys. Rev. **D75**, 124024 (2007)
T. Baumgarte, P. Brady, J.D.E. Creighton, L. Lehner, F. Pretorius et al., Learning about compact binary merger: the Interplay between numerical relativity and gravitational-wave astronomy. Phys. Rev. **D77**, 084009 (2008)
E. Berti, V. Cardoso, M. Casals, Eigenvalues and eigenfunctions of spin-weighted spheroidal harmonics in four and higher dimensions. Phys. Rev. D **73**, 024013 (2006)
E. Berti, V. Cardoso, J.A. Gonzalez, U. Sperhake, M. Hannam, S. Husa, B. Brügmann, Inspiral, merger, and ringdown of unequal mass black hole binaries: a multipolar analysis. Phys. Rev. D **76**, 064034 (2007)
P.R. Brady, J.D.E. Creighton, K.S. Thorne, Computing the merger of black hole binaries: the IBBH problem. Phys. Rev. **D58**, 061501 (1998)
S. Brandt, B. Bruegmann, A simple construction of initial data for multiple black holes. Phys. Rev. Lett. **78**, 3606–3609 (1997)
S. Brandt, R. Correll, R. Gomez, M. Huq, P. Laguna et al., Grazing collisions of black holes via the excision of singularities. Phys. Rev. Lett. **85**, 5496–5499 (2000)
B. Bruegmann, Binary black hole mergers in 3-d numerical relativity. Int. J. Mod. Phys. **D8**, 85 (1999)
B. Bruegmann, W. Tichy, N. Jansen, Numerical simulation of orbiting black holes. Phys. Rev. Lett. **92**, 211101 (2004)
A. Buonanno, T. Damour, Effective one-body approach to general relativistic two-body dynamics. Phys. Rev. **D59**, 084006 (1999)
A. Buonanno, G.B. Cook, F. Pretorius, Inspiral, merger and ring-down of equal-mass black-hole binaries. Phys. Rev. **D75**, 124018 (2007)
M. Campanelli et al., Accurate evolutions of orbiting black-hole binaries without excision. Phys. Rev. Lett. **96**, 111101 (2006)
C. Capano, Y. Pan, A. Buonanno, Impact of higher harmonics in searching for gravitational waves from non-spinning binary black holes. Phys. Rev. D **89**, 102003 (2014)
J. Clark, L. Cadonati, J. Healy, I.S. Heng, J. Logue, N. Mangini, L. London, L. Pekowsky, D. Shoemaker, Investigating binary black hole mergers with principal component analysis. In this volume (2014). arXiv:1406.5426v2
G.B. Cook et al., Boosted three-dimensional black hole evolutions with singularity excision. Phys. Rev. Lett. **80**, 2512–2516 (1998)
N. Cornish. April APS Meeting (2014)

C. Cutler, T.A. Apostolatos, L. Bildsten, L.S. Finn, E.E. Flanagan et al., The last three minutes: issues in gravitational wave measurements of coalescing compact binaries. Phys. Rev. Lett. **70**, 2984–2987 (1993)

M. Dominik, E. Berti, R. O'Shaughnessy, I. Mandel, K. Belczynski et al., Double compact objects III: gravitational wave detection rates (2014). arXiv:1405.7016

O. Dreyer, B. Krishnan, D. Shoemaker, E. Schnetter, Introduction to isolated horizons in numerical relativity. Phys. Rev. **D67**, 024018 (2003)

S. Fischetti, J. Healy, L. Cadonati, L. London, S.R.P. Mohapatra et al., Exploring the use of numerical relativity waveforms in burst analysis of precessing black hole mergers. Phys. Rev. **D83**, 044019 (2011)

E.E. Flanagan, S.A. Hughes, Measuring gravitational waves from binary black hole coalescences: 1. Signal-to-noise for inspiral, merger, and ringdown. Phys. Rev. **D57**, 4535–4565 (1998a)

E.E. Flanagan, S.A. Hughes, Measuring gravitational waves from binary black hole coalescences: 2. The Waves' information and its extraction, with and without templates. Phys. Rev. **D57**, 4566–4587 (1998b)

F. Foucart, L. Buchman, M.D. Duez, M. Grudich, L.E. Kidder et al., First direct comparison of nondisrupting neutron star-black hole and binary black hole merger simulations. Phys. Rev. **D88**(6), 064017 (2013)

D. Gerosa, M. Kesden, E. Berti, R. O'Shaughnessy, U. Sperhake, Resonant-plane locking and spin alignment in stellar-mass black-hole binaries: a diagnostic of compact-binary formation. Phys. Rev. **D87**(10), 104028 (2013)

R. Gomez, L. Lehner, R.L. Marsa, J. Winicour, A.M. Abrahams et al., Stable characteristic evolution of generic three-dimensional single black hole space-times. Phys. Rev. Lett. **80**, 3915–3918 (1998)

C. Gundlach, J.M. Martin-Garcia, G. Calabrese, I. Hinder, Constraint damping in the Z4 formulation and harmonic gauge. Class. Quantum Grav. **22**, 3767–3774 (2005)

M. Hannam, Modelling gravitational waves from precessing black-hole binaries: progress, challenges and prospects. Gen. Rel. Grav. **46**, 1767 (2014)

M. Hannam, S. Husa, J.G. Baker, M. Boyle, B. Bruegmann et al., The Samurai Project: verifying the consistency of black-hole-binary waveforms for gravitational-wave detection. Phys. Rev. **D79**, 084025 (2009)

M. Hannam, D.A. Brown, S. Fairhurst, C.L. Fryer, I.W. Harry, When can gravitational-wave observations distinguish between black holes and neutron stars? Astrophys. J. **766**, L14 (2013)

G.M. Harry, Advanced LIGO: the next generation of gravitational wave detectors. Class. Quantum Grav. **27**, 084006 (2010)

J. Healy et al., Template mode hierarchies for binary black hole mergers. Phys. Rev. D **88**(2), 024034 (2013)

I. Hinder et al., Error-analysis and comparison to analytical models of numerical waveforms produced by the NRAR Collaboration. Class. Quantum Grav. **31**, 025012 (2014)

I. Kamaretsos, M. Hannam, S. Husa, B.S. Sathyaprakash, Black-hole hair loss: learning about binary progenitors from ringdown signals. Phys. Rev. **D85**, 024018 (2012a)

I. Kamaretsos, M. Hannam, B. Sathyaprakash, Is black-hole ringdown a memory of its progenitor? Phys. Rev. Lett. **109**, 141102 (2012b)

B.J. Kelly, J.G. Baker, Decoding mode mixing in black-hole merger ringdown. Phys. Rev. **D87**, 084004 (2013)

M. Kesden, G. Lockhart, E. Sterl Phinney, Maximum black-hole spin from quasi-circular binary mergers. Phys. Rev. **D82**, 124045 (2010)

G. Khanna, J.G. Baker, R.J. Gleiser, P. Laguna, C.O. Nicasio et al., Inspiralling black holes: The Close limit. Phys. Rev. Lett. **83**, 3581–3584 (1999)

S, Klimenko, April APS Meeting (2014)

P. Kumar, I. MacDonald, D.A. Brown, H.P. Pfeiffer, K. Cannon et al., Template banks for binary black hole searches with numerical relativity waveforms. Phys. Rev. **D89**, 042002 (2014)

E.W. Leaver, An analytic representation for the quasi normal modes of Kerr black holes. Proc. R. Soc. Lond. **A402**, 285–298 (1985)

L. London, J. Healy, D. Shoemaker, Modeling ringdown: beyond the fundamental quasi-normal modes (2014). arXiv:1404.3197

C.O. Lousto, Y. Zlochower, Orbital evolution of extreme-mass-ratio black-hole binaries with numerical relativity. Phys. Rev. Lett. **106**, 041101 (2011)

C.O. Lousto, H. Nakano, Y. Zlochower, M. Campanelli, Statistical studies of Spinning Black-Hole Binaries. Phys. Rev. **D81**, 084023 (2010)

G. Lovelace, Private communication (2014)

G. Lovelace et al. Accurate gravitational waveforms for binary-black-hole mergers with nearly extremal spins. Class. Quantum Grav. **29**, 045003 (2012)

I. MacDonald, A.H. Mroue, H.P. Pfeiffer, M. Boyle, L.E. Kidder et al., Suitability of hybrid gravitational waveforms for unequal-mass binaries. Phys. Rev. **D87**, 024009 (2013)

C. Misner, *Role of Gravitation in Physics - Report from the 1957 Chapel Hill Conference*. Max Planck Research Library for the History and Development of Knowledge (1957)

A.H. Mroue et al., A catalog of 174 binary black-hole simulations for gravitational-wave astronomy. Phys. Rev. Lett. **111**, 241104 (2013)

F. Ohme, M. Hannam, S. Husa, Reliability of complete gravitational waveform models for compact binary coalescences. Phys. Rev. **D84**, 064029 (2011)

Y. Pan, A. Buonanno, J.G. Baker, J. Centrella, B.J. Kelly et al., A data-analysis driven comparison of analytic and numerical coalescing binary waveforms: nonspinning case. Phys. Rev. **D77**, 024014 (2008)

L. Pekowsky, J. Healy, D. Shoemaker, P. Laguna, Impact of higher-order modes on the detection of binary black hole coalescences. Phys. Rev. D **87**, 084008 (2013a)

L. Pekowsky et al., Comparing gravitational waves from nonprecessing and precessing black hole binaries in the corotating frame. Phys. Rev. D **88**(2), 024040 (2013b)

L. Pekowsky, J. Healy, D. Shoemaker, Impact of higher-order modes on parameter recovery from binary black hole coalescences (2014, in preparation)

W.H. Press, S.A. Teukolsky, Perturbations of a rotating black hole. II. Dynamical stability of the kerr metric. Astrophys. J. **185**, 649–674 (1973)

F. Pretorius, Evolution of binary black hole spacetimes. Phys. Rev. Lett. **95**, 121101 (2005)

S. Privitera, S.R.P. Mohapatra, P. Ajith, K. Cannon, N. Fotopoulos et al., Improving the sensitivity of a search for coalescing binary black holes with non-precessing spins in gravitational wave data. Phys. Rev. **D89**, 024003 (2014)

J.S. Read, L. Baiotti, J.D.E. Creighton, J.L. Friedman, B. Giacomazzo et al., Matter effects on binary neutron star waveforms. Phys. Rev. **D88**, 044042 (2013)

J.D. Schnittman, Spin-orbit resonance and the evolution of compact binary systems. Phys. Rev. **D70**, 124020 (2004)

K. Somiya, Detector configuration of KAGRA: the Japanese cryogenic gravitational-wave detector. Class. Quantum Grav. **29**, 124007 (2012)

K. Thorne, in *Three Hundred Years of Gravitation*, ed. by S.W. Hawking, W. Israel (Cambridge University Press, Cambridge, 1987)

B. Vaishnav et al., Matched filtering of numerical relativity templates of spinning binary black holes. Phys. Rev. D **76**, 084020 (2007)

Memory Effect from Spinning Unbound Binaries

Lorenzo De Vittori, Achamveedu Gopakumar, Anuradha Gupta, and Philippe Jetzer

Abstract We present a recently developed prescription to obtain ready-to-use gravitational wave (GW) polarization states for spinning compact binaries on hyperbolic orbits. We include leading order spin-orbit interactions, invoking 1.5PN-accurate quasi-Keplerian parametrization for the radial part of the orbital dynamics. We also include radiation reaction effects on h_+ and $h_\times$ during the interaction. In the GW signals from spinning binaries there is evidence of the memory effect in both polarizations, in contrast to the non-spinning case, where only the cross polarizations exhibits non-vanishing amplitudes at infinite time. We also compute 1PN-accurate GW polarization states for non-spinning compact binaries in unbound orbits in a fully parametric way, and compare them with existing waveforms.

1 Introduction

Among the plausible sources of GWs for ground and space based detectors, compact binaries moving on hyperbolic orbits could represent an interesting fraction of the observable events. In the last decades there has been much progress in the description of signals emitted from unbound systems. Some of the most important works providing crucial inputs required to construct polarization states are the computation of the quadrupolar order radiation field from non-spinning compact objects in Newtonian hyperbolic orbits by Turner (1977), and its extension to 1PN order by Junker and Schäfer (1992), who made use of the quasi-Keplerian parametrization approach by Damour and Deruelle (1985) describing 1PN-accurate motion on bound and unbound orbits. Explicit GW polarization states for non-spinning compact binaries on hyperbolic orbits were computed for the first time by Majár and Vasúth (2008) and Majár et al. (2010). Many other works provided interesting investigations on different aspects of non-spinning binaries on hyperbolic orbits,

L. De Vittori (✉) • P. Jetzer
Physik-Institut, Universität Zürich, Winterthurerstrasse 190, 8057 Zürich, Switzerland
e-mail: lorenzo@physik.uzh.ch

A. Gopakumar • A. Gupta
Department of Astrophysics, Tata Institute of Fundamental Research, Mumbai 400005, India

C.F. Sopuerta (ed.), *Gravitational Wave Astrophysics*, Astrophysics and Space Science Proceedings, DOI 10.1007/978-3-319-10488-1_22

such as the energy released during the interactions and the power spectrum of the GW emission (Wagoner and Will 1976; De Vittori et al. 2012), or the event rate estimate of such flybys (Capozziello et al. (2008)).

Here we summarize the work done recently by De Vittori et al. (to be published), where the GW polarization states for spinning compact binaries on hyperbolic orbits were computed, up to 1.5PN order including leading order spin-orbit coupling. We show the presence of the so-called memory effect, which consists in the non-zero amplitude of the waveform at $t = +\infty$, for both polarization states.

Taking a look at the non-spinning limit, we compute the 1PN-accurate waveforms in a fully parametric manner and show that only the cross polarization exhibits memory effects. Finally, we compare our signals in this limit case with those in Junker and Schäfer (1992) and Favata (2011) and those obtained through a true anomaly parametrization in Majár et al. (2010).

2 Waveforms for Spinning Compact Binaries on Hyperbolic Orbits

It is customary to express GWs polarization states at Newtonian order through the following expression in terms of the velocity vector $\boldsymbol{v}$, the unit relative separation vector $\boldsymbol{n}$ and the plane-of-the-sky unit vectors $\boldsymbol{p}$ and $\boldsymbol{q}$:

$$h_+ = \frac{2G\mu}{c^4R}\left\{(\boldsymbol{p}\cdot\boldsymbol{v})^2 - (\boldsymbol{q}\cdot\boldsymbol{v})^2 - \frac{Gm}{r}\left[(\boldsymbol{p}\cdot\boldsymbol{n})^2 - (\boldsymbol{q}\cdot\boldsymbol{n})^2\right]\right\}, \tag{1}$$

$$h_\times = \frac{4G\mu}{c^4R}\left\{(\boldsymbol{p}\cdot\boldsymbol{v})\,(\boldsymbol{q}\cdot\boldsymbol{v}) - \frac{Gm}{r}\,(\boldsymbol{p}\cdot\boldsymbol{n})\,(\boldsymbol{q}\cdot\boldsymbol{n})\right\}, \tag{2}$$

where m is the total mass of the system, μ the reduced mass, R the distance to the observer and r the relative separation. It is possible to express the polarization states for general orbits including leading order spin-order coupling contributions up to 1PN, as given e.g. in Gopakumar and Schäfer (2011). Choosing the reference frames in an appropriate way and performing some straightforward computations, one can express the above vectors and dot products in terms of the dynamics, i.e. as a function of the relative radius r, the radial velocity $\dot{r}$, the phasing angle Φ and its derivative $\dot{\Phi}$, and the precession angles α and ι, obtaining

$$h_+|_Q(t) = h_+(r,\dot{r},\Phi,\dot{\Phi},\alpha,\iota) \quad \text{and} \quad h_\times|_Q(t) = h_\times(r,\dot{r},\Phi,\dot{\Phi},\alpha,\iota)\,. \tag{3}$$

Explicit expressions can be found in Sect. II of De Vittori et al. (to be published).

Since we are interested in temporally evolving $h_+|_Q(t)$ and $h_\times|_Q(t)$, we need to specify how r, $\dot{r}$, Φ, $\dot{\Phi}$, α and ι vary in time along PN-accurate hyperbolic orbits. In the case of spinning binaries, we have to divide this analysis into two parts, the radial and the angular solution of the dynamics. The former can be solved in a customary parametric way, while the latter has to be tackled numerically.

Using expressions for non-spinning binaries on eccentric PN-accurate orbits from Damour and Deruelle (1985), results including spin effects were found by Gopakumar and Schäfer (2011). They solve the radial part of the dynamics through the usual Kepler equations

$$r = a_\mathrm{r}\,(1 - e_\mathrm{r}\cos u)\,, \tag{4}$$

$$l = n\,(t - t_0) = u - e_\mathrm{t}\sin u\,, \tag{5}$$

where a_r, e_r, e_t and n are the PN extensions of the semi-major axis, the radial eccentricity, the time eccentricity and the mean motion, respectively. Afterwards, they find expressions for the orbital parameters a_r, e_t and e_r for eccentric orbits, as functions of the dimensionless parameter $\xi = Gmn/c^3$, the reduced angular momentum L and energy E, the symmetric mass ratio η, and the spin-orbit interaction parameter Σ, describing the spin misalignment in terms of $\boldsymbol{k}\cdot\boldsymbol{s}_1$ and $\boldsymbol{k}\cdot\boldsymbol{s}_2$. Finally, they can write down the solution to Eq. (4) as a series in $1/c$ up to 1.5PN order.

We obtain the hyperbolic counterpart of the Kepler equations and their solutions invoking the analytic continuation technique from Damour and Deruelle (1985), i.e. we let $u = \imath v$ and $n = -\imath\bar{n}$. The analog of Eqs. (4) and (5) for unbound orbits is

$$r = a_\mathrm{r}\,(e_\mathrm{r}\cosh v - 1)\,, \tag{6}$$

$$l = \bar{n}\,(t - t_0) = e_\mathrm{t}\sinh v - v\,, \tag{7}$$

and computing the hyperbolic version of the orbital parameters, the radial solution can be expressed through $\bar{\xi} = Gm\bar{n}/c^3$ as

$$r = \frac{Gm}{c^2}\frac{1}{\bar{\xi}^{2/3}}\left\{e_\mathrm{t}\cosh v - 1 - \bar{\xi}^{2/3}\frac{18 - 2\eta + (6 - 7\eta)\,e_\mathrm{t}\cosh v}{6} + \frac{\bar{\xi}\,\Sigma}{\sqrt{e_\mathrm{t}^2 - 1}}\right\}, \tag{8}$$

$$\dot{r} = \bar{\xi}^{1/3}\,\frac{c\,e_\mathrm{t}\sinh v}{e_\mathrm{t}\cosh v - 1}\left\{1 - \bar{\xi}^{2/3}\,\frac{6 - 7\eta}{6}\right\}. \tag{9}$$

For the angular part of the dynamics, we employ again the results in Gopakumar and Schäfer (2011) for eccentric orbits. Using the previous arguments to find the hyperbolic counterpart, we find the differential equations driving the temporal evolution:

$$\dot{s_1} = \frac{c^3}{Gm} \frac{\bar{\xi}^{5/3} \sqrt{e_t^2 - 1}}{(e_t \cosh v - 1)^3} \delta_1 \, \boldsymbol{k} \times \boldsymbol{s}_1 \, , \tag{10}$$

$$\dot{s_2} = \frac{c^3}{Gm} \frac{\bar{\xi}^{5/3} \sqrt{e_t^2 - 1}}{(e_t \cosh v - 1)^3} \delta_2 \, \boldsymbol{k} \times \boldsymbol{s}_2 \, , \tag{11}$$

$$\dot{\boldsymbol{k}} = \frac{c^3}{Gm} \frac{\bar{\xi}^2 \, (\delta_1 \chi_1 q \, \boldsymbol{s}_1 + \delta_2 \chi_2 / q \, \boldsymbol{s}_2) \times \boldsymbol{k}}{(e_t \cosh v - 1)^3} \, , \tag{12}$$

$$\dot{\Phi} = \frac{\bar{n} \sqrt{e_t^2 - 1}}{(e_t \cosh v - 1)^2} \left\{ 1 + \bar{\xi}^{2/3} \left(\frac{4 - \eta}{e_t \cosh v - 1} + \frac{\eta - 1}{e_t^2 - 1} \right) - \bar{\xi} \frac{\Sigma}{\sqrt{e_t^2 - 1}} \left(\frac{1}{e_t \cosh v - 1} + \frac{1}{e_t^2 - 1} \right) \right\} - \cos \iota \, \dot{\alpha} \, , \tag{13}$$

where the derivative in the last term of Eq. (13) reads $\dot{\alpha} = (k_x \dot{k}_y - k_y \dot{k}_x)/(k_x^2 + k_y^2)$. Additionally, it is possible to incorporate numerically the effects of radiation reaction during the hyperbolic passage. To the above set, we add the following 2.5PN-accurate coupled differential equations for $\bar{n}$ and e_t, adapted from Damour et al. (2004):

$$\frac{d\bar{n}}{dt} = -\frac{c^6}{G^2 m^2} \frac{\bar{\xi}^{11/3} \, 8 \, \eta}{5 \, \beta^7} \left(-49\beta^2 - 32\beta^3 + 35(e_t^2 - 1)\beta - 6\beta^4 + 9e_t^2\beta^2 \right) \, , \tag{14}$$

$$\frac{de_t}{dt} = -\frac{c^3}{Gm} \frac{\bar{\xi}^{8/3} 8\eta (e_t^2 - 1)}{15 \, \beta^7 \, e_t} \left(-49\beta^2 - 17\beta^3 + 35(e_t^2 - 1)\beta - 3\beta^4 + 9e_t^2\beta^2 \right) \, , \tag{15}$$

where for simplicity $\beta = e_t \cosh v - 1$. Specifying the initial binary configuration in terms of the masses m_1, m_2, the Kerr parameters χ_1, χ_2, the initial values for $\bar{n}$ and e_t and the initial spin orientations through the angles (θ_1^i, θ_2^i) and (ϕ_1^i, ϕ_2^i), we implement a numerical method to evolve the set of 12 differential equations (10)–(15), and combine it to the results for the radial part in Eqs. (8) and (9). Finally, we impose these variations in the expressions for $h_+|_Q(t)$ and $h_\times|_Q(t)$, given in (3), and obtain the desired waveforms for spinning compact binaries on hyperbolic orbits.

In Fig. 1 we display the waveform for both polarization states, resulting from such an implementation. We compare the signals for two different initial eccentricities $e_t = 1.5$ and $e_t = 1.3$, and for two different mass ratios $q = 1$ and $q = 4$.

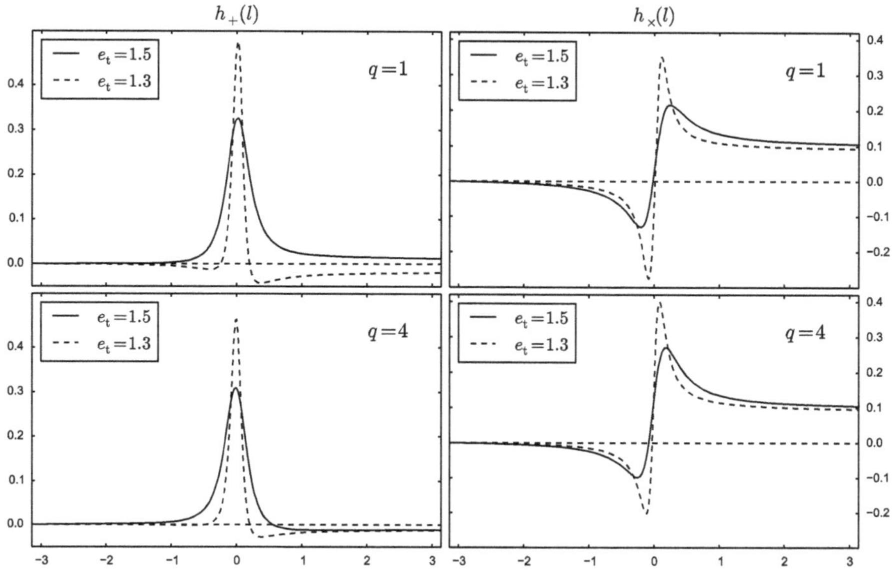

Fig. 1 We plot $h_+|_Q$ and $h_\times|_Q$ for mass ratios $q = 1$ and $q = 4$. As one can see, both polarizations show a memory effect: the amplitude of the waveform at $t = +\infty$ is different from the initial. The hyperbolic passage generates a burst and leaves a trace of the interaction in the metric. We see the influence of the eccentricity on the waveform by comparing the $e_t = 1.5$ and $e_t = 1.3$ cases

3 Waveforms for Non-spinning Binaries on Hyperbolic Orbits

In the limit case of non-spinning compact binaries it is easier to obtain PN-accurate waveforms since there is no spin-orbit coupling and therefore no precession of the orbital plane. It is then enough to find the evolution to the radius r and its derivative $\dot{r}$, as well as the phasing angle ϕ and its derivative $\dot{\phi}$ to fully describe the dynamics of the binary. This can be done in a parametric manner, including 1PN correction terms, as done in Damour and Deruelle (1985). Results for r, $\dot{r}$, ϕ, $\dot{\phi}$ can be found in Sect. III of De Vittori et al. (to be published).

The expressions for the polarization states we use are the same as for the spinning case, whereas the dot products in (1)–(2) and in their 1PN generalization are much simpler. As usual, we can write them as a PN series:

$$h_+ = -\frac{G\mu}{c^4 R}\left(h_+^N + \frac{1}{c}h_+^{0.5} + \frac{1}{c^2}h_+^1\right), \quad h_\times = -\frac{G\mu}{c^4 R}\left(h_\times^N + \frac{1}{c}h_\times^{0.5} + \frac{1}{c^2}h_\times^1\right). \tag{16}$$

In Sect. III of De Vittori et al. (to be published) we show the full 1PN expressions in terms of r, $\dot{r}$, ϕ and $\dot{\phi}$.

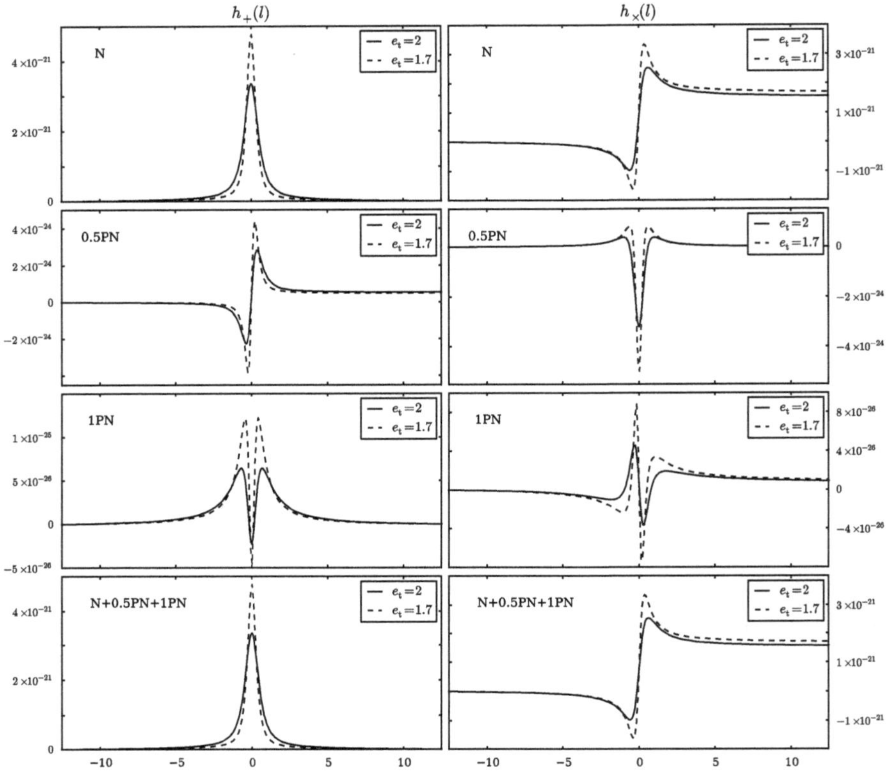

Fig. 2 We plot here $h_+|_Q$ and $h_\times|_Q$ for non-spinning binaries, at Newtonian, 0.5PN and 1PN order. In contrast to the spinning case, here only the cross polarization shows a memory effect. Also, one can see that the signal is symmetric with respect to the $t = 0$ axis at each PN order. This is due to the fact that the orbital plane is not precessing. The effect of the eccentricity on the waveform is clearly visible. Low eccentricities lead to violent interactions and therefore sharp and high peaks

Specifying the initial configuration of the binary and imposing the evolution equations in the above expressions, we obtain waveforms for non-spinning compact binaries on hyperbolic orbits.

In Fig. 2 we plot some results for a binary system with masses $m_1 = 8M_\odot$ and $m_2 = 13M_\odot$, for $e_t = 2$ and $e_t = 1.7$, showing the waveform at different PN orders.

We can compare our results for non-spinning binaries along unbound orbits with existing waveforms. We have reproduced temporal evolution for the real and the imaginary parts of appropriate time derivatives of various mass and current multipole moments that are displayed in Fig. 8 in Junker and Schäfer (1992), so that we can be sure of the correctness of our approach. Also, results depicted in Fig. 2 of Favata (2011) for the (2,2) GW mode coincide with ours.

In Majár and Vasúth (2008), a generalized true anomaly parametrization approach to obtain 1PN-accurate polarization states is developed. The prescription is carried out in Majár et al. (2010) and some results are shown. A visual comparison of our $e_t = 2$ plots in Fig. 2 with Figs. 6, 7 and 10 of Majár et al. (2010) reveals considerable differences. The memory effect exhibited in the cross polarization is clearly different, and none of the GW signals depicted seems to be symmetric with respect to the $t = 0$ axis, something we would expect. We suspect that the observed differences may be due to the way temporal evolution is implemented in Majár et al. (2010).

Conclusions

We provide a new prescription to obtain ready-to-use PN-accurate GW templates for spinning compact binaries on hyperbolic orbits, including leading order spin-orbit coupling contributions and radiation reaction effects. We are able to reproduce previous results for the non-spinning limit case, and find some interesting new feature arising from the spin-orbit coupling. In particular, we notice that for spinning binaries both the plus and the cross polarization states exhibit a memory effect, i.e. a non-vanishing amplitude at time $t = +\infty$, whereas for non-spinning systems only the cross polarization shows such a behaviour. In this last case, we can write

$$\lim_{t\to+\infty} h_+|_Q(t) = \lim_{t\to-\infty} h_+|_Q(t)\,, \qquad \lim_{t\to+\infty} h_\times|_Q(t) = -\lim_{t\to-\infty} h_\times|_Q(t)\,, \tag{17}$$

while for spinning binaries not only the sign but also the magnitude of $h_{+,\times}$ changes from $t = -\infty$ to $t = +\infty$. Since the waveform is strongly influenced by spin effects, the structure of such an asymmetric burst-like signal carries many informations about the system which could be useful for parameter estimation. It will be interesting to incorporate the 2PN order non-spinning contributions to our 1.5PN-accurate orbital dynamics, as well as dominant order spin-spin interactions.

References

S. Capozziello, M. de Laurentis, F. de Paolis, G. Ingrosso, A. Nucita, Mod. Phys. Lett. A **23**, 99 (2008)
T. Damour, N. Deruelle, Ann. Inst. Henri Poincaré Phys. Théor. **43**, 107 (1985)
T. Damour, A. Gopakumar, B.R. Iyer, Phys. Rev. D **70**, 064028 (2004)
L. De Vittori, A. Klein, P. Jetzer, Phys. Rev. D **86**, 044017 (2012)
L. De Vittori, A. Gopakumar, A. Gupta, P. Jetzer, Phys. Rev. D. (to be published)
M. Favata, Phys. Rev. D **84**, 124013 (2011)
A. Gopakumar, G. Schäfer, Phys. Rev. D **84**, 124007 (2011)

W. Junker, G. Schäfer, Mon. Not. R. Astron. Soc. **254**, 146 (1992)
J. Majár, M. Vasúth, Phys. Rev. D **77**, 104005 (2008)
J. Majár, P. Forgács, M. Vasúth, Phys. Rev. D **82**, 064041 (2010)
M. Turner, Astrophys. J. **216**, 914 (1977)
R.V. Wagoner, C.M. Will, Astrophys. J. **210**, 764 (1976)

The Challenges in Gravitational Wave Astronomy for Space-Based Detectors

Edward K. Porter

Abstract The GW universe contains a wealth of sources which, with the proper treatment, will open up the universe as never before. By observing massive black hole binaries to high redshifts, we should begin to explore the formation process of seed black holes and track galactic evolution to the present day. Observations of extreme mass ratio inspirals will allow us to explore galactic centers in the local universe, as well as providing tests of General Relativity and constraining the value of Hubble's constant. The detection of compact binaries in our own galaxy may allow us to model stellar evolution in the Milky Way. Finally, the detection of cosmic (super)strings and a stochastic background would help us constrain cosmological models. However, all of this depends on our ability to not only resolve sources and carry out parameter estimation, but also on our ability to define an optimal data analysis strategy. In this presentation I will examine the challenges that lie ahead in GW astronomy for the ESA L3 Cosmic Vision mission, eLISA.

1 Introduction

Gravitational waves (GWs) are a prediction of Einstein's theory of General Relativity and can be thought of as ripples of spacetime itself, which are caused by the high acceleration of massive objects. This can be from the motion of a binary of compact objects, a supernova or the creation of the Universe itself. GWs share many common points with electromagnetic (EM) waves. They have two polarisations, which instead of a 90° rotation as is seen with photons, are rotated to each other by 45°. Rather than being a dipolar radiation, GWs are a quadrupolar radiation. They obey the inverse square law and have a propogation speed of that of light. The difference is that whereas photons are oscillations of the EM field, GWs are oscillations of the fabric of spacetime itself. The other main difference is that photons are scattered and diffracted as they make their way towards the Earth. As gravity is the weakest of the four fundamental forces, GWs interact very weakly with matter, meaning

E.K. Porter (✉)
François Arago Center, APC, Université Paris Diderot, CNRS/IN2P3, CEA/Ifru, Obs. de Paris, Sorbonne Paris Cité, 10, rue Alice Domon et Léonie Duquet, 75205 Paris Cedex 13, France
e-mail: porter@apc.univ-paris7.fr

C.F. Sopuerta (ed.), *Gravitational Wave Astrophysics*, Astrophysics and Space Science Proceedings, DOI 10.1007/978-3-319-10488-1_23

there is almost no scattering or diffraction.The weakness of the GWs makes their detection a very difficult task. However, due to their low environmental interaction, they travel from source to detector with virtually no information loss. This allows us to investigate the sources of GWs with unprecedented accuracy

In the next two decades, GWs will open a new window on the universe, providing the field of astrophysics with an unprecedented richness of sources. The ground-based interferometers Advanced Virgo and Advanced LIGO will (http://wwwcascina.virgo.infn.it/advirgo/, https://www.advancedligo.mit.edu/) come on-line in 2015 and operating at frequencies greater than one Hertz, will be sensitive to stellar mass compact objects in the $(1 - 10)M_{\odot}$ mass range. In the nanohertz regime, researchers are investigating the detection of GWs from supermassive black hole binaries (SMBHBs) with masses of between $(10^8 - 10^{10})M_{\odot}$ via pulsar timing measurements (www.ipta4gw.org/). It is highly possible that both methods will make the first direct GW detection of astrophysical sources before the end of the current decade. In addition, ESA has recently chosen the theme of the "Gravitational Wave Universe" for the ESA Cosmic Vision L3 mission (Amaro-Seoane et al. 2012) which will explore the millihertz regime. This frequency band, which is inaccessible to the other experiments, will explore SMBHBs with masses of $(10^3 - 10^7)M_{\odot}$.

An efficient way to detect GWs is by using interferometry. The passage of a GW through an interferometer of length L changes the distance between space-craft by an amount ΔL. The strain amplitude induced by the passage of the GW is given by $h \sim \Delta L/L$. In general, nature provides a natural strain on the order of $h \sim 10^{-21}$. As the GW changes the distance between space-craft by approximately 20 pm, we can see that a space-craft separation of $L \sim 10^{-12}/10^{-21} \sim 10^6$ kms is needed to detect the GW.

The main detection technique for GWs is based on optimal or Weiner matched filtering. The matched filter is the optimal linear filter in the presence of stochastic noise. In most cases, we assume that the GW source is buried in instrumental noise and is not visible in the time domain. Matched filtering works perfectly in this case as it allows us to detect the source by transforming to the Fourier domain and using a theoretical waveform model, or template, to correlate with the data. By adjusting the parameters of the template, we can optimise the signal to noise ratio between the template and a possible signal in the data. The downside of matched filtering is that it requires a theoretical model of the possible signal that we are trying to detect. Furthermore, it is mostly dependent on the phase matching between a template and signal, so the technique requires very accurate phase modeling of the waveforms.

In this lecture, we will investigate the main GW sources of interest for a space-based interferometer, listing the current state of the art and outstanding problems in both astrophysics and algorithmic development.

2 Galactic Binaries

The eLISA data stream will be source dominated with approximately 60 million GW sources. Approximately 99 % of these sources will be Milky Way binaries composed of white dwarf, neutron star and stellar mass black holes. While made up of compact objects which are described by General Relativity, these binaries can be thought of as being Newtonian in terms of their orbits (i.e. the orbital velocity of the binary components is very small compared to the speed of light). Due to the wide separation of the system components, these binaries are essentially monochromatic with simple waveform polarization models.

While most of these galactic binaries will be monochromatic, we do expect a certain number of interacting binaries where we have mass transfer due to Roche lobe overflow. In these cases, we expect to be able to measure the first frequency derivative for a certain number of sources, and for a smaller number of sources, possibly even the second frequency derivative. A measurement of the frequency derivatives would allow us to break certain parameter correlations and improve parameter estimation.

For the LISA mission it was estimated that we should be able to individually resolve approximately 25×10^3 sources (Timpano et al. 2006). For the eLISA mission this number falls to around 3,500 sources with 1 year of observation (Amaro-Seoane et al. 2012), with an additional 10^3 sources per year (Littenberg 2014). This is due to the fact that the frequency resolution Δf goes as the inverse of the observation time T_{obs}. Therefore as T_{obs} gets larger, the width of each frequency bin Δf gets smaller allowing us to resolve more sources. While the total number of resolved binaries decreases with the eLISA mission, one of the benefits is the lack of a confusion problem. This arises from the fact that many thousands of galactic binaries occupy the same Fourier bin. Only the brightest sources in each bin can be resolved, leaving behind many unresolved sources per frequency bin. With LISA, even after the subtraction of 25×10^3, the background of unresolved sources remained above the noise curve. This excess galactic noise could then effect the parameter estimation of other sources in the data. For eLISA, subtracting the 3,500 galactic binaries reduces the galaxy to the instrumental noise level, eliminating the galactic confusion problem.

2.1 Verification Binaries

These are compact binaries in the Milky Way for which we already have EM information. These are guaranteed sources for an ESA L3 mission. Due to the fact that we already have information on these binaries, they should also be the first sources we detect. As well as the fact that they are guaranteed sources, these binaries will also allow us to recombine the data sets in a coherent manner in case of a disturbance or disruption of the data, due to the inherent knowledge of the frequency of the binary.

At present there are less than ten known verification binaries. However, in 2011 a new white dwarf binary was discovered with an orbital period of 39 min (Kilic et al. 2011). This binary, labelled J0106-1000, has a GW frequency right in the middle of the eLISA band. It is also hoped that GAIA (http://sci.esa.int/gaia/) will discover a number of additional verification binaries. While GAIA will not have the resolution to provide full parameter estimation, it is possible that follow-up observations with other EM telescopes will provide many tens, if not hundreds, of new verification binaries by time a mission like eLISA launches.

2.2 Current Status of Existing Algorithms

Only a small number of algorithms exist worldwide to search for galactic binaries (Crowder and Cornish 2007; Blaut et al. 2010; Littenberg and Cornish 2010). These algorithms were based on Markov Chain Monte Carlo or stochastic template bank algorithms. The first of these algorithms uses a stochastic search process to improve the parameter estimation of the source by moving to points of higher likelihood in the parameter space. The algorithm can be accelerated by using techniques such as simulated annealing (i.e. the likelihood surface is heated to remove all but the highest peaks, and is then slowly cooled back to a normal state) and by the use of source specific proposals when making the jump to the next point in the parameter space. The second algorithm is more of a brute force method where the parameter space is covered in a grid of templates, and each template is correlated with the data set. Points of high correlation are then investigated further. However, since 2010 there has not been much further development in galactic binary algorithms. While the algorithms were applied to Mock Data Challenges (The MLDC Taskforce and Participants 2007) with varying levels of success, they were never applied to the reformulated eLISA mission. However, as the number of resolvable binaries has decreased, if anything, the existing algorithms should be more performant.

2.3 Outstanding Problems

As with all the sources, it is possible to split the outstanding problems into two branches: astrophysical and algorithmic.

1. We still do not have a solid grasp on stellar evolution in the Milky Way. In a standard evolutionary picture, we know that the compact remnant of a supernova is imparted with a spatial kick velocity. If big enough, this kick velocity could disrupt the binary. For a compact binary to form, the system would need to

survive the kick velocities from two supernovae. Furthermore, as we do not fully understand the initial mass distribution of stars in the Milky Way, it is very difficult to estimate the distribution of binaries in terms of those composed of white dwarves, neutron stars and stellar mass black holes. We would also expect there to be a certain fraction of mixed systems.
2. On the algorithmic side, the development of search and resolution algorithms needs attention. While it has been shown that there is no confusion between a single galactic binary and an other source type, such as a massive black binary, we have seen that the galaxy of white dwarf binaries can fool an algorithm into believing it has found a massive black hole binary (this is the so-called white dwarf transform). However, this can be circumvented using a Bayesian model analysis. The big question that needs to be answered is can we search for galactic binaries without removing the other brighter sources first? This should be a priority line of study in the coming years.
3. Finally, the current algorithms model galactic binary sources as circular binaries. However, we know that the famous Hulse-Taylor binary B1913+16 has an eccentricity of $e \sim 0.6$ (Hulse and Taylor 1975). While it is well known that GWs are very efficient in circularising a binary, it may be that a large fraction of the galactic binary sources will have a non-negligible eccentricity.

3 Supermassive Black Hole Binaries

A principal GW source for a space-based mission is the merger of supermassive black hole binaries (SMBHBs). The luminosity of a SMBH merger in GWs is around $10^{26} L_{\odot}$, making them the most violent and brightest events in the Universe (In comparison, a supernova releases $\sim 10^{14} L_{\odot}$ in photons and generally outshines the host galaxy). As the waves travel towards the detector almost untouched, it allows us to see the source almost as it was and estimate the parameters of the binary system with incredible accuracy. It is estimated that we will be able to measure the component masses to less than 1 % error (see for example Fig. 3 of Amaro-Seoane et al. (2012)—a measurement of astrophysical masses on a level never seen before), the luminosity distance with an error between 1 and 50 % and estimate the time of merger to within a few minutes. The expected event rate for these sources is ~100/year (Sesana et al. 2007, 2011). Detection of these sources would allow us to map the black hole mass function and the spin history of the binary as a function of cosmic time, observe the mechanisms for the formation of massive black holes during the dark ages of the Universe and test alternative theories of gravity during the inspiral, merger and ringdown of the binary.

There is almost unquestionable evidence that SMBHs lie at the center of each galaxy. There is also EM evidence for the existence of massive black hole binaries, although we should mention that resolution of these sources is incredibly difficult, so a note of caution should be aired. What is certain is that we know that galaxies collide and that the SMBHs that we see today did not form as they are observed.

The working hypothesis for black hole growth is that the seeds of SMBHs formed at high redshifts ($10 \leq z \leq 17$), and via subsequent galactic mergers, grew to the size they are observed at today. These merger tree models(see Volonteri 2010 for example) seem to give a reasonable explanation of the Universe we observe today. However, one of the main questions in observational cosmology and astrophysics is how did the first massive black holes form.

At present there are two main hypotheses (Volonteri 2010; Begelman et al. 2006; Devecchi et al. 2012; Madau and Rees 2001; Mayer et al. 2006; Tegmark et al. 1997): the first is that the first black hole seeds came from the remnants of Population III stars. These were very massive, low metal stars with very short lifetimes ($\sim$ million years). The resulting supernova would produce black holes with masses of between $(10-100)M_{\odot}$. The second hypothesis is that some protogalaxies underwent direct gravitational collapse forming black holes with masses of between $(10^3 - 10^4)M_{\odot}$. However, no observational evidence exists for either scenario.

3.1 Algorithmic Development

As they are a priority source, the area of development of search algorithms for SMBHBs has probably seen the highest activity in the last decade. A number of algorithms of different types have been developed to search for SMBHBs. Each algorithm was successful at a certain level, but through the MLDCs it was realised that many of the algorithms had fundamental limitations. The developed algorithms followed a number of different strategies. Some groups tried to apply the template grid method that was developed in the ground-based community. These groups either tried a multi-stage algorithm (Brown et al. 2007) or a stochastic template bank (Babak 2008). Both of these algorithms detected the source and had varying success in estimating the parameters of the source. It is not clear how well these algorithms will work in the future when applied to realistic sources in a full sized parameter space. In general, the number of templates for a template bank grows geometrically with the dimensionality of the parameter space. If we assume that one day we will search for spinning, eccentric binaries, then the bank needs to cover a 17 dimensional parameter space.

Other groups applied a time-frequency algorithm (see The MLDC Taskforce and Participants 2007 for example) where short duration Fourier transforms of the data are used to identify coherent tracks in the time-frequency plane. As the phase information of the wave is lost, it makes it very difficult to carry out parameter estimation using this technique and imposes a fundamental limit of the usefulness of the algorithm.

A third group of algorithms were developed with the idea of conducting the matched filtering in a more sophisticated manner. These were based on adapted Metropolis-Hasting Markov Chains (Cornish and Porter 2005, 2006, 2007a, b), Nested Sampling (Feroz et al. 2009), Genetic Algorithms (Petiteau et al. 2010) or Evolutionary Algorithms (Gair and Porter 2009). These algorithms were very successful in the detection of SMBHBs, and in most cases estimating the parameters of the source. The main difference between the algorithms was that the Metropolis-Hastings algorithm searches for single sources in series, while the Nested Sampling, Evolutionary and Genetic Algorithms search for multiple sources at the same time.

3.2 Outstanding Problems and Questions

While a lot of progress has been made in astrophysical modeling, GW theory and GW astronomy in the last decade, a number of outstanding questions still remain.

1. There still is no solution to the question of how the sources should be searched for. It could be that is possible to search for the SMBHBs without having to worry about the other source types, in which case any of the above style of algorithms will work. However, if this is not true, and we have to search for sources in terms of their signal to noise ratio, then it maybe makes more sense to develop algorithms that only search for single sources.
2. The current studies have yet to come close to an astrophysically realistic data set in terms of source numbers. It is estimated that there could be anything up to 100 SMBHBs per year in the data stream. For the cases already studied, a solution exists for the restricted case of a small number of Schwarzschild SMBHBs. The community has begun to investigate the case of Kerr SMBHBs, but a solution does not yet exist.
3. In the previous studies, the priors on the parameter values have been quite narrow. It is unclear how the current algorithms would perform on open priors.
4. While we would like to detect every merger possible, the question arises as to do we need to do this in real time? If we believe that there may be EM counterparts to the merger, then it is clear that we would like to inform the wider astrophysical community. However, there may be a luminosity cutoff which means that not every merger will have a detectable EM signature. These systems could then be treated at a more relaxed pace.
5. Do we expect to see EM counterparts for SMBHBs? It could be that even if these processes are super-Eddington in nature, the source is simply too distant to observe the counterpart.
6. How do gas and accretion disks effect the evolution of spin in SMBHBs? We know that in gas rich environments, the spins of the black holes align with the total angular momentum of the system. In this case, we could use a simplified template to search for SMBHBs.
7. What are the formation processes of massive black holes at high redshift? It is possible that nature will present us with a mixture of high and low mass seeds. However, none of the current astrophysical models explain the presence of hypermassive quasars at high redshift. Because of these uncertainties in the models, the actual expected event rate is very difficult to calculate.
8. Do we have realistic templates? I will discuss this below.

3.3 Comparable Mass Waveforms

As stated earlier, the main detection technique in the search for GWs is matched filtering. When developing a template, we need to have a strong theoretical knowledge of the waveform model if we are to carry out an accurate estimation of parameters. Up to now, pretty much all algorithmic development and parameter estimation studies were conducted using circular Schwarzshild templates.

In reality we know that these waveform models will be good enough to announce a detection, but will not be good enough for parameter estimation. It was demonstrated that the effects of spin and spin precession break correlations between parameters and improve parameter estimation (Lang and Hughes 2006). There is a further breaking of parameter correlations when one includes higher harmonic corrections to the waveforms. It was shown that these corrections provide a massive improvement in both the luminosity distance and the sky position of the source (Porter and Cornish 2008; Trias and Sintes 2008; Arun et al. 2007). While we know that GWs are very efficient at circularising binaries, it was shown that three body interactions in stellar clusters and galactic centers can push the eccentricity of the system so high that it enters the detection band with a non-negligible value. In this case, it is extremely difficult to properly estimate the parameters of the source using circular templates (Amaro-Seoane et al. 2010; Porter and Sesana 2010).

In many cases, the merger and ringdown of the waveform have also been neglected. However, with advances in Numerical Relativity, not only are longer waveforms being produced, but analytic waveforms that are calibrated to the numerical results are also being produced (Damour et al. 2013; Pan et al. 2014).

Ultimately, when a space-based GW observatory launches, we would like to have templates that include eccentricity, spins, higher harmonic corrections, merger and ringdown. This presents a major theoretical challenge for the next decade.

4 Extreme Mass Ratio Inspirals

Another major source of GWs are the Extreme Mass Ratio Inspirals or EMRIs. This is a system composed of a stellar mass compact object orbiting a central SMBH. This object could be anything from an ordinary star to a stellar mass black hole. It is assumed that before reaching the true strong field regime close to the central black hole, a stellar mass star would be tidally disrupted, and compact objects such as white dwarves and neutron stars may undergo a thermonuclear detonation from tidal heating due to the eccentric orbits. It is therefore believed that the main observable systems will be composed of a stellar mass black hole orbiting a SMBH.

EMRIs are considered to be a very exciting source. It is expected that we will see approximately 50 per year out to a redshift of $z \sim 0.7$ (Amaro-Seoane et al. 2012; Gair and Porter 2013). As the stellar mass object spends the last 1–2 years of its life in the strong field regime of the central black hole (1–3 Schwarzschild radii), these objects are an incredible opportunity to conduct spacetime cartography. In this final 1–2 year period, we expect the smaller body to trace out between $10^4 - 10^6$ orbits in the vicinity of the central black hole, allowing us to confirm that it is a Kerr black hole as predicted by General Relativity, and not something more exotic like a Boson star.

4.1 Algorithmic Development

The EMRIs in general have received the least amount of attention in terms of development. Two main algorithms exist. These algorithms were adapted from the SMBHB search codes and are essentially based on a Markov Chain Monte Carlo type algorithm. During a mock data challenge, four out of a possible five EMRIs were detected with good parameter estimation. However, once again, this is nowhere near the expected astrophysical rate of 50/year. Furthermore, the searches took place with very narrow priors.

The main problem with the development of algorithms for EMRIs is the run-time with current computational power. The other sources are easier to develop for, as an algorithm can be set running and the user knows within a couple of hours if things are going well or not. This is a luxury we do not have with the EMRIs. It usually takes a few days before one can say whether or not there is a problem. This slows down the algorithmic development dramatically.

4.2 Outstanding Issues

While the list is shorter, there is no doubt that the following problems present an immense technical difficulty.

1. The question of whether or not we have realistic templates arises once more. All of the studies and challenges conducted for EMRIs have used a phenomenological waveform that contains the necessary complexity (i.e. the correct number of parameters). However, EMRIs are very difficult to model. They have three fundamental frequencies and have very complex orbits. While progress has been made on including self-force calculations (see for example Barack 2009), it is not clear how close we are to having realistic EMRI templates.
2. As the current algorithms are based on phenomenological waveforms, their success was based on the analytic knowledge of the separation between harmonics and sub-harmonics. This is something that will not be known with realistic waveforms and might affect the performance of our algorithms.
3. Is there an EMRI confusion problem? For those systems with little frequency evolution in each harmonic, a real data set may contain multiple systems that

contain a harmonic in the same Fourier bin. This could cause problems in identifying the source.

4. Will imperfect removal of a galactic binary affect the detection of an EMRI? Again for those sources with very little frequency evolution, an EMRI can theoretically be modeled by a collection of monochromatic sources. It is not inconceivable that an EMRI harmonic could be mistaken for a galactic binary and accidentally removed.
5. What are the EMRI formation channels? Do we require purely centrophilic orbits or can we expect the birth of compact objects in an accretion disk around the central black hole? One can also approach this question from the other direction. If we detect EMRIs, what information can we provide on the formation channel?
6. Can we deal with wide priors and astrophysically realistic event rates? In theory, there is no reason to conduct an EMRI analysis in real time as we do not expect to have an EM counterpart. Further research is necessary to estimate how many EMRI detections are needed to answer any scientific questions that we deem important.

5 Cosmological Sources

With the recent BICEP 2 result (Ade et al. 2014), there is a large current interest in GWs from cosmological sources. While the GWs discovered by BICEP 2 are not in the same frequency band as eLISA, there are many possible sources that will be. Two such sources are first order phase transitions and networks of cosmic (super)strings. In a previous MLDC (The MLDC Taskforce and Participants 2007) the community investigated the possibility of detecting bursts of GWs from cusps of cosmic (super)strings. While each source was detected, the parameter estimation, and in particular the localisation of the source on the sky, was not very good.

This was mostly due to how a space-based observatory attains a sky position for the source. For long lived sources, such as SMBHBs, the motion of the detector around the sun induces a Doppler motion effect into the waveform. Because the detector sees the source differently at different points in the orbit, it allows us to fix the source in the sky. However, for short transient events like GW bursts from cusps, the detector has hardly moved during the lifetime of the source. This means that to get the sky position, we need to use triangulation between the space-craft. This is much less accurate than using the Doppler motion and results in an inferior estimation of the distance.

5.1 *Outstanding Issues*

1. Very little development has gone into algorithms to detect a stochastic cosmological background. It is unclear if one needs a three-arm detector to detect such a background, or if it is possible with a two-arm detector.
2. There is a prediction that a network of cosmic (super)strings will produce a background that would be the dominant source of GWs for a mission such as eLISA (Binetruy et al. 2012). In this case the stochastic background would become the foreground source. The question then arises if it is still possible to detect the astrophysical sources that would lie below such a foreground?
3. It is hoped that cosmological theories will be more constrained in the next decade. This theoretical work would allow a better prediction of the strength of a possible stochastic background, as well as the possible event rates for transient bursts of cosmological origin.

Conclusion

This is an exciting time in the field of GW astronomy. The recent BICEP 2 announcement of the discovery of GWs from the early universe has brought the subject to the forefront of both the research and public eye. With both Advanced LIGO and Virgo due to come on-line in the next couple of years and the advance in pulsar timing arrays, there is a real possibility of the first detection of GWs from astrophysical origins before the end of the current decade. The recent decision by ESA to chose the Gravitational Wave Universe as the theme of the Cosmic Vision L3 mission would allow the community to cover the GW spectrum from 10^{-9} Hz up to ~kHz.

The last decade has seen a massive improvement in the modelling of astrophysical sources of GWs. From advanced computer modeling of accretion disks and gas interactions, to merger tree models, to three body interactions, our understanding of the astrophysical universe has never been better. The discovery of new binary systems with EM parameter estimation has now begun to increase the number of expected verification sources in our own galaxy. In the next decade we hope to see the discovery of many more of these systems and possibly even the concrete detection of SMBHBs with sub-milliparsec separations.

Since 2004, the field has also witnessed giant leaps in the development of search algorithms and statistical techniques for space-based GW sources. For some sources, the limit to our development is simply the computational power available to us today. The community has developed sophisticated techniques that now give us confidence in our ability to carry out GW astronomy for objects such as SMBHBs, EMRIs and galactic binaries.

(continued)

However, a lot of work still lies ahead. A constant improvement in the efficiency and accuracy of our algorithms is needed. More sophisticated templates incorporating the maximum of physical effects need to be developed for all sources. Better constraints on astrophysical event rates and cosmological models will be needed to guide the development of the field of GW astronomy.

While the next decade will be challenging, it will also be very rewarding. A new field of astronomy is opening up before us, and promises to provide us with a view of the universe never seen before.

Acknowledgements The author would like to thank the organisers of the Sant Cugat Forum on Astrophysics for their invitation to give this lecture.

References

P.A.R. Ade et al., arXiv:1403.3985 (2014)
P. Amaro-Seoane, C. Eichhorn, E.K. Porter, R. Spurzem, MNRAS **401**, 2268 (2010)
P. Amaro-Seoane et al., Class. Quantum Grav. **29**, 124016 (2012)
K.G. Arun et al., Phys. Rev. D **76**, 104016 (2007)
S. Babak, Class. Quantum Grav. **25**, 195011 (2008)
L. Barack, Class. Quantum Grav. **26**, 213001 (2009)
M.C. Begelman et al., MNRAS **370**, 289 (2006)
P. Binetruy, A. Bohé, C. Caprini, J.-F. Dufaux, JCAP **1206**, 027 (2012)
A. Blaut et al., Phys. Rev. D **81**, 063008 (2010)
D. Brown et al., Class. Quantum Grav. **24**, S595 (2007)
N.J. Cornish, E.K. Porter, Class. Quantum Grav. **22**, S927 (2005)
N.J. Cornish, E.K. Porter, Class. Quantum Grav. **23**, S761 (2006)
N.J. Cornish, E.K. Porter, Phys. Rev. D **75**, 021301 (2007a)
N.J. Cornish, E.K. Porter, Class. Quantum Grav. **24**, 5729 (2007b)
J. Crowder, N.J. Cornish, Phys. Rev. D **75**, 043008 (2007)
T. Damour, A. Nagar, S. Bernuzzi, Phys.Rev. D **87** 8, 084035 (2013)
B. Devecchi et al., MNRAS **421**, 1465 (2012)
F. Feroz, J.R. Gair, M.A. Hobson, E.K. Porter, Class. Quantum Grav. **26**, 215003 (2009)
J.R. Gair, E.K. Porter, Class. Quantum Grav. **26**, 225004 (2009)
J.R. Gair, E.K. Porter, *Proceedings of the 9th LISA Symposium*, Paris. ASP Conference Series, Vol. 467. San Francisco: Astronomical Society of the Pacific **173**, (2013)
R.A. Hulse, J.H. Taylor, ApJ **195**, L51 (1975)
M. Kilic et al., MNRAS **413**, L101 (2011)
R. Lang, S.A. Hughes, Phys. Rev. D **74**, 122001 (2006)
T. Littenberg, N.J. Cornish, Phys. Rev. D **82**, 103007 (2010)
T. Littenberg, (Private Communication) (2014)
P. Madau, M. Rees, ApJ **551**, L27 (2001)
L. Mayer et al., Nature **433**, 604 (2006)
Y. Pan et al., Phys. Rev. D **89**, 084006 (2014)
A. Petiteau, Y. Shang, S. Babak, F. Feroz, Phys. Rev. D **81**, 104016 (2010)
E.K. Porter, N.J. Cornish, Phys. Rev. D **78**, 064005 (2008)
E.K. Porter, S. Sesana, arXiv:1005.5296 (2010)

A. Sesana et al., MNRAS **377**, 1711 (2007)
A. Sesana et al., Phys. Rev. D **83**, 44036 (2011)
M. Tegmark et al., ApJ **474**, 1 (1997)
The MLDC Taskforce and Participants, Class. Quantum Grav. **24**, S529 (2007); **24**, S551 (2007); **25**, 114037 (2008); **25**, 184026 (2008); **27**, 084009 (2010)
S. Timpano, L. Rubbo, N.J. Cornish, Phys. Rev. D **73**, 122001 (2006)
M. Trias, A. Sintes, Phys. Rev. D **77**, 024030 (2008)
M. Volonteri, A&ARv **18**, 279 (2010)

Investigating Binary Black Hole Mergers with Principal Component Analysis

J. Clark, L. Cadonati, J. Healy, I.S. Heng, J. Logue, N. Mangini, L. London, L. Pekowsky, and D. Shoemaker

Abstract Despite recent progress in numerical simulations of the coalescence of binary black hole systems, highly asymmetric spinning systems and the construction of accurate physical templates remain challenging and computationally expensive. We explore the feasibility of a prompt and robust test of whether the signals exhibit evidence for generic features that can educate new simulations. We form catalogs of numerical relativity waveforms with distinct physical effects and compute the relative probability that a gravitational wave signal belongs to each catalog. We introduce an algorithm designed to perform this task for coalescence signals using principal component analysis of waveform catalogs and Bayesian model selection and demonstrate its effectiveness.

1 Introduction

The coalescence of two black holes is arguably the most powerful source of GWs detectable by the second generation of ground based detectors: Advanced LIGO (Harry 2010), Advanced Virgo (Acernese et al. 2009), and KAGRA (Somiya 2012). The discovery of these signatures, forecast within the next few years (Aasi et al. 2013b), will open a new era of gravitational wave astrophysics, where the GW signature will provide insights on the physics of the source.

J. Clark (✉) • N. Mangini
University of Massachusetts Amherst, Amherst, MA 01003, USA
e-mail: clark@physics.umass.edu

L. Cadonati
University of Massachusetts Amherst, Amherst, MA 01003, USA

Cardiff University, Cardiff CF24 3AA, UK

I.S. Heng • J. Logue
SUPA, School of Physics and Astronomy, University of Glasgow, Glasgow G12 8QQ, Scotland, UK

J. Healy • L. London • L. Pekowsky • D. Shoemaker
Center for Relativistic Astrophysics, Georgia Institute of Technology, Atlanta, GA 30332, USA

C.F. Sopuerta (ed.), *Gravitational Wave Astrophysics*, Astrophysics and Space Science Proceedings, DOI 10.1007/978-3-319-10488-1_24

To decode the information in the GW waveform, we need a careful mapping with the masses and the spin magnitude and orientation of the black holes; this is the charge of NR. While available NR waveforms span an increasing portion of the physical parameter space of unequal mass, spin and precessing BBHs (Ajith et al. 2012; Hinder et al. 2014), each simulation takes a week or more to run. A complete coverage of the full parameter space remains a slow but important endeavor to enable GW matched filtering and parameter estimation (Thorne 1987; Aasi et al. 2013a).

The LIGO and Virgo Collaborations have refined techniques for the search of generic GW transients, or *bursts*, which don't assume a specific waveform but rely on a coherent GW in multiple detectors for a variety of plausible sources (Abadie et al. 2012; Andersson et al. 2013). The work presented here aims to answer the question of how a transient detected by a template-less burst search can trigger new NR simulations in interesting regions of the BBH parameter space. We introduce a proof-of-concept study, which uses the method of Principal Component Analysis (PCA) to compare a plausible signal to catalogs of NR waveforms, which represent certain regions of the BBH physical parameter space.

2 Binary Black Hole Merger Simulations

The GW waveform produced by solar and intermediate mass BBH systems spans the sensitive band of ground based detectors through the inspiral, merger and ringdown phases. While post-Newtonian and perturbation theories adequately describe the inspiral and ringdown, numerical relativity is necessary to capture the physics of the merger. NR has been probing the parameter space of binary black hole mergers since the breakthrough of 2005 (Pretorius 2005) achieving extreme mass ratios (Lousto and Zlochower 2011), extreme spin magnitudes (Lovelace et al. 2012) and many precessing runs (Mroue et al. 2013; Pekowsky et al. 2013).

The NR waveforms used in this paper were produced by the MAYA code of the Georgia Institute of Technology (Vaishnav et al. 2007) The MAYA code uses the `Einstein Toolkit`,[1] which is based on the CACTUS[2] infrastructure and CARPET mesh refinement (Schnetter et al. 2004). The output of all simulations is the Weyl Scalar, Ψ_4, decomposed into spin-weighted spherical harmonics that is then converted to strain (Reisswig and Pollney 2011).

For this work we use 48 NR runs, listed in Table 1 without hybridization with post-Newtonian waveforms. The Q-series contains 13 non-spinning, unequal-mass simulations. We use 15 runs from the HR-series, a set of unequal-mass, equal spin simulations, with initial spin parallel to the initial angular momentum. The RO3-series is a set of 20 unequal-mass simulations with the lighter black hole

[1]http://www.einsteintoolkit.org.

[2]http://www.cactuscode.org.

Table 1 Physical parameters for the three catalogs used in this study

Name	Q	HR	RO3
Mass ratio, $q = m_1/m_2$	1–2.5	1–4	1.5–4
Spin magnitude, a	0.0	$0.0 - 0.9$	0.4, 0.6
Tilt angle, θ	0.0	0.0	45°–270°
N waveforms	13	15	20

spin aligned to the initial angular momentum (z-axis) and the other black hole at a tilt angle θ with the z-axis in the xz-plane; these systems are precessing and the tilt-angles are defined at a specific separation of the black holes at one instant in the evolution of the binary system and change in time. While the runs are tabulated with initial parameters, there is no functional form to relate one waveform to the next; we use a Principal Component Analysis to determine the main features of each catalog.

3 Principal Component Analysis and Bayesian Model Selection

We parametrize the NR waveform catalogs of Sect. 2 with an orthonormal set of principal components (PCs), obtained with a standard singular value decomposition (Heng 2009; Röver et al. 2009). For a catalog of n waveforms $\{h_i\}_{i=1,\dots,n}$ with m samples, we create a matrix $\mathbf{H}$ whose columns corresponds to each waveform. We then factorize the resulting $m \times n$ matrix $\mathbf{H}$ so that:

$$\mathbf{H} = \mathbf{U}\mathbf{S}\mathbf{V}^T, \tag{1}$$

where $\mathbf{U}$ is an $m \times m$ matrix whose columns are the eigenvectors of $\mathbf{H}\mathbf{H}^T$ and $\mathbf{V}$ is an $n \times n$ matrix whose columns are eigenvectors of $\mathbf{H}^T\mathbf{H}$. The $m \times n$ matrix $\mathbf{S}$ will have all zeros, except for the $\{S_{jj}\}_{j=1,\dots,n}$ terms, which correspond to the square root of the jth eigenvalue. $\mathbf{U}$ contains the catalog's PCs, ranked by their corresponding eigenvalue: the first column is the first PC, which encapsulates the most significant features common to all waveforms in the catalog, the second column, corresponding to the second largest eigenvalue, describes the second most significant common features in the catalog, and so on. The waveforms in $\mathbf{H}$ can be reconstructed as a linear combination of PCs:

$$h_i \approx \sum_{j=1}^{k} U_j \beta_j \ , \tag{2}$$

where h_i is the catalog waveform, u_j is the jth PC and β_j is the corresponding coefficient, obtained by projecting h_i onto u_j. The sum over k PCs is an approximation

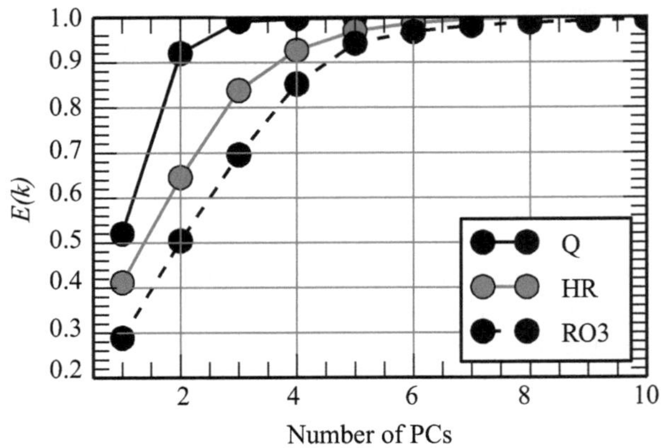

Fig. 1 Cumulative eigenvector energy as a function of the number of principal components for the three catalogs in this study. We use the number of PCs that provides 90 % of the energy: 2 PCs for set Q, 4 PCs for set HR and 5 PCs for set RO3

of the desired waveform, since in general $k < n$. In this analysis, the choice of k is determined by the *cumulative eigenvalue energy*, $E(k)$, shown in Fig. 1:

$$E(k) = \frac{\sum_{i=1}^{k} S_{ii}}{\sum_{j=1}^{n} S_{jj}} \tag{3}$$

In this analysis we use k PCs, so that $E(k) \geq 0.9$. This corresponds to 2, 4 and 5 PCs for the Q, HR and RO3 catalogs respectively. A selection of the waveforms from the HR catalog and corresponding PCs are shown in Fig. 2.

Following the seminal work on Burst signals in Clark et al. (2007), Logue et al. (2012), the PCs can be used to identify generic features for a measured waveform through the posterior odds ratio, which is widely used in GW data analysis to compare the probabilities of two competing models M_i and M_j. Given data D, the odds ratio $\mathscr{O}_{ij}$ is the ratio of posterior probabilities for each model:

$$\mathscr{O}_{ij} = \frac{p(M_i)}{p(M_j)} \frac{p(D|M_i)}{p(D|M_j)} = \pi_{ij} \frac{Z_i}{Z_j} \ , \tag{4}$$

where π_{ij} is the *prior odds ratio* which reflects any bias one has for the models. Z_i is the evidence for model M_i. The evidence ratio Z_i/Z_j is referred to as the Bayes' factor B_{ij} and reflects the influence of the data. To demonstrate the efficiency of our algorithm, we assume here $\pi_{ij} = 1$. In this context, the models are the waveform catalogs and the evidences are obtained by marginalizing over all model parameters which are the $\{\beta_j\}$ coefficients used to construct the signal model in Eq. (2) from the catalog's PCs. We adopt a uniform prior for $\{\beta_j\}$, in a range obtained by projecting the waveforms from each catalog onto its corresponding PCs. As in Logue et al.

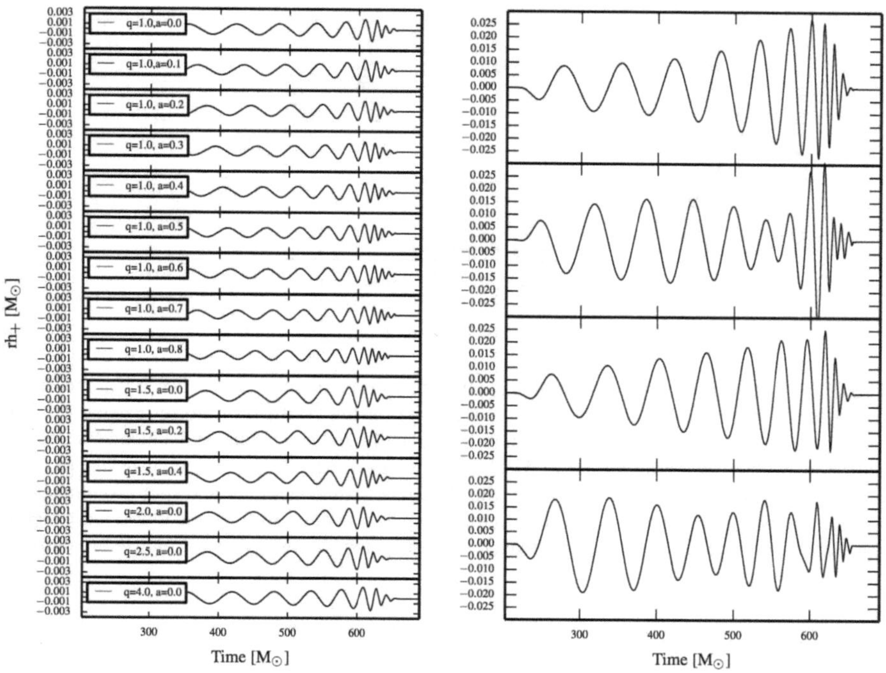

Fig. 2 *Left*: The waveforms in the HR catalog of spinning, non precessing waveforms used in this study. *Right*: The principle component decomposition of the HR catalog

(2012), the likelihood and corresponding evidences are computed with a nested sampling algorithm. The model evidence is largest for the most parsimonious model that best explains the data; $B_{ij} > 1$ indicates M_i is preferred over M_j.

4 Identifying Binary Black Hole Merger Phenomenology

We demonstrate the efficacy of the PCA-based Bayesian model selection with a Monte-Carlo analysis where simulated GW signals from each catalog are added to colored, Gaussian noise, which is generated as in Logue et al. (2012). For this proof-of-principle study we assume a single aLIGO detector operating at design sensitivity in the "zero-detuned, high-power" configuration (Harry 2010). We make the further assumptions that the time of peak amplitude of the signal is known, that the source is optimally oriented and located on the sky with respect to the detector and, finally, that the total mass of the system is 250 $M_\odot$. This choice of mass ensures that the signals "switch on" below the minimum sensitive frequency of the aLIGO noise spectrum (10 Hz). The physical distance of the simulated signal is scaled such that the injections have SNR = 50. The GW signals from our catalogs are injected into

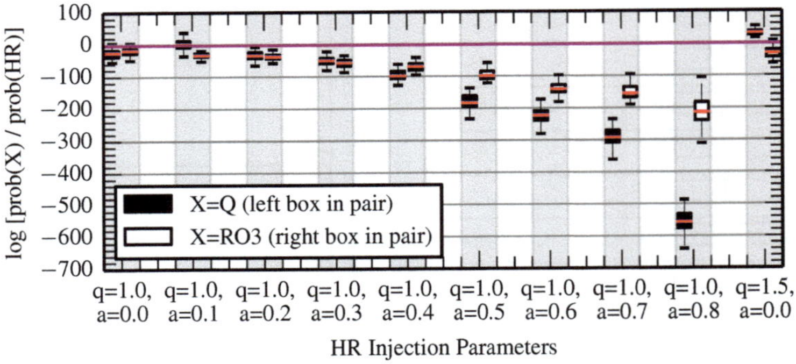

Fig. 3 Distribution of Bayes factors for HR waveforms. Each pair of boxes in the figure corresponds to the sample of Bayes factors (Eq. (5)) for the 50 different noise realizations. The *boxes* denote the interquartile range of the distribution, the *red lines* indicate the median value and the *whiskers* show the outliers within 1.5× the interquartile range. The x-axis indicates the physical parameters of the injection performed; for the HR catalog, the mass ratio and spin magnitudes are varied. The two $a = 0$ systems are seen to be difficult to distinguish from the Q catalog which is not surprising since Q catalog contain waveforms for non-spinning systems (Color figure online)

50 independent noise realizations. Thus, for each waveform we obtain 50 evidence values for the waveform to belong to one of the catalogs: $Z_{\rm Q}$, $Z_{\rm HR}$ and $Z_{\rm RO3}$.

To demonstrate that model selection can correctly identifying which catalog a given injection originated from, for an injection from a given catalog C we form the Bayes factors between the other catalogs and the model M_C. For example, if an injection is performed from the HR catalog, we compute the log Bayes factors:

$$\log B_{\rm Q,HR} = \log Z_{\rm Q} - \log Z_{\rm HR} \quad \text{and} \quad \log_e B_{\rm RO3,HR} = \log Z_{\rm RO3} - \log Z_{\rm HR} \,. \tag{5}$$

If the algorithm correctly discriminates between the waveform catalogs, both $B_{\rm Q,HR}$ and $B_{\rm RO3,HR}$ will be less than unity.

Figure 3 summarizes the distribution of Bayes factors for HR waveforms. The majority of the boxes lie well below zero, indicating that the algorithm correctly identifies the HR catalog as the most probable for these simulations, with $Z_{\rm HR} > \max(Z_{\rm Q}, Z_{\rm RO3})$. Qualitatively similar results are found when analyzing signals from the Q and RO3 catalogs and will be explored more fully in a follow-up publication.

Acknowledgements This work was supported by NSF grants PHY-0955773 and 0955825, SUPA and STFC UK. Simulations were supported by NSF XSEDE PHY120016 andPHY090030, and CRA Cygnus cluster.

References

J. Aasi et al., Parameter estimation for compact binary coalescence signals with the first generation gravitational-wave detector network. Phys. Rev. D **88**, 062001 (2013a)

J. Aasi et al., Prospects for Localization of Gravitational Wave Transients by the Advanced LIGO and Advanced Virgo Observatories. Preprint arXiv:1304.0670 (2013b)

J. Abadie et al., All-sky search for gravitational-wave bursts in the second joint LIGO-Virgo run. Phys. Rev. D **85**, 122007 (2012)

F. Acernese et al., Virgo Technical Report VIR-0027A-09. https://tds.ego-gw.it/ql/?c=6589 (2009)

P. Ajith et al., The NINJA-2 catalog of hybrid post-Newtonian/numerical-relativity waveforms for non-precessing black-hole binaries. Class. Quantum Grav. **29**, 124001 (2012)

N. Andersson et al. The Transient Gravitational-Wave Sky. Class. Quantum Grav. **30**, 193002 (2013)

J. Clark et al., An evidence based search method for gravitational waves from neutron star ring-downs. Phys. Rev. D **76**, 043003 (2007)

G.M. Harry, Advanced LIGO: the next generation of gravitational wave detectors. Class. Quantum Grav. **27**, 084006 (2010)

I.S. Heng, Rotating stellar core-collapse waveform decomposition: a principal component analysis approach. Class. Quantum Grav. **26**, 105005 (2009)

I. Hinder et al., Error-analysis and comparison to analytical models of numerical waveforms produced by the NRAR Collaboration. Class. Quantum Grav. **31**, 025012 (2014)

J. Logue et al., Inferring core-collapse supernova physics with gravitational waves. Phys. Rev. D **86**, 044023 (2012)

C.O. Lousto, Y. Zlochower, Orbital evolution of extreme-mass-ratio black-hole binaries with numerical relativity. Phys. Rev. Lett. **106**, 041101 (2011)

G. Lovelace et al., Accurate gravitational waveforms for binary-black-hole mergers with nearly extremal spins. Class. Quantum Grav. **29**, 045003 (2012)

A.H. Mroue et al., A catalog of 174 binary black-hole simulations for gravitational-wave astronomy. Phys. Rev. Lett. **111**, 241104 (2013)

L. Pekowsky et al., Comparing gravitational waves from nonprecessing and precessing black hole binaries in the corotating frame. Phys. Rev. D **88**(2), 024040 (2013)

F. Pretorius, Evolution of binary black hole spacetimes. Phys. Rev. Lett. **95**, 121101 (2005)

C. Reisswig, D. Pollney, Notes on the integration of numerical relativity waveforms. Class. Quantum Grav. **28**, 195015 (2011)

C. Röver et al., Bayesian reconstruction of gravitational wave burst signals from simulations of rotating stellar core collapse and bounce. Phys. Rev. D **80**, 102004 (2009)

E. Schnetter et al., Evolutions in 3D numerical relativity using fixed mesh refinement. Class. Quantum Grav. **21**, 1465–1488 (2004)

K. Somiya, Detector configuration of KAGRA: The Japanese cryogenic gravitational-wave detector. Class. Quantum Grav. **29**, 124007 (2012)

K.S Thorne, in *300 Years of Gravitation*, ed. by S. W. Hawking, W. Israel (Cambridge University Press, Cambridge, 1987)

B. Vaishnav et al., Matched filtering of numerical relativity templates of spinning binary black holes. Phys. Rev. D **76**, 084020 (2007)

Split Bregman Method for Gravitational Wave Denoising

Alejandro Torres, Antonio Marquina, José A. Font, and José M. Ibáñez

Abstract This paper presents a progress report in our aim to develop a Total Variation algorithm for denoising of gravitational waves. These algorithms, are routinely employed in the context of image processing and they do not need any a priori information on the signals. We apply our method to two different types of numerically-simulated gravitational wave signals, namely burst produced from the core collapse of rotating stars and waveforms from binary black hole mergers, and present a preliminary assessment of its capabilities.

1 Introduction

Obtaining useful information from a dataset is one of the most important goals in data analysis. In this sense, reducing the unwanted modifications and interferences produced by noise is a major challenge due to the different statistical properties of each variety of noise in the emerging field of gravitational wave (GW) astronomy. The signals to be measured by GW detectors can be very distorted as a consequence of the large number of noise sources present. For this reason, a great effort has been made to develop specific techniques of noise removal and signal detection. Even with the next generation of detectors, with their enhanced sensitivity, Advanced Ligo (Harry 2010), Advanced Virgo (Advanced Virgo Baseline Design 2009) and KAGRA (Aso et al. 2013), these methods are absolutely necessary to achieve detection.

In this paper we present preliminary results in our project of developing a data analysis technique to remove noise from a measured signal from a GW detector. This method is based on Total Variation (TV) denoising algorithms and

A. Torres (✉) • J.A. Font • J.M. Ibáñez
Departamento de Astronomía y Astrofísica, Universitat de València, Dr. Moliner 50, 46100 Burjassot, València, Spain
e-mail: Alejandro.Torres@uv.es; J.Antonio.Font@uv.es; Jose.M.Ibanez@uv.es

A. Marquina
Departamento de Matemática Aplicada, Universitat de València, Dr. Moliner 50, 46100 Burjassot, València, Spain
e-mail: Antonio.Marquina@uv.es

C.F. Sopuerta (ed.), *Gravitational Wave Astrophysics*, Astrophysics and Space Science Proceedings, DOI 10.1007/978-3-319-10488-1_25

is complementary to existing data analysis procedures such as Matched Filtering or Bayesian Reconstruction (Röver et al. 2009). TV-norm was introduced in 1992 in Rudin et al. (1992). However, it was not until the last decade, with the publication of compressed sensing (CS) methods (Donoho 2006), that these type of methods began to have a great development. They have been successfully tested for image processing, medical imaging and magnetic resonance imaging. TV denoising is considered to be one of the best denoising models, but also one of the most difficult to implement and use.

We show here how to adapt one of these techniques to GW signals. More precisely, we apply the Split-Bregman (SB) method to waveforms from two catalogs of numerically simulated gravitational waves. It must be pointed out that the SB algorithm does not use any information about the source or the morphology of the signal, and the catalogs have been only employed to assess the method. As algorithms do not make any distinction about the features of the signal, the methods can be applied both in the time and in the frequency domain, as desired. The interested reader is addressed to Torres et al. (2014) for further details on this ongoing research.

2 Mathematical Background

We assume the usual linear degradation model that is set as,

$$f = u + n \,, \tag{1}$$

where f is the observed signal, n is the noise and u is the signal to be recovered. We assume that n is Gaussian white noise, (i.e., n is a square integrable function with zero mean).

Solving this problem involves finding a signal u whose distance, given by the L_2-norm, to the measured signal f, is the standard deviation of noise n, σ, i.e. the constraint

$$||u - f||_{L_2}^2 = \sigma^2 \,, \tag{2}$$

is satisfied.

This problem is classically solved with algorithms based on least squares. However, solutions obtained by these procedures are contaminated by Gibbs' phenomena (ringing) and often the linear system can be large, close to singular and even indeterminate. To avoid these problems, it is necessary to minimize the energy,

$$E(u) = \int |\nabla u|_{L_2}^2 \,, \tag{3}$$

to ensure that u is regular. The denoising problem turns into a variational problem where the energy of u must be minimized subject to so called fidelity term given by Eq. (2).

Applying Tikhonov regularization, the constrained variational problem can be formulated as an unconstrained variational problem by adding to the regularizing functional the constraint weighted by a positive Lagrange multiplier λ which is also an unknown of the problem.

$$u = \min_u \left\{ \int |\nabla u|^2_{L_2} + \frac{\lambda}{2} ||f - u||^2_{L_2} \right\} . \tag{4}$$

The regularization parameter $\lambda > 0$ is the scale parameter. Finer scales are recovered for large values of λ. However, the presence of noise amplifies high frequencies of the solutions and λ cannot be chosen large enough.

The Rudin-Osher-Fatemi (ROF) model (Rudin et al. 1992) replaces L_2 norm in Eq. (4) by the L_1 norm,

$$u = \min_u \left\{ \int_\Omega |\nabla u|_{L_1} + \frac{\lambda}{2} ||f - u||^2_{L_2} \right\} . \tag{5}$$

Although the L_1 norm is usually avoided because it produces singular distributions, it avoids the oscillations produced by Gibbs effect and preserves the large gradients promoting zeros for the smaller ones.

The Split-Bregman method (Goldstein and Osher 2009) is an efficient implementation to solve the ROF model. It is an iterative procedure that decouples the variational problem into L_1 and L_2 portions of energy to be minimized separately.

3 Gravitational Wave Catalogs

The first catalog we employ to evaluate our SB algorithm corresponds to burst type signals from rotating core collapse. This catalog was obtained by Dimmelmeier et al. (2008) and contains 128 waveforms. The span of parameters covered by the initial models include non-rotating progenitors with distinct mass at zero age main sequence, several values of precollapse central angular velocity, three parametrizations of the degree of differential rotation and two equations of state. The second catalog is composed by waveforms obtained from simulations of binary black hole (BBH) mergers built by Mroué et al. (2013) very recently. This catalog is the result of 174 numerical simulations of which 167 cover more than 12 orbits, and 91 represent precessing binaries. It also extends previous simulations by incorporating models with larger mass ratios, and includes simulations with the first systematic sampling of eccentric BBH waveforms.

4 Results

All the signals we analyze have been resampled to the LIGO/VIRGO sampling rate (16,384 Hz). The non-white Gaussian noise is generated applying the methods implemented in the LIGO Scientific Collaboration algorithm libraries (LAL) (Creighton et al. 2007). The one-sided detector noise power spectral density $\sqrt{S(f)}$ corresponds to Advanced LIGO in the proposed broadband configuration. The signal-to-noise ratio (SNR) has been computed as,

$$SNR = \sqrt{4\Delta t^2 \Delta f \sum_{k=1}^{N_f} \frac{|\tilde{h}(f)|^2}{S(f_k)}}, \tag{6}$$

where $\tilde{h}$ indicates Fourier Transform of signal h.

For the sake of illustrative purposes we choose signal "s20a3o15_LS" from the core collapse catalog and signal "0001" from the BBH catalog. The SNR is set to 20, and we assume that there is a signal detected in the dataset. Although the chosen scenario is simple, it allows us to illustrate the performance of our denoising algorithm. The results from applying our method to the burst signal is shown in Fig. 1. The positive and negative peaks in the waveform are properly captured. However, the damped oscillations of the proto neutron star are indistinguishable from the noise due to their low amplitude and small gradients. Correspondingly, for the BBH signal the results are shown in Fig. 2. The frequency of the merger part

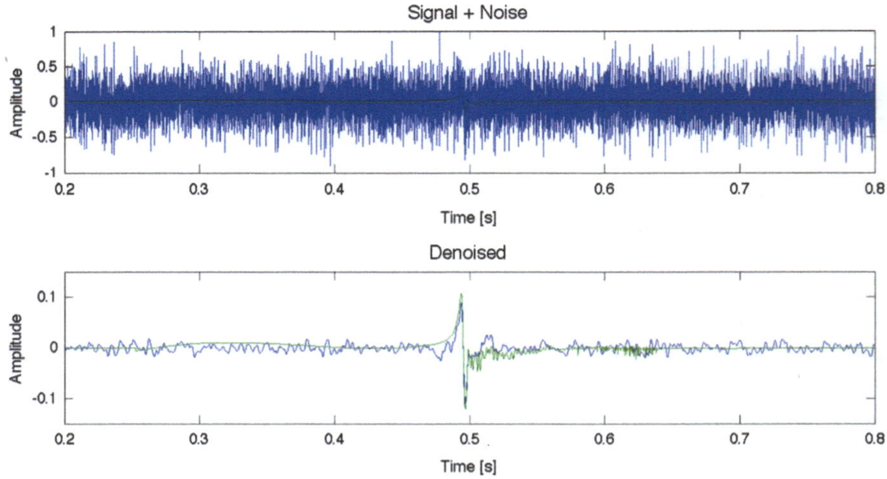

Fig. 1 Results of applying the Split Bregman method to signal "s20a3o15_LS" of the core collapse catalog. The upper panel contains the original measured signal (*blue*) embedded in non-white Gaussian noise. The *lower panel* shows the obtained denoised signal (*blue*). In both panels the original signal from the catalog is displayed for comparison reasons (Color figure online)

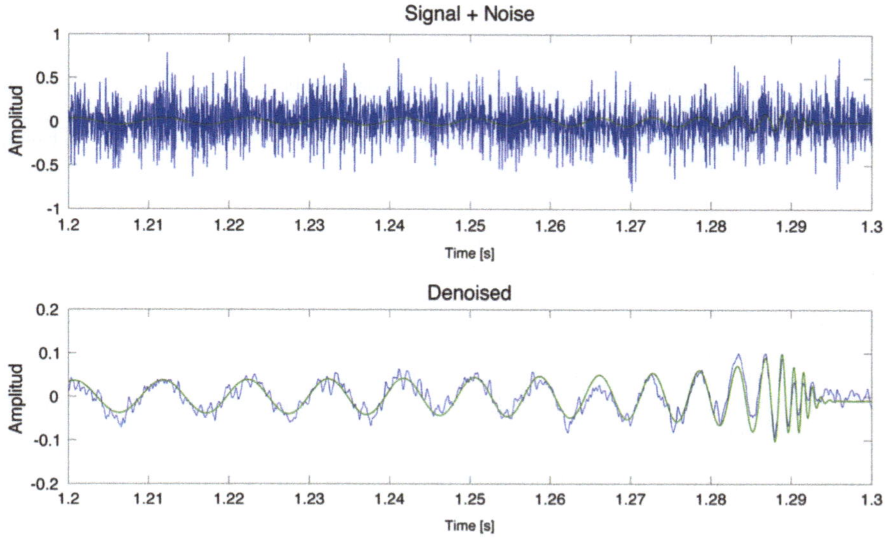

Fig. 2 Results of applying the Split Bregman method to signal "0001" of the BBH catalog. The upper panel contains the original measured signal (*blue*) embedded in non-white Gaussian noise. The lower panel shows the obtained denoised signal (*blue*). The time span shown corresponds to the last few cycles before the merger. In both panels the original signal from the catalog is displayed for comparison reasons (Color figure online)

around $t \approx 1.29$ s is recovered properly, although its amplitude is attenuated. The reason lies in the value of the regularization parameter λ, since a fixed value cannot recover all the signal scales. Despite this fact, the main features of the signals are well obtained in both examples.

5 Summary

We have presented preliminary results of our goal of developing a total variation algorithm for gravitational wave denoising. We have applied the method to two gravitational wave signals with different morphology, and we have not used any a priori information about the source. The method has proven to be useful to remove non-white Gaussian noise and to obtain a recognizable waveform. Further technical details about our method will be presented in Torres et al. (2014).

Acknowledgements This work was supported by the Spanish MICINN (AYA 2010-21097-C03-01), by the Generalitat Valenciana (PROMETEO-2009-103) and by Spanish MIMECO (MTM2011-28043).

References

Advanced Virgo Baseline Design, The Virgo Collaboration, VIR-0027A-09 (2009), available from https://pub3.ego-gw.it/itf/tds/

Y. Aso, Y. Michimura, K. Somiya, M. Ando, O. Miyakawa, T. Sekiguchi, D. Tatsumi, H. Yamamoto, The KAGRA Collaboration, Phys. Rev. D **88**, 043007 (2013)

J. Creighton et al., LAL software documentation (2007), http://www.lsc-group.phys.uwm.edu/daswg/projects/lal.html.

H. Dimmelmeier, C.D. Ott, A. Marek, H.-T. Janka, Phys. Rev. D **78**, 064056 (2008)

D.L. Donoho, IEEE Trans. Inf. Theory **52**(4), 1289–1306 (2006)

T. Goldstein, S. Osher, J. Imag. Sci. **2**, 323–43 (2009)

G.M. Harry, Advanced LIGO: the next generation of gravitational wave detectors. Class. Quantum Grav. **27**, 084006 (2010)

A.H. Mroué, M.A. Scheel et al., Phys. Rev. Lett. **111**, 241104 (2013)

C. Röver, M.A. Bizouard, N. Christensen, H. Dimmelmeier, I.S. Heng, R. Meyer. Phys. Rev. D **80**, 102004 (2009)

L.I. Rudin, S. Osher, E. Fatemi, Physica D **60**, 259–268 (1992)

A. Torres, A. Marquina, J.A. Font, J.M. Ibáñez, Phys. Rev. D (2014) (Submitted)

Interaction of Gravitational Waves with Charged Particles

Thulsi Wickramasinghe, Will Rhodes, and Mitchell Revalski

Abstract It is shown here that a cloud of charged particles could in principle absorb energy from gravitational waves (GWs) incident upon it, resulting in wave attenuation. This could in turn have implications for the interpretation of future data from early universe GWs.

1 Introduction

GWs emanating from the inflationary era of the cosmos en route to us must have also traveled through plasma before the decoupling era. We attempt to show here that these GWs traveling through clouds of charged particles could suffer attenuation. The strain satisfies

$$\Box^2 h^{\mu\nu} = -2\kappa T^{\mu\nu} = 0; \quad h << 1 \tag{1}$$

with plane wave solutions in *empty space* given by

$$h^{\mu\nu} = A^{\mu\nu} exp(i\kappa_\rho x^\rho) \quad \text{with} \quad A^{\mu\nu}\kappa_\nu = 0$$

These waves are commonly thought of as having little or no interaction with matter, reaching us unattenuated, making GWs indispensable in detecting events during and after the inflationary era.

A gauge in general relativity refers to a specific observer. In the TT gauge, the observer is exactly on a particle interacting with a GW, and for a wave propagating in the $\hat{z}$ direction, we have

$$[h^{TT}_{\mu\nu}(t,z)] = \begin{pmatrix} h_+ & h_\times \\ h_\times & -h_+ \end{pmatrix} \cos\left[\omega(t - \frac{z}{c})\right]$$

T. Wickramasinghe (✉) • W. Rhodes • M. Revalski
The College of New Jersey, 2000 Pennington Road, Ewing, NJ 08628, USA
e-mail: wick@tcnj.edu; rhodesw1@tcnj.edu; revalsm1@tcnj.edu

C.F. Sopuerta (ed.), *Gravitational Wave Astrophysics*, Astrophysics and Space Science Proceedings, DOI 10.1007/978-3-319-10488-1_26

In the TT gauge the coordinate positions of free particles remain unchanged since $h_{00}, h_{0i} \rightarrow 0$ due to the imposed gauge condition. In addition $\Gamma^i_{00} = 0$, which is only valid if h^2 terms are neglected. Despite this, in the proper detector frame, particles will move with respect to one another. On Earth, the TT gauge is approximately the same as the proper detector frame since $\Gamma^i_{00} \approx 0$ and $h \sim 10^{-21}$ (expected) in our frame. The motion of free particles in the proper detector frame can be described by a *Newtonian type of force*

$$F_i = \frac{1}{2} m \, \ddot{h}^{TT}_{ij} \, \xi^j$$

Therefore, we have the ability to describe the effects of GWs on particles in the proper detector frame purely in a Newtonian way, without referring to GR, remembering that h and ξ are the values measured in the TT gauge (Maggiore 2008) However, this is possible if and only if Eq. (1) is satisfied; that is, matter at the observing point does not add significantly to the curvature of spacetime. However, to use the geodesic equation to first order, we need to make sure that the size of the detector is smaller than the wavelength of the GW.

2 Interaction of GWs with Charged Particles

For an astrophysical source, the GW strain at a distance r is $h \sim 2GMv^2/rc^4$. To understand how such a wave would interact with a cloud of charged particles of appropriately small size, we need to look at the Einstein field equations (Heintzmann 1968)

$$R_{\mu\nu} - \frac{1}{2} g_{\mu\nu} R = -\kappa (T_{\mu\nu} + F_{\mu\nu}[EM]) \tag{2}$$

with

$$g^{\mu\nu} \nabla_\mu \nabla_\mu A_\alpha = \mu_0 j_\alpha \tag{3}$$

In addition, the following Lorenz gauge condition should also be met.

$$\nabla_\alpha A^\alpha = 0 \quad \text{and} \quad [A^\mu] = \left(\frac{\phi}{c}, \mathbf{A} \right)$$

It is clear from Eq. (1) that a TT gauge cannot be chosen inside a matter distribution as $T^{\mu\nu} \neq 0$. However, here we assume that motionless (no currents are inside the cloud and $\mathbf{B} \neq 0$) charges do not contribute significantly to the curvature of spacetime. This can be assumed at least *initially*. Then (1) and (2) indicate that a *TT frame, and thus a detector frame, may be used even inside a charged cloud under the influence of a GW.* Under these assumptions, Eq. (3) gives the vector potential to be

$$\Box^2 A_i = 0 \quad \text{(since there are no currents, at least initially)}$$

and

$$\Box^2 A_0 = c\mu_0\rho + h^{\mu\nu}\nabla_\mu\nabla_\nu A_0 \tag{4}$$

When no GWs are present solutions to the inhomogeneous wave equation (4), which satisfies the Lorenz gauge automatically, are well-known and are given by

$$A_0 = \frac{\mu_0}{4\pi}\int d^3y\,\frac{j_0(ct - |\mathbf{x}-\mathbf{y}|, \mathbf{y})}{|\mathbf{x}-\mathbf{y}|} = \frac{1}{4\pi\epsilon_0 c}\int d^3y\frac{\rho}{|\mathbf{x}-\mathbf{y}|} \tag{5}$$

where $\mathbf{x}$ is the field point while $\mathbf{y}$ is a point inside the charged distribution measured from the center of a spherical cloud. Integrals are evaluated at the retarded time $t_r = t - |\mathbf{x}-\mathbf{y}|/c$. Equation (5) gives the zeroth order solution for A_0 in (4). A first order solution to (4) can be written as $A_0^{(1)} = \epsilon + A_0$. Upon substitution into (4), this yields to first order (neglecting $\partial_\mu\partial_\nu\epsilon$)

$$\Box^2\epsilon \approx h^{\mu\nu}\partial_\mu\partial_\nu A_0$$

Thus, the first order solution becomes

$$A_0^{(1)} = \frac{1}{4\pi\epsilon_0 c}\int d^3y\frac{\rho}{|\mathbf{x}-\mathbf{y}|} + \frac{1}{4\pi}\int d^3y\,\frac{h^{\mu\nu}\partial_\mu\partial_\nu A_0}{|\mathbf{x}-\mathbf{y}|} \tag{6}$$

While the first term of the foregoing gives the electrostatic potential due to charges, the second term emerging from the derivatives of the charge density within the cloud [see Eq. (5)] results in a time-varying vector potential giving rise to an electromagnetic (EM) field at the observation point P. Since this field is produced due to the GWs ($h^{\mu\nu}$) some amount of energy of the incident waves is lost in transit via the cloud.

2.1 GW of a Specific Simple Form

Consider the following specific GW form propagating along the $\hat{r}$ direction at a distance r from a source, as an example.

$$h^{\mu\nu} = \delta^\mu_\nu\,\frac{\lambda}{r}\cos\omega\left(t - \frac{r}{c}\right) \quad \text{with} \quad h^{00} = h^{11} = 0$$

Then, Eq. (6) becomes

$$A_0^{(1)} = \frac{1}{4\pi\epsilon_0 c}\int d^3y \frac{\rho}{|\mathbf{x}-\mathbf{y}|} + \frac{\lambda}{4\pi r}\cos(\omega(t-r/c))\int d^3y \frac{(\partial_z\partial_z A_0 - \partial_2\partial_2 A_0)}{|\mathbf{x}-\mathbf{y}|} \tag{7}$$

The EM fields may be easily calculated via

$$\mathbf{E} = -\nabla c A^0 - \partial_t \mathbf{A} \quad \text{and} \quad B = \nabla \times \mathbf{A} \quad \text{with} \quad \partial_0 A^0 - \nabla \cdot \mathbf{A} = 0$$

3 Results and Discussion

We see from Eq. (7) that in the asymptotic region far away from the charge distribution the Poynting flux is proportional to $1/|\mathbf{x}-\mathbf{y}|^2$. Thus, the cloud of charges absorbs a *finite amount* of energy provided that those gradients of the charge density are nonzero. From Eq. (7), we see that the amount of absorbed energy depends on the inhomogeneity of the charge cloud and how far it is from the source of GWs.

For an EM wave, a charged particle oscillating in the transverse directions stays at rest even after the wave passes showing that the particle does not absorb energy. This is not the case in the direction of propagation (Marsh 2011). The displacement of the charged particle in the direction of propagation of the wave is proportional to $1/2(B_0^2 - A_0^2)$. Then it is clear that unless the incident radiation is circularly polarized, a charged particle will oscillate in the direction of propagation, absorbing energy.

But for a GW, longitudinal proper oscillations in the direction of propagation of the wave will still not vanish even if the incoming GW is circularly polarized ($h_+ = \pm h_\times$) (Marsh 2011). Thus as time goes on, GWs sweeping through a charged cloud will modify the proper charge density and the gradients in Eq. (7) will be established. The result is that the GW will be attenuated, losing energy absorbed by the charges in the cloud. If magnetic fields are present initially (Kleidis et al. 1993), the attenuation will be even more pronounced as $A_i \neq 0$ in Eq. (4).

Our analysis shows that GWs incident upon an inhomogeneous cloud of charged particles will be attenuated even if the charges are not subjected to any magnetic field. Since waves change the proper charge density at least in the direction of the propagation of the wave, even a completely homogeneous cloud will eventually become inhomogeneous and the gradients of the charge density will be nonzero. When this occurs, we see from Eq. (7) that an EM field will be produced and GW energy will be absorbed. It is clear that near a source, attenuation can be rather significant. Therefore, it is vital for us to consider these effects especially when trying to understand GWs from the inflationary era. This could have additional implications requiring a reevaluation of the magnitudes of h from various sources expected here on Earth.

References

H. Heintzmann, Detection of gravitational waves by electromagnetic radiation. Zeitschrift für Physik. **210**, 380–390 (1968)

K. Kleidis, H. Varvoglis, D. Papadopoulos, Interaction of gravitational waves with charged particles of various polarizations and directions of propagation. Astron. Astrophys. **275**, 309–317 (1993)

M. Maggiore, *Gravitational Waves. Theory and Experiments*, vol. 1 (Oxford university Press, Oxford, 2008)

G.E. Marsh, Electromagnetic and gravitational waves: the third dimension. Can. J. Phys. **89**(2), 1187–1194 (2011)

The Emission of Electromagnetic Radiation from Charges Accelerated by Gravitational Waves and Its Astrophysical Implications

Mitchell Revalski, Will Rhodes, and Thulsi Wickramasinghe

Abstract We provide calculations and theoretical arguments supporting the emission of electromagnetic radiation from charged particles accelerated by gravitational waves (GWs). These waves have significant indirect evidence to support their existence, yet they interact weakly with ordinary matter. We show that the induced oscillations of charged particles interacting with a GW, which lead to the emission of electromagnetic radiation, will also result in wave attenuation. These ideas are supported by a small body of literature, as well as additional arguments for particle acceleration based on GW memory effects. We derive order of magnitude power calculations for various initial charge distributions accelerated by GWs. The resulting power emission is extremely small for all but very strong GWs interacting with large quantities of charge. If the results here are confirmed and supplemented, significant consequences such as attenuation of early universe GWs could result. Additionally, this effect could extend GW detection techniques into the electromagnetic regime. These explorations are worthy of study to determine the presence of such radiation, as it is extremely important to refine our theoretical framework in an era of active GW astrophysics.

1 Introduction

Gravitational waves (GWs) are ripples in space which propagate at the speed of light, conveying the changing gravitational field of a system. Systems undergoing anisotropic acceleration will have changing gravitational fields and emit GWs (Schutz 2000). The theoretical framework for GWs has been well developed and clarified making the field of study more accessible (Thorne 1995), with many astrophysical applications explored (Sathyaprakash and Schutz 2000). Despite this theoretical competency, gravitational radiation is the last prediction of general relativity which has not yet been detected directly.

M. Revalski (✉) • W. Rhodes • T. Wickramasinghe
The College of New Jersey, 2000 Pennington Road, Ewing, NJ 08628, USA
e-mail: revalsm1@tcnj.edu; rhodesw1@tcnj.edu; wick@tcnj.edu

C.F. Sopuerta (ed.), *Gravitational Wave Astrophysics*, Astrophysics and Space Science Proceedings, DOI 10.1007/978-3-319-10488-1_27

Significant indirect evidence has accumulated through monitoring orbital decay in binary star systems as predicted by general relativity (Weisberg et al. 2010; Hermes et al. 2012). This decay is attributed to orbital energy radiated in the form of GWs. Recently, B-mode polarization has been measured in the Cosmic Microwave Background (CMB), predicted if stochastic GWs in the early universe induced a curling polarization in the CMB (Ade et al. 2014). If confirmed, this will provide significant evidence for the existence of GWs.

The interaction of GWs with normal matter is extremely weak, with the graviton cross-section $\sim$80 orders of magnitude smaller than the Thomson cross-section (Maggiore 2008)! Also, as compared to electromagnetic waves, the wavelengths of GWs are determined by the bulk movement of a source rather than atomic transitions. Thus GWs cannot be used to resolve sources in the way electromagnetic radiation is employed (Flanagan and Hughes 2005). Regardless, there is interest in GWs due to their pristine nature, propagating nearly unperturbed by galactic or interstellar media. Despite their weak interaction, it has been well established that GWs should cause a ring of particles to oscillate. This toy model is often the prototype for studying the interaction of particles with GWs (Hobson et al. 2006).

An interesting situation arises when considering a distribution of *charged* test particles. If a GW causes a distribution of charged particles to oscillate, then they could experience an acceleration and radiate energy in the form of electromagnetic radiation. If this occurs, significant astrophysical consequences could result including attenuation of GWs and new avenues for GW detection.

We aim here to show that charges interacting with a GW will effectively accelerate, absorb GW energy, and radiate that energy in the form of electromagnetic radiation, leading to wave attenuation. The meaning of *effective* acceleration is detailed in these proceedings in the contribution by Thulsi Wickramasinghe where it is shown that a charged gas cloud under the influence of a passing gravitational wave displays the same time varying terms in the calculated magnetic potential as those produced by accelerating charges, leading to measurable radiation. Thus we are able to forgo an extensive investigation of whether or not charges experience the acceleration, as we find it leads to the same physics. We discuss the previous work on this topic in Sect. 2, and present our calculations for the radiated power in Sect. 3. We give additional arguments for acceleration in Sect. 4, and devote Sect. 5 to discussing the astrophysical implications. Our conclusions are given in the final section.

2 Previous Work

The concept of electromagnetic emission from charged particles accelerated by GWs is not new; however, it has been sparsely explored. Early work concluded that charged particles could extract energy from GWs and radiate it in the form of electromagnetic radiation (Heintzmann 1968). Further work agrees, with the condition that a pre-existing magnetic field be present (Papadopoulos and Esposito

1981). Additionally, significant wave attenuation could occur under certain resonant conditions, allowing particles to extract large amounts of energy from the GW (Voyatzis et al. 2006; Kleidis et al. 1993). More recent work has examined these resonant and nonlinear interactions in magnetic fields, finding that charges are accelerated (Kleidis et al. 1995). In examining astrophysical situations in which these conditions may occur, it was shown that charges could be accelerated to high energies by the GWs emitted during the collapse of massive magnetized stars (Vlahos et al. 2004). Preliminary calculations for general distributions of particles show that while charged particles can absorb large amounts of energy from GWs, it would constitute at most 1 part in $\sim 10^9$ of the wave's overall energy (Voyatzis et al. 2006).

With this literature it is more firmly grounded that charged particles will, at least under some conditions, absorb energy from GWs, and emit electromagnetic radiation. The primary constraint in previous calculations is the requirement of a pre-existing magnetic field. We aim to show that charged particles will be accelerated by GWs and radiate in general without an external magnetic field, as further developed employing Maxwell's equations in the parallel paper led by Thulsi Wickramasinghe.

An important detail to carry through from this brief literature survey is that we expect charged particles will be accelerated by the GW with respect to their electrostatic fields, and not freely falling. Additionally, debates on if uniformly accelerating charges radiate do not apply here, as the direction of the acceleration is not constant for particles under the influence of a GW, meaning $\dddot{a}(t)$ will be nonzero.

3 Radiation from Accelerating Charges: Calculations

We turn now to calculating the power radiated by these charges, constructing an order of magnitude calculation from the non-relativistic Larmor formula. When the equations of motion for particles under the influence of a GW are examined in the transverse-traceless (TT) gauge, we may dispense with a relativistic treatment and work with Newtonian equations of motion as described in Maggiore (2008); Buonanno (2006).

First, equations describing the acceleration for each particle in a ring distribution are derived. Following this we employ the Larmor formula to find a general expression for the total power radiated by all charges in a given distribution. We consider the total charge of the system to be constant in time. Using the Larmor formula, the power radiated from a single non-relativistic charge is

$$P_e = \frac{1}{4\pi\epsilon_0}\frac{2}{3}\frac{e^2 a^2}{c^3} \tag{1}$$

where e is the particle's charge and a is its acceleration. The acceleration can be calculated based on the particle's motion in the proper detector frame as derived in

Hobson et al. (2006). Considering motion in the plane perpendicular to the direction of the wave propagation vector, the particle's position for one wave polarization is given by

$$X(t) = x\left(1 + \frac{h}{2} sin(\omega t)\right) \tag{2}$$

where h and ω are the amplitude and angular frequency of the GW, respectively. $X(t)$ is the position of the particle in the proper detector frame and x is its initial coordinate position. Differentiating twice with respect to time we obtain

$$\ddot{X}(t) = -\frac{1}{2} x h \omega^2 sin(\omega t) \tag{3}$$

Substituting Eq. (3) into the Larmor formula (1), we obtain the power emitted as a function of time.

$$P_e(t) = \frac{1}{4\pi\epsilon_0} \frac{e^2}{6c^3} x^2 h^2 \omega^4 sin^2(\omega t) \tag{4}$$

The emitted power which is sinusoidal in time will have a continuous frequency distribution which may be obtained using Fourier techniques (Jackson 1998).

To generalize our solution and make it astrophysically applicable, we extend this result to rings, disks, spheres, and cylinders of charges. Considering a ring in 2D space, we extend Eq. (3) for both $\ddot{X}(t)$ and $\ddot{Y}(t)$ and obtain the proper acceleration, $\sqrt{\ddot{X}^2 + \ddot{Y}^2} = r\omega^2 h/2\ sin(\omega t)$, which is independent of the coordinate system. As the ring oscillates the charge density remains constant to first order in h, since the circumference of the ellipse, $4r(1 + h/2\ sin\omega t)E(2h\ sin\omega t)$, is still $2\pi r$. For a ring with total charge $Q = Ne$ consisting of N point charges the total power radiated is

$$P_{ring}(t) = \frac{1}{4\pi\epsilon_0} \frac{2e}{3c^3} Q \ddot{r}^2 = \frac{1}{4\pi\epsilon_0} \frac{e h^2 \omega^4 Q}{6c^3} r^2 sin^2(\omega t) \tag{5}$$

Expanding this to a disk of charges, we consider a series of concentric annuli with charge density σ, which is constant in time to first order in h. Integrating over these annuli yields a total charge on the disk of $Q_{disk} = (\pi r_2^2 - \pi r_1^2)\sigma$, where r_1 and r_2 are the inner and outer radii, respectively. The charge of each infinitesimally thin ring is $dQ = 2\pi r dr \sigma = 2Q_{disk} r dr/(r_2^2 - r_1^2)$. From Eq. (5) we can write the power emitted by dQ as

$$dP = \frac{1}{4\pi\epsilon_0} \frac{e h^2 \omega^4}{6c^3} r^2 sin^2(\omega t)\ dQ$$

Substituting in dQ and integrating over r, allowing $r_1 \to 0$ for a uniform disk, we find

$$P_{disk}(t) = \frac{1}{4\pi\epsilon_0}\frac{eh^2\omega^4 Q_{disk}}{12c^3} r^2 sin^2(\omega t) \tag{6}$$

where r is the proper radius of the disk at $t = 0$ and Q_{disk} is the total charge.

Considering a spherical distribution, we sum our result from Eq. (6) for different cross-sections of a sphere. At $t = 0$, the proper radius of the sphere, R, is related to the cross-sectional radius by $R^2 = z^2 + r^2$. The total charge of the sphere is $Q_{sph} = 4\pi R^3\rho/3$, where the charge density ρ is constant with time to first order in h. The charge for each infinitesimal cross-section is then $dQ = 4\pi r^2\rho dz = 3Q_{sph}r^2dz/4R^3$. The power associated with this cross section from Eq. (6) is

$$dP = \frac{1}{4\pi\epsilon_0}\frac{eh^2\omega^4}{12c^3} r^2 sin^2(\omega t)dQ$$

Substituting in dQ, changing variables, and integrating from $z = -R \rightarrow R$, we find

$$P_{sph}(t) = \frac{1}{4\pi\epsilon_0}\frac{eh^2\omega^4 Q_{sph}}{15c^3} R^2 sin^2(\omega t) \tag{7}$$

Far from a source, GWs may be treated as plane waves, propagating across a distribution of matter cylindrically. Thus we also consider a cylindrical distribution of charge with a base perpendicular to the GW propagation vector, $\mathbf{z}$. Using Eq. (6) with amplitude $h(z) = h_1/z$, and integrating over the cylinder length, $z_1 \rightarrow z_2$, we obtain

$$P_{cyl}(t) = \frac{1}{4\pi\epsilon_0}\frac{eh_1^2\omega^4 Q_{cyl}}{12c^3} r^2 sin^2(\omega t)\ \frac{1}{z_1 z_2} \tag{8}$$

Using the equations above, the power radiated from astrophysical systems is calculated. We consider a small spherical cloud of 10^{12} electrons; so that our approximations hold. Examples are given in the following table.

h	ω	N	Power (J/s)	h	ω	N	Power (J/s)
10^{-20}	10^6	10^{12}	$\mathcal{O}(10^{-51})$	10^{-5}	10^6	10^{12}	$\mathcal{O}(10^{-21})$
10^{-12}	10^2	10^{12}	$\mathcal{O}(10^{-43})$	10^{-12}	10^{10}	10^{12}	$\mathcal{O}(10^{-27})$

Integrating over the duration of the wave will increase the amount of energy absorbed and re-emitted during the process. It is important to note that the GW wavelength limits the value of r, determining the maximum separation that two particles could have and still be treated with simplified Newtonian equations of motion.

For our first-order treatment here, we have considered the equations of motion to be linear, which holds for small amplitude GWs ($h << 1$). The amplitude is considered to be constant over the distance which the charge distribution occupies, which is reasonable far from the source. An exception is Eq. (8), where we have done a more complete treatment. As mentioned earlier, the charge density is constant in time to first order in h. Additionally, to calculate the total power from any distribution with a significant length in the direction of wave propagation as compared with the wavelength of the GW, the phase of the wave plays an important role (Marsh 2011). For simplicity, we neglect this minor effect which when averaged over only contributes a factor of $\sim 1/2$ to the power radiated.

4 Additional Arguments for Radiation: Memory Effects

To make a stronger case for charges experiencing acceleration, we briefly examine the various *memory* effects of GWs. It has been established that all sources of gravitational radiation will produce some degree of memory effect (Favata 2010).

The Velocity-Coded Memory effect is a permanent relative velocity between two bodies following the passage of a GW. This effect should present itself when there is an asymmetric rate in the rise and fall of the GW amplitude, as shown in figure 1 of Grishchuk and Polnarev (1989). This work showed the effect would be very small, but could be of experimental importance.

The Linear Memory effect permanently changes the separation of two bodies following the passage of a GW (Braginskii and Thorne 1987). If the separation changes permanently, the particles should have accelerated despite the initial and final velocities being zero.

The Nonlinear Memory component results in permanent displacements which may be only one order of magnitude weaker than the oscillatory portion of strong GWs (Christodoulou 1991). This memory effect surprisingly shows up at leading quadrupole order in post-Newtonian expansions (Favata 2010), and is due to waves produced by the energy carried in the initial GWs, in the form of the emitted gravitons (Thorne 1992).

While these memory effects could be used in detection, the resulting displacements are very small. The possible exception being velocity-coded memory, as the particles continue to increase in separation even after the wave has passed. These effects are likely undetectable in ground based interferometers as the test masses are not truly free falling, relying on damping systems to maintain a delicate equilibrium.

A final motion to consider is the acceleration of particles in the direction of wave propagation. While almost always neglected when studying particle interactions with GWs, as the displacement is exceedingly small, it is calculated to be nonzero (Braginskii and Grishchuck 1985).

The results to exploit here are not the memory effects themselves, but the corresponding acceleration that particles interacting with GWs should experience due to these memory effects.

5 Astrophysical Implications

The orders of magnitude given in Sect. 3 are very small, even when GWs with very large amplitudes and high frequencies are involved. For average parameters the energy radiated per second by the charge distribution is on the order of a CMB photon; however, peak values yield $\sim 10^{-10}\ J \cdot s^{-1}$. In terms of the overall power of the GW as derived from its flux in Hobson et al. (2006), we find the average fractional power extracted from a disk of charges to be

$$\frac{P_{disk}}{P_{incident}} \approx 1.8 \times 10^{-70} \omega^2 Q_{disk} \tag{9}$$

Currently, the likelihood of detecting such weak radiation is highly unlikely.

Current resources invested in GW detection focus primarily on instruments like the Laser Interferometer Gravitational Wave Observatory (LIGO). If emission of electromagnetic radiation is induced by GWs through the process explored above, it would open a new regime for GW detection. These new detection methods, along with the use of charged particle storage rings (Dong and Huang 2003) and atom interferometers (Dimopoulos et al. 2009), could allow GWs to be exploited in new studies of the universe.

Furthermore, if charges are able to attenuate GWs more than is currently understood, significant consequences would result for early universe cosmology. GWs produced in the initial expansion of the universe were likely very powerful, and many free charges would have been present, possibly contributing to significant attenuation of these waves. Further cases of attenuation would result during supernovae explosions, for example. Calculations regarding the significance of this attenuation are a current focus in extending this work.

Conclusions

We have discussed how GWs should accelerate charged particles, leading to the emission of electromagnetic radiation and wave attenuation. Due to the complexities of analyzing particle motion in a general relativistic framework, there may be arguments to oppose this which we have not yet considered. Despite this, an examination using Maxwell's equations yield radiation from a gas cloud under the influence of a GW, even if the particles are not experiencing the tidal accelerations. By further exploring the memory effects of GWs, we lent additional evidence to the ability of GWs to accelerate charged particles.

Our preliminary calculations show that even under idealistic circumstances in which systems produce strong GWs, the emitted radiation would be feeble.

(continued)

This follows since if high quantities of radiation were being produced, the effect would have likely been detected by now.

The most important conclusion is the theoretical support that this effect may occur. While it likely contributes only in a minor way to wave attenuation, it is significant in forming a coherent and complete pedagogy for GW interactions in our universe.

In the future we hope to collaborate with others and challenge the ideas presented here in order to reach a more secure conclusion. During this exciting time in GW astronomy, explorations such as this are important in completing our understanding of GW astrophysics and for explaining future observational phenomena.

Acknowledgements We would like to thank the School of Science and the Department of Physics at The College of New Jersey for funding to present these results at the Sant Cugat Forum on Astrophysics held in Barcelona during 22–25 of April, 2014. In addition, we are very grateful to the forum organizers for partially subsidizing our registration. This research has made use of NASA's Astrophysics Data System (ADS).

References

P.A.R. Ade, R.W. Aikin, D. Barkats et al., [BICEP2 Collaboration]: BICEP2 I: detection Of B-mode Polarization at Degree Angular Scales. In Press. arXiv:1403.3985 (2014)

V.B. Braginskii, L.P. Grishchuck, Kinematic resonance and memory effects in free-mass gravitational antennas. Zh. Eksp. Teor. Fiz. **89**, 744–750 (1985)

V.B. Braginskii, K.S. Thorne, Gravitational-wave bursts with memory and experimental prospects. Nature. **327**, 123–125 (1987)

A. Buonanno, Gravitational waves, in *Proceedings of Les Houches Summer School. Particle Physics and Cosmology*, Les Houches, 31 July–25 August 2006

D. Christodoulou, Nonlinear nature of gravitation and gravitational-wave experiments. Phys. Rev. Lett. **67**, 1486–1489 (1991)

S. Dimopoulos, P.W. Graham, J.M. Hogan et al., Gravitational wave detection with atom interferometry. Phys. Lett. B. **678**(1), 37–40 (2009)

D. Dong, C. Huang, Response of charged particles in a storage ring to gravitational waves. Commun. Theor. Phys. **40**, 299–300 (2003)

M. Favata, The gravitational-wave memory effect. Class. Quantum Grav. **27**(8) (2010). doi:10.1088/0264-9381/27/8/084036

É.É. Flanagan, S.A. Hughes, The basics of gravitational wave theory. New J. Phys. **7**(1), 30 (2005)

L.P. Grishchuk, A.G. Polnarev, Gravitational wave pulses with "velocity-coded memory". Zh. Eksp. Teor. Fiz. **96**, 1153–1160 (1989)

H. Heintzmann, Detection of gravitational waves by electromagnetic radiation. Zeitschrift für Physik. **210**(4), 380–390 (1968)

J.J. Hermes, M. Kilic, W.R. Brown et al., Rapid orbital decay in the 12.75-minute binary white dwarf J0651+2844. ApJL **757**(2), L21 (2012)

M.P. Hobson, G. Efstathiou, A.N. Lasenby, *General Relativity An Introduction for Physicists* (Cambridge University Press, Cambridge, 2006)

D.J. Jackson, *Classical Electrodynamics* (Wiley, New York, 1998)

K. Kleidis, H. Varvoglis, D. Papadopoulos, Interaction of charged particles with gravitational waves of various polarizations and directions of propagation. Astron. Astrophys. **275**, 309–317 (1993)

K. Kleidis, H. Varvoglis, D. Papadopoulos et al., Non-linear interaction of a gravitational wave with a distribution of particles. Astron. Astrophys. **294**, 313–321 (1995)

M. Maggiore, *Gravitational Waves. Theory and Experiments*, vol. 1 (Oxford University Press, Oxford, 2008)

G.E. Marsh, Electromagnetic and gravitational waves: the third dimension. Can. J. Phys. **89**(2), 1187–1194 (2011)

D. Papadopoulos, F.P. Esposito, On the transformation of gravitational radiation into electromagnetic radiation. ApJ **248**(1), 783–789 (1981)

B.S., Sathyaprakash, B.F. Schutz, Physics, astrophysics and cosmology with gravitational waves. Living Rev. Relativity **12**(2), 1–141 (2000). doi:10.12942/lrr-2009-2

B. Schutz, Gravitational radiation, in *Encyclopedia of astronomy and astrophysics*, ed. by P. Murdin (Institute of Physics Publishing, Bristol, 2000)

K.S. Thorne, Gravitational-wave bursts with memory: the Christodoulou effect. Phys. Rev. D. **45**(2), 520–524 (1992)

K.S. Thorne, Gravitational waves, in *Singapore Proceedings of Snowmass Summer Study, 29 June - 14 July, 1994*, ed. by E.W. Kolb, R.D. Peccei. Particle and Nuclear Astrophysics and Cosmology in the Next Millennium (World Scientific, Singapore, 1995)

L. Vlahos, G. Voyatzis, D. Papadopoulos, Impulsive electron acceleration by gravitational waves. ApJ **604**(1), 297–305 (2004)

G. Voyatzis, L. Vlahos, S. Ichtiaroglou, D. Papadopoulos, Acceleration of low energy charged particles by gravitational waves. Phys. Lett. A. **352**(4–5), 261–266 (2006)

J.M. Weisberg, D.J. Nice, J.H. Taylor, Timing measurements of the relativistic binary pulsar PSR B1913+16. ApJ **722**(2), 1030–1034 (2010)

MIX
Papier aus verantwortungsvollen Quellen
Paper from responsible sources
FSC® C105338
FSC
www.fsc.org

If you have any concerns about our products, you can contact us on
ProductSafety@springernature.com

In case Publisher is established outside the EU, the EU authorized representative is:
Springer Nature Customer Service Center GmbH
Europaplatz 3, 69115 Heidelberg, Germany

Printed by Libri Plureos GmbH
in Hamburg, Germany